Texts and Readings in Physical Sciences

Volume 21

The **Texts and Readings in Physical Sciences** series publishes high-quality textbooks, research-level monographs, lecture notes and contributed volumes. Undergraduate and graduate students of physical sciences and applied mathematics, research scholars and teachers would find this book series useful. The volumes are carefully written as teaching aids and highlight characteristic features of the theory. Books in this series are co-published with Hindustan Book Agency, New Delhi, India.

More information about this series at https://link.springer.com/bookseries/15139

Amal Kumar Raychaudhuri

Classical Theory of Electricity and Magnetism

A Course of Lectures

Amal Kumar Raychaudhuri (Deceased)
Department of Physics
Presidency University
Kolkata, West Bengal, India

ISSN 2366-8849 ISSN 2366-8857 (electronic)
Texts and Readings in Physical Sciences
ISBN 978-981-16-8141-7 ISBN 978-981-16-8139-4 (eBook)
https://doi.org/10.1007/978-981-16-8139-4

This work is a co-publication with Hindustan Book Agency, New Delhi, licensed for sale in all countries in electronic form, in print form only outside of India. Sold and distributed in print within India by Hindustan Book Agency, P-19 Green Park Extension, New Delhi 110016, India. ISBN:978-81-951961-5-9

This Springer imprint is published by the registered company Springer Nature Singapore Pte Ltd.
The registered company address is: 152 Beach Road, #21-01/04 Gateway East, Singapore 189721, Singapore

Foreword to the Revised Edition

A. K. Raychaudhuri (1923–2005) (in short, AKR) was well known for his research in general relativity and cosmology. D.Sc. from Calcutta University as well as D.Sc. (Honoris Causa) Burdwan University, he was Professor at his *alma mater*, Presidency College, Calcutta, from 1961 to 1986. Subsequent to his retirement from Presidency College, he was Senior Scientist, INSA and Honorary Visiting Professor, Jadavpur University. His most important contribution was the deduction of the eponymous Raychaudhuri equations in 1955; this provided the basis for subsequent fundamental work on gravitational singularities by Roger Penrose and Stephen Hawking that led to the award in 2020 of the Nobel Prize in Physics to Sir Roger Penrose.

The first edition of AKR's *Classical Theory of Electricity and Magnetism*, published by Oxford University Press in 1990, was instantly recognized as a masterful pedagogical exposition of the subject. AKR taught generations of physics graduates who passed through the portals of Presidency College, and many of the leading physicists of India have benefited from his instruction, in taste and style, and especially in substance.

This revised edition is occasioned firstly by a desire to see the book made available to new generations of students of physics. AKR's treatment of the subject has its own whimsy and is valuable in its own right. In addition, there were some errors in the first edition that he (and others) had noted, and it was important that these be corrected. Fortunately, we had his personal copy on which he had made handwritten notes, and the changes that he wanted have now been made here. We have (very occasionally) changed some more archaic phrases that he had used, and have also taken this opportunity to redraw and amend some of the figures.

The most significant change, however, has been in the modernization of the units and the notation. In the past few decades, SI units are standard in instruction, and although the Gaussian units are more "natural" in some contexts, we felt that in order to address contemporary students, it was necessary to change the units in most of the chapters (exceptions have been indicated clearly). This required both care and diligence, and we would like to thank our colleagues, especially Prof. Anindya Datta (University of Calcutta) for his thorough reading of the entire text. Our thanks are also due to Profs. S. Uma Sankar and B. Nandi of IIT Bombay for going through some of the chapters, and to Prof. Debashis Ghoshal of the JNU for his careful editing of the final manuscript and for redrafting some of the figures.

It is our pleasure to thank Oxford University Press for their generosity with regard to the copyright, and the Hindustan Book Agency for bringing out the TRiPS series, this volume in particular. Our special thanks go out to Prof. Ram Ramaswamy for his initiative and his enthusiastic participation in seeing the revival of this textbook, which serves to remind us once more of the great scientific and pedagogic contributions of AKR.

Mumbai, India
Kolkata, India
May 2021

Dipan Ghosh
Parongama Sen

Preface

It is customary to present every book with a preface. Usually, it speaks of the motivation or the background which has led to the writing of the book, spells out the topics that have been covered, indicating in particular the type of readers that the author has in view and not often adds a word claiming some special merit of the book. So far as the present book is concerned, the author does not feel the necessity of such a detailed preface but would like to write just two sentences. Firstly, he believes that the book will meet the needs of the average student studying for his B.Sc. or M.Sc. degree in physics (or mathematics) if classical electromagnetism forms a part of his course. Secondly, the author has been guided by the story of a teacher in ancient India who was called upon to educate some princes within a rather short time. His reaction was expressed in a verse, which, translated feely, runs as follows:

> The sea of knowledge knows no bounds, but life is short and obstacles are many. So rejecting non-essentials, the fundamentals are to be grasped just as swans select milk out of a mixture with water.

The author hardly believes that swans can indeed do this magic, nevertheless, concentrating on the 'fundamentals', and pushing what the author considers to be 'non-essentials' to the background, the author has completed the book in some odd three hundred pages, while standard books on the subject have more than double the volume. He leaves the judgement on the wisdom of this procedure to his readers.

Kolkata, India
1990

Amal Kumar Raychaudhuri

Preface

Contents

About the Author

Amal Kumar Raychaudhuri (1923–2005) taught at the Department of Physics, Presidency College, Kolkata between 1961 and 1986. An extremely popular teacher, Prof. Raychaudhuri is known for the eponymous "Raychaudhuri Equations" of the General Theory of Relativity.

During his long research and teaching career, Professor Raychaudhuri received a number of awards, most notably the Vainu Bappu Memorial Award of the Indian National Science Academy in 1991. He was elected Fellow of the Indian Academy of Sciences in 1982 and Fellow of the Indian National Science Academy in 1987. He was also Fellow of the National Academy of Sciences as well as an Honorary Fellow of the Astrophysical Society of India and of the Inter-University Centre of Astronomy and Astrophysics, Pune.

Subsequent to retirement from Presidency College, he held the position of UGC Emeritus Fellow from 1986 to 1988 and INSA Senior Scientist from 1988 to 1991. In addition to being awarded the Doctor of Science degree in 1960 from the University of Calcutta, Professor Raychaudhuri also received doctorates (honoris causa) from Burdwan, Kalyani and Vidyasagar Universities.

About the Author

Chapter 1
Empirical Basis of Electrostatics

We begin with the study of electrostatics, i.e., of phenomena associated with 'electric charges' at rest. The quotation marks indicate that we must define, or at least form some idea of, what we mean by the term electric charge. The empirical basis of electrostatics is essentially the following.

1.1 The Idea of Electric Charge

When some materials undergo some special treatment (e.g., when a glass rod is rubbed with a piece of silk cloth), the materials sometimes exhibit the power of attracting or repelling other such materials. The causative agency of these forces is what we call electric charge. The fact that they exhibit both types of forces (unlike the gravitational phenomenon where there is only one type of force, namely attraction) makes it evident that charges are at least of two types. Again, as the two types may annul each other's effects, one is led to associate positive and negative signs to their magnitudes. However, the particular adjectives, positive and negative, are purely conventional—one might as well reverse the nomenclature throughout. Our present knowledge tells us that electrical phenomena are mainly due to some particles which are called electrons—their charge is taken to be negative while the other universal constituent of matter is called the proton and it has a charge of the opposite sign to the electron. The proton charge is thus taken as positive.

1.2 The Law of Interaction of Point Charges

The practice in classical physics is to consider matter as an assembly of discrete particles, to enunciate the laws obeyed by such particles and then to develop the consequences for matter in bulk. This will be clear if one recalls the case of mechanics—

A. K. Raychaudhuri, *Classical Theory of Electricity and Magnetism*, Texts and Readings in Physical Sciences 21, https://doi.org/10.1007/978-981-16-8139-4_1

Newton's laws are stated for point particles and from these one develops the mechanics of continua like rigid bodies or fluids. Thus, a classical physicist introduced the idea of point charges and arrived at the law of their interaction—called Coulomb's law. Today with all our sophisticated experiments, we are not able to assign any finite magnitude to the dimension of the electron, and we do consider it as a genuine point body. We can break up Coulomb's law into several parts:

(a) The force between any two point charges is directed along the line joining the charges, i.e., the force is central. Now if the sources of interaction have no intrinsic directional property (i.e., if sources be describable as scalars), simple symmetry consideration shows that the force must be central. Conversely, from the central nature of the force, we may conclude that electrical charge is a scalar quantity.

(b) The force between two point charges *in vacuo* varies inversely as the square of the distance between them. The direct verification of this law was based on the torsion balance method of measuring the force between two charged bodies of finite size. However, such a method is of very limited accuracy. There are indirect methods based on a consequence of the inverse square law, that the charge resides entirely on the outer surface of a conductor or equivalently that the hollow space within a charged closed conductor is devoid of any electrical influences, and these methods are capable of a very high degree of accuracy.
The inverse square law has very far-reaching implications. The electrostatic interaction is usually given a field-theoretic interpretation, i.e., the interaction is broken up into two factors: (i) something called the *field* is assumed to exist in space surrounding any electrical charge, and (ii) this field is considered responsible for the force on any other charge in the region. The important thing is that the field is assumed to have a reality even if there is no second charge so that there is no observable interaction. Now to use some modern terminology, the quantization of this field leads to the idea of photons, and the electrostatic interaction is said to arise from the exchange of photons. The standard field theory formulations show that the interaction involves an exponential term of the form e^{-r/r_0} where r_0 is related with the mass of the exchanged particle. Such a term makes the interaction fall off sharply with distance, being effectively negligible for $r \gg r_0$. If the interaction is really of Coulomb form, there is no such exponential fall and we are to conclude that the rest mass of the photon is zero.

(c) The force between like charges is one of repulsion and that between unlike charges is of attraction, and force varies directly as the strength of either charge. One part of this, namely the symmetry between the 'source' and the 'object' charge follows from Newton's third law. What we mean is this—suppose the force on charge e_A due to charge e_B is of the form $f(e_A)\phi(e_B)$, apart from the dependence on distance, then the force on B due to A will be of the form $f(e_B)\phi(e_A)$. From the equality of action and reaction, we have $f(e_A)\phi(e_B) = f(e_B)\phi(e_A)$. If this is to be true for arbitrary e_A and e_B, then f and ϕ must be the same function.

The question arises whether the stated proportionality of the force to the charge is merely a prescription as to how different charges are to be measured (or rather compared) or does it contain something more?

That it contains something more may be seen from the following consideration. We may give a prescription for preparing a charge, but then it is natural to assume that charges so prepared are of equal magnitude. Now put in two such point charges in contact, if the force on the combination of these two charges due to a third constant charge be double of that on a single charge due to the third charge, our idea of proportionality will be supported.

(d) The resultant force on a given point charge is the vector sum of the force due to each other charge, the force between any two charges being reckoned as if the other charges did not exist.

The above is referred to as the principle of linear superposition. Let us examine it a little. Suppose there are charges B, C, D, E, etc. Consider the force on charge A. This is

$$\mathbf{F} = \mathbf{F}_{AB} + \mathbf{F}_{AC} + \mathbf{F}_{AD} + \dots \tag{1.1}$$

where $\mathbf{F}_{AB}$ represents the force on A due to B alone and is independent of the presence of C, D, etc. This may seem at first sight to be fairly obvious and trivial, but it is really not so. This is the case in electrostatics, in the Newtonian law of gravitation, but it is not the case in Einstein's theory of gravitation (called the general theory of relativity). Mathematically, this means that the differential equations of electrostatics or Newtonian gravitation are linear while those in general relativity are not. The linear differential equations are much easier to handle, but there is an intrinsic logical incompleteness in linear field theories. Thus, it is sometimes speculated that the ultimate equations of electrostatics (or rather the equations of electromagnetism) would also be non-linear. If that be so, then this empirical principle of linear superposition must have limitations.

We have so far talked about the interaction between charges *in vacuo*. The law would require modification in case there are material media. We shall not go into that right here but note that empirical observations lead to a classification of materials into two types. In the first type, which we call conductors, the smallest conceivable electrical force would cause a motion of the charges (i.e., the electrons). This motion will go on till the electrical forces in the medium are completely annulled. Thus, in static conditions there cannot be any electrical force in a conductor. We shall take this as the definition of a conductor. Of course, all these statements are true only on a macroscopic level—there are in general electrical forces on the atomic scale even in a conductor. There may also be forces so long as the conditions are not static. The other type is called insulators or dielectrics and in them there may be electrical forces on a macroscopic scale even in the static condition. We shall return to the discussion of dielectrics later.

For the force between two point charges separated by a distance r *in vacuo*, we may thus write

$$\mathbf{F}_2 = C\frac{e_1 e_2}{r^3}\mathbf{r} \tag{1.2}$$

where e_1, e_2 are the magnitude of the two charges, $\mathbf{r}$ the vector joining them directed from e_1 to e_2, $\mathbf{F}_2$ is the force on e_2 and C is a constant which depends on the system of units used. The simplest is to take the constant C to be unity. This is, in fact, used in the Gaussian system of units. The modern practice is to use a system in which the unit of electric charge is specified. In this system, known as the *Système International* (SI), the electric charge is measured in coulomb, distance r in metres and the force $\mathbf{F}$ in newtons, so that the constant C has the units of N-m^2/C^2. In SI units, the constant of proportionality is taken as $\frac{1}{4\pi\epsilon_0}$ where ϵ_0 is known as the permittivity of the free space, and has a value $\epsilon_0 = 8.854 \times 10^{-12}$ C^2 N^{-1} m^{-2}. Later, we will see how this constant is related to other fundamental constants such as the velocity of light in vacuum and permeability of free space. We may split up the above equation into two parts:

$$\mathbf{E} = \frac{1}{4\pi\epsilon_0}\frac{e_1}{r^3}\mathbf{r} \quad \text{and} \quad \mathbf{F}_2 = e_2\mathbf{E} \tag{1.3}$$

We call $\mathbf{E}$ the electric field or the electric intensity—it may thus be defined as

$$\mathbf{E} = \lim_{e\to 0}\frac{\mathbf{F}}{e} \tag{1.4}$$

The reason for going to the limit is that if the charge e which responds to the field be finite, then it may lead to a change of $\mathbf{E}$ by causing a redistribution of the charges on the conductors that may be present in the neighbourhood. We shall give a simple example of this. Suppose there is a charged spherical conductor; in the absence of other charges, the charge will distribute itself uniformly over the surface of the sphere and the force on an infinitesimal charge at a distance r from the centre of the sphere will be

$$\mathbf{F} = \frac{1}{4\pi\epsilon_0}\frac{Qe}{r^3}\mathbf{r} \tag{1.5}$$

where Q is the charge on the sphere and e is an infinitesimal charge. If, however, e be finite, the distribution of charge over the conducting sphere will be non-uniform—if e and Q are of the same sign, the charges will flow to the farther side due to repulsion. As we shall see later, the force now can be expressed as the sum of two terms, the Coulomb-type term $Qe\mathbf{r}/r^3$ plus a term proportional to the square of e. Thus, we get the undistorted field by going over to the limit $e \to 0$. (Those who care for rigour would object that the limit $e \to 0$ is in reality unattainable for the smallest charge that we can have is the electron charge which is a finite quantity. Quarks are considered to have charges equal to one-third and two-third of the electron charge but have not been observed in free state. However, our study is restricted to macroscopic physics in which the electron charge is physically infinitesimal.)

Equation (1.5) allows us to write the electric intensity as a gradient

$$\mathbf{E} = -\nabla\phi, \qquad \phi = \frac{1}{4\pi\epsilon_0}\frac{e_1}{r} + \phi_0 \tag{1.6}$$

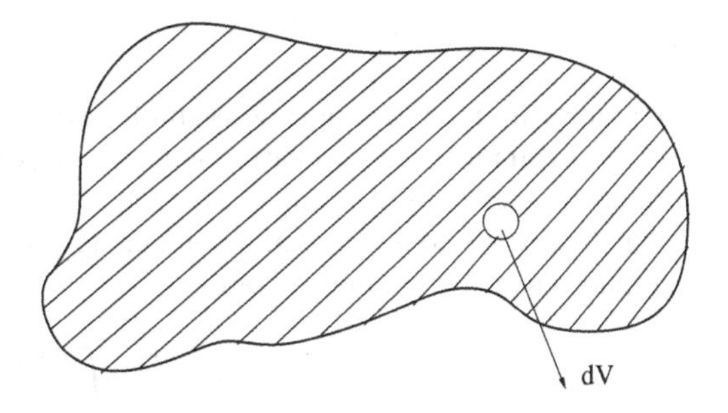

Fig. 1.1 A volume element within a continuous charge distribution

where ϕ is called the electrostatic potential. In fact, a potential function exists because of the central nature of the field irrespective of the inverse square form. Equation (1.6) determines the potential function up to an arbitrary constant ϕ_0. For cases where the charges are situated within a bounded region, it is usual to normalize the potential by the condition that it vanishes at infinity. If, however, the distribution be unbounded, e.g., an infinitely long charged cylinder or an infinite plate, this normalization cannot be used. It is readily seen that the potential function exists not only in the case of a single point source but also for any arbitrary distribution of charges—discrete or continuous. Using the linear superposition principle, we may write for the field in case of discrete charges:

$$\begin{aligned}\mathbf{E}(\mathbf{r}) &= \mathbf{E}_1 + \mathbf{E}_2 + \mathbf{E}_3 + \ldots \\ &= \frac{e_1}{4\pi\epsilon_0}\frac{(\mathbf{r}-\mathbf{r}_1)}{|\mathbf{r}_1-\mathbf{r}|^3} + \frac{e_2}{4\pi\epsilon_0}\frac{(\mathbf{r}-\mathbf{r}_2)}{|\mathbf{r}_2-\mathbf{r}|^3} + \ldots\end{aligned} \tag{1.7}$$

In case there is a continuous distribution of charge as shown in Fig. 1.1, we can divide the region into infinitesimal parts dv' in which the charge is $\rho(\mathbf{r}')dv'$. The field due to this charge at the point $\mathbf{r}$ is

$$d\mathbf{E}(\mathbf{r}) = \frac{\rho(\mathbf{r}')dv'}{4\pi\epsilon_0}\frac{(\mathbf{r}-\mathbf{r}')}{|\mathbf{r}-\mathbf{r}'|^3} \tag{1.8}$$

The total field will be obtained by integrating the above expression over entire space, i.e.,

$$\mathbf{E}(\mathbf{r}) = \int d\mathbf{E}(\mathbf{r}) = \frac{1}{4\pi\epsilon_0}\int \frac{(\mathbf{r}-\mathbf{r}')}{|\mathbf{r}-\mathbf{r}'|^3}\rho(\mathbf{r}')dv' \tag{1.9}$$

$\rho(\mathbf{r}')$ being the charge density, i.e., charge per unit volume at position $\mathbf{r}'$.

Equations (1.7) and (1.9) correspond, respectively, to distributions of discrete and continuous charges. Of course, one may conceive of situations where there are some discrete charges along with a continuous distribution. In that case, we would have an expression involving discrete terms as in (1.7) as also integrals as in (1.9). In any case, as each term in the right of (1.7) represents a central field, it can be expressed as the gradient of a scalar. The sum of such gradients will also be a gradient:

$$\mathbf{E} = -\nabla\phi_1 - \nabla\phi_2 \cdots = -\nabla(\phi_1 + \phi_2 + \ldots) = -\nabla\phi$$

The argument in the case of (1.9) is similar. Any element of the integral is a central field due to an infinitesimal charge element, and so can be written as a gradient of an infinitesimal scalar. Integrating, we get the resultant field as the gradient of a (finite) scalar. Analogous to the case of a point charge, we may write the potential due to a continuous distribution in the form

$$\phi(\mathbf{r}) = \frac{1}{4\pi\epsilon_0}\int \frac{\rho(\mathbf{r}')dv'}{|\mathbf{r} - \mathbf{r}'|}$$

However, this form is meaningful only if the integral converges. In case the distribution is not bounded, we may not use the normalizing condition $\phi \to 0$ at infinity.

Problems

1. Calculate the electric intensity due to a uniformly charged infinite wire. Show that this can be expressed as $-\nabla V'$ but the potential does not agree with $\frac{1}{4\pi\epsilon_0}\int \frac{\sigma dl}{|\mathbf{r} - \mathbf{r}'|}$, where σ is the charge per unit length of the wire and dl an element of the wire at $\mathbf{r}'$.
2. The potential at a point is sometimes defined as the work done in bringing unit positive charge from infinity to that point. Examine this definition critically.
3. Does the following represent a possible electrostatic field?

$$\mathbf{E} = \left(E_x = \frac{1}{4\pi\epsilon_0}\left(\frac{2x}{r^3} - \frac{3x^3}{r^5}\right),\ E_y = -\frac{1}{4\pi\epsilon_0}\frac{3x^2y}{r^5},\ E_z = -\frac{1}{4\pi\epsilon_0}\frac{3xyz}{r^5}\right)$$

Chapter 2
Direct Calculation of Field in Some Cases

A point charge is called a monopole. The next in simplicity is an arrangement of two equal and opposite charges separated by a distance. We shall consider the separation between the charges to be small in comparison to the distance of the field point (i.e., the point at which we calculate the field) from the charges. In a more mathematical language, we shall consider the separation ℓ between the charges to lend towards zero and the charges q to tend towards infinity in such a way that the product $q\ell$ remains finite and constant. Such an arrangement of two charges is called a dipole.

Before going to calculate the field due to a dipole, we establish a simple but useful result. The scalar r, the distance between the field point and the source point, is given by

$$r^2 = (x - x_s)^2 + (y - y_s)^2 + (z - z_s)^2$$

where (x, y, z) are the coordinates of the field point and (x_s, y_s, z_s) are those of the source point. We now have by direct differentiation

$$\frac{\partial r}{\partial x} = \frac{x - x_s}{r}, \qquad \frac{\partial r}{\partial x_s} = -\frac{x - x_s}{r}, \text{ etc.}$$

so that

$$\frac{\partial r}{\partial x} = -\frac{\partial r}{\partial x_s} \quad \text{or} \quad \nabla(r) = -\nabla_s(r)$$

Consequently, for any function $F(r)$ of r, we have

$$\frac{\partial F(r)}{\partial x_s} = \frac{dF(r)}{dr}\frac{\partial r}{\partial x_s} = -\frac{dF(r)}{dr}\frac{\partial r}{\partial x} = -\frac{\partial F(r)}{\partial x}$$

Consider now a dipole AB as in Fig. 2.1. The potential at P is given by

A. K. Raychaudhuri, *Classical Theory of Electricity and Magnetism*, Texts and Readings in Physical Sciences 21, https://doi.org/10.1007/978-981-16-8139-4_2

Fig. 2.1 An electric dipole

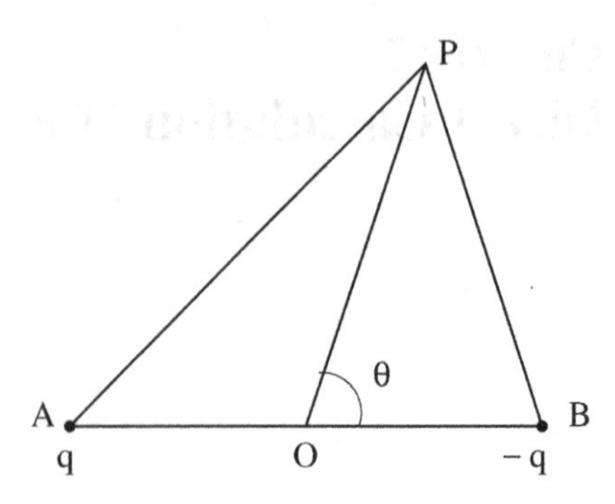

$$\phi_P = \frac{1}{4\pi\epsilon_0}\left(\frac{q}{AP} + \frac{q}{BP}\right) = \frac{q}{4\pi\epsilon_0}\mathbf{AB}\cdot\nabla_s\left(\frac{1}{r}\right)$$
$$= \frac{q}{4\pi\epsilon_0}\frac{\boldsymbol{l}\cdot\mathbf{r}}{r^3}$$

where r is the distance of the field point P from O and $\boldsymbol{l}$ is the fixed vector **AB**. (Note that in the limit $|\boldsymbol{l}| \to 0$, A, B and O coincide.) We now define $q\boldsymbol{l}$ as the dipole moment $\boldsymbol{\mu}$. It is a vector with its direction reckoned from the negative to the positive charge. So finally

$$\phi_P = \frac{1}{4\pi\epsilon_0}\frac{\boldsymbol{\mu}\cdot\mathbf{r}}{r^3} = \frac{1}{4\pi\epsilon_0}\frac{\mu\cos\theta}{r^2} \tag{2.1}$$

The electric intensity is therefore

$$\mathbf{E} = -\nabla\phi_P = \frac{1}{4\pi\epsilon_0}\left(-\frac{\boldsymbol{\mu}}{r^3} + \frac{3(\boldsymbol{\mu}\cdot\mathbf{r})}{r^5}\right) \tag{2.2}$$

While the monopole field falls off as r^{-2}, the dipole field falls off as r^{-3}, and besides, it is direction-dependent whereas the monopole field has spherical symmetry about the source. Obviously, this last difference arises from the fact that while the monopole strength is a scalar, the dipole source strength, i.e., the dipole moment, is a vector. Later, we shall consider fields due to sources whose strengths are tensors of higher rank.

2.1 Interaction Between Dipoles

The potential ϕ gives the potential energy of a unit positive charge (provided the potential vanishes at infinity—we are taking the potential energy for the unit positive charge at infinity to be zero). Thus, in an external field the potential energy of the dipole is

$$-q\phi + q\,(\phi + (\boldsymbol{l}\cdot\nabla)\phi) = \boldsymbol{\mu}\cdot\nabla\phi = -\boldsymbol{\mu}\cdot\mathbf{E}$$

It is easy to calculate the force on the dipole

$$\mathbf{F} = -q\mathbf{E} + q\,(\mathbf{E} + (\boldsymbol{l}\cdot\nabla)\mathbf{E}) = (\boldsymbol{\mu}\cdot\nabla)\mathbf{E}$$

The torque about the centre of the dipole is

$$-\frac{1}{2}\boldsymbol{l}\times(-q\mathbf{E}) + \frac{1}{2}\boldsymbol{l}\times(q(\mathbf{E}+d\mathbf{E})) = \boldsymbol{\mu}\times\mathbf{E} \tag{2.3}$$

where in the limit $l \to 0$, $d\mathbf{E}$ vanishes.

We can now have an expression for the potential energy of one dipole in the field of another dipole. This is called the interaction energy of the two dipoles.

$$\begin{aligned}\boldsymbol{\mu}_2\cdot\mathbf{E}_1 &= -\boldsymbol{\mu}_2\cdot\left(-\frac{\boldsymbol{\mu}_1}{r^3} + \frac{3(\boldsymbol{\mu}_1\cdot\mathbf{r})\mathbf{r}}{r^5}\right)\\ &= \frac{\boldsymbol{\mu}_1\cdot\boldsymbol{\mu}_2}{r^3} - 3\frac{(\boldsymbol{\mu}_1\cdot\mathbf{r})(\boldsymbol{\mu}_2\cdot\mathbf{r})}{r^5}\end{aligned} \tag{2.4}$$

The expression is symmetric in 1 and 2, as indeed it should be as it is a mutual effect.

2.2 Potential due to Surface Distribution of Charges and Dipoles

If the charge be situated on a surface, the potential due to an element dS of the surface is $\frac{\sigma dS}{4\pi\epsilon_0 r}$, where σ is the charge per unit area at dS and r is the distance of the field point from dS. The total potential is obtained by integrating over the entire charged surface. (Again, all this is valid provided the integral converges; if the charge distribution extends to infinity, this calculation of potential may not hold good but we may still have a finite electric intensity and thereby calculate a finite potential function.)

We consider the simple case of a charged plane disc of radius a and calculate the potential at a point lying on the perpendicular to the disc at its centre (see Fig. 2.2). The potential is

$$\begin{aligned}\phi &= \frac{\sigma}{2\epsilon_0}\int_0^a \frac{r\tan\theta\, d(r\tan\theta)}{r\sec\theta} = \frac{r\sigma}{2\epsilon_0}(\sec\theta - 1)\\ &= \frac{\sigma}{2\epsilon_0}\left(\sqrt{r^2+a^2} - r\right)\end{aligned}$$

As $r \to 0$, the potential tends to $\frac{\sigma a}{2\epsilon_0}$, irrespective of whether we approach the disc from the left or right. There is, therefore, no discontinuity of the potential as we

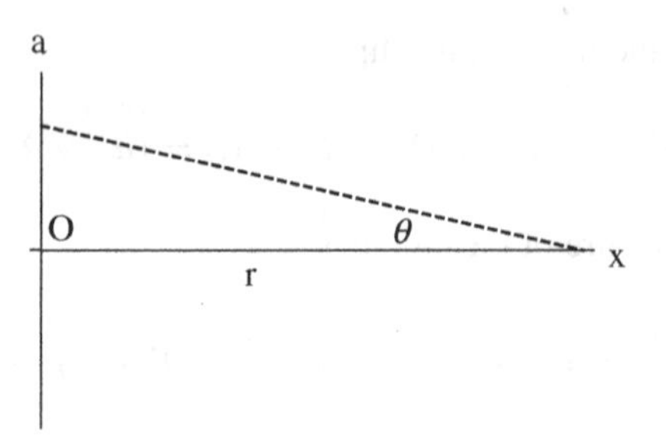

Fig. 2.2 A charged disc

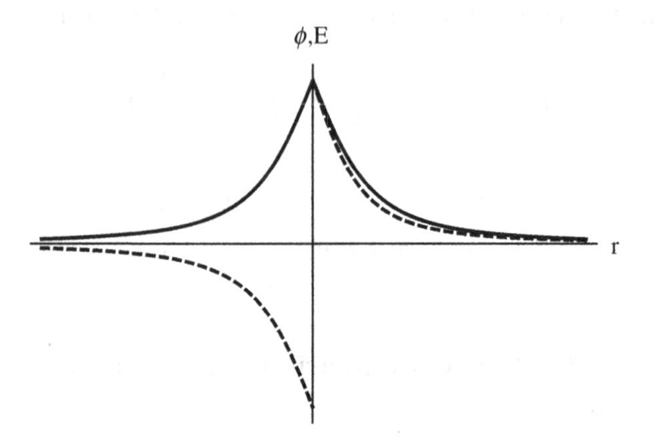

Fig. 2.3 Potential (dashed line) and electric Intensity (solid line) due to a charged disc

cross the disc. The intensity is normal to the plane and if the charges on the disc be positive, the intensity is directed away from the plane. Its magnitude is

$$\frac{\sigma}{2\epsilon_0}\int_0^{\theta}\frac{r\tan\theta d(r\tan\theta)}{r^2\sec^2\theta}\cos\theta=\frac{\sigma}{2\epsilon_0}(1-\cos\theta)=\frac{\sigma}{2\epsilon_0}\left(1-\frac{r}{\sqrt{r^2+a^2}}\right)$$

As the point X approaches the disc, the magnitude of the intensity tends to $\frac{\sigma}{2\epsilon_0}$ on either side but the direction of the intensity is opposite on two sides of the disc. Hence, there is a discontinuity in the intensity of magnitude σ/ϵ_0. This means a discontinuity in the gradient of ϕ as shown in Fig. 2.3. One may ask "What is the value of the intensity just at the centre of the disc?" The two limiting values are different, as we have just seen. However, from symmetry it seems legitimate to say that the intensity vanishes there. (The symmetry consideration is that if we consider a reflection at the disc, $x \to -x$, but there is no change in the physical situation, so the intensity should not change.)

Next, we consider the case of a distribution of dipoles. One may imagine opposite charges of equal density distributed over two parallel surfaces separated by a short

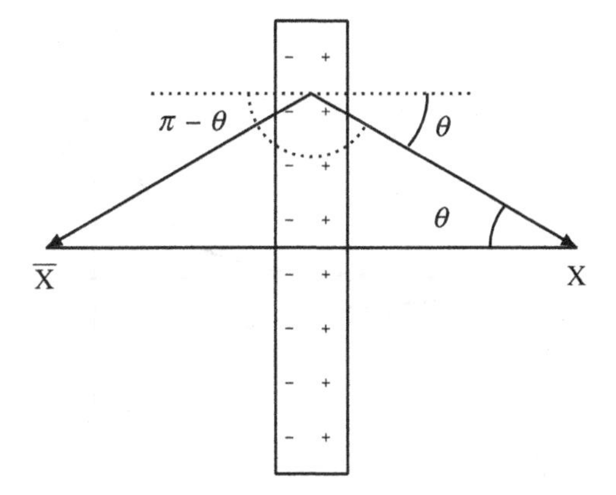

Fig. 2.4 Potential due to a distribution of dipoles

distance, e.g., over the two sides of a very thin disc. Of course, mathematically, we are interested in the limiting situation where the charge densities increase indefinitely and the separation of the surfaces tend to vanish keeping their product a finite constant. Thus, we introduce a dipole moment Ω per unit area. We have for the potential due to the annular region of the disc lying between the angles θ and $\theta + d\theta$ (see Fig. 2.4)

$$d\phi = \frac{1}{4\pi\epsilon_0}\frac{\Omega\cos\theta\, 2\pi(r\tan\theta)d(r\tan\theta)}{r^2\sec^2\theta}$$

where we have used Eq. (2.1) for the dipole potential. Therefore, the total potential at X is

$$\begin{aligned}\phi &= \frac{\Omega}{2\epsilon_0}\int_0^{\theta_0}\sin\theta\, d\theta = \frac{\Omega}{2\epsilon_0}(1-\cos\theta_0)\\ &= \frac{\Omega}{2\epsilon_0}\left(1-\frac{r}{\sqrt{r^2+a^2}}\right)\end{aligned} \tag{2.5}$$

If we had considered a point $\overline{X}$ on the opposite side of the disc, all the terms in the potential calculation would have been the same except that the angle between the dipole moment and the vector joining the element to $\overline{X}$ would be $\pi - \theta$ in place of θ. Hence, the potential would be

$$\phi = \frac{\Omega}{2\epsilon_0}\int_0^{\theta_0}\sin(\pi-\theta)d(\pi-\theta) = -\frac{\Omega}{2\epsilon_0}(1-\cos\theta_0) \tag{2.6}$$

As X and $\overline{X}$ approach the disc indefinitely, $\theta \to \pi/2$ and the jump in the potential is Ω/ϵ_0 (see Fig. 2.5a). We may calculate the intensity directly, or obtain it by differentiating the potential. Thus,

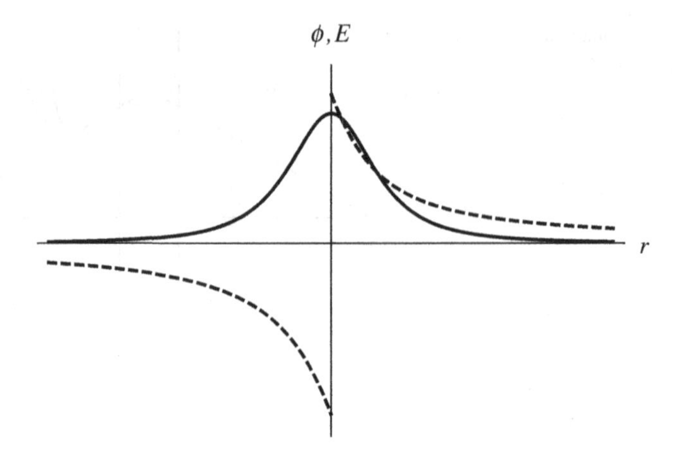

Fig. 2.5 The potential (dashed line) and the electric field (solid line) due to a dipole layer

$$E_r = -\frac{\partial \phi}{\partial r} = \frac{\Omega}{2\epsilon_0} \frac{a^2}{(a^2 + r^2)^{3/2}} \tag{2.7}$$

Apparently, therefore, the intensity does not suffer any discontinuity (see Fig. 2.5b). This, however, makes the jump in potential rather strange. We may consider the field within two layers of charge of opposite signs and separated by a finite but small distance. If the distance be sufficiently small, then from the result that we have obtained in the case of a charged disc, the field due to each layer would be $\frac{\sigma}{2\epsilon_0}$, and both directed from the positively charged layer to the negatively charged one. Hence, the resultant field would be σ/ϵ_0, and consequently, the potential jump comes out as $\frac{\sigma l}{\epsilon_0} = \frac{\Omega}{\epsilon_0}$.

The potential due to a dipole layer can be given an elegant form valid for the arbitrary shape of the dipole layer surface. Let Ω be the dipole moment per unit area directed normally to the surface element ds, then the potential at X at a distance r due to the element ds is (see Fig. 2.6)

$$d\phi = \frac{\Omega\, ds \cos\theta}{4\pi\epsilon_0 r^2}, \tag{2.8}$$

where θ is the angle between the normal to ds in the direction of Ω and r. Thus, the potential is $\Omega d\omega$, where $d\omega$ is the solid angle subtended by ds at X, $d\omega$ being reckoned positive if θ be acute. Thus, the resultant potential due to a finite surface is $\frac{1}{4\pi\epsilon_0}\int \Omega d\omega$ and in case X is indefinitely close to the surface this becomes $\frac{\Omega}{2\epsilon_0}$, whereas if X be on the opposite side, the potential is $-\frac{\Omega}{2\epsilon_0}$. Thus, quite generally, we get back the result that in crossing a dipole layer the potential undergoes a jump of Ω/ϵ_0.

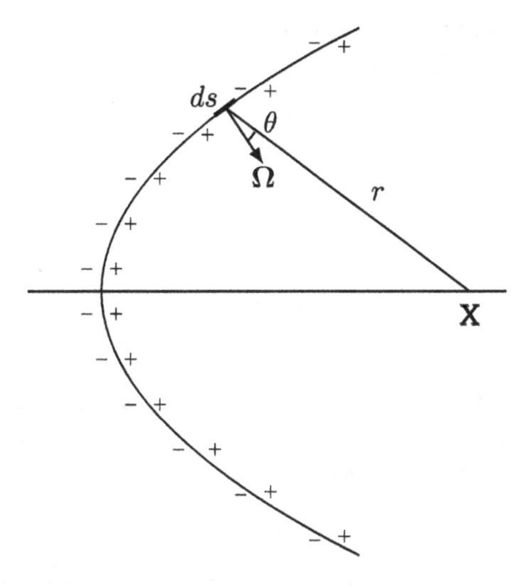

Fig. 2.6 Potential due to a surface layer of arbitrary shape

2.3 The Dirac δ-Function

We now introduce the elementary idea of the Dirac δ-function, which we shall use in the next chapter. Consider a function of a single variable x defined as follows:

$$F(x) = \begin{cases} 0 & \text{if } \mid x \mid > a/2 \\ h & \text{if } \mid x \mid < a/2 \end{cases}$$

As the region $|x| < \frac{a}{2}$ shrinks and the value of $F(x)$ in this region increases so that the product ha remains constant. Ultimately when $a \to 0$ and $h \to \infty$ but still $ha = 1$, the resulting function $F(x)$ is called the δ-function. Obviously, the δ-function vanishes everywhere except at a certain unique value of the argument where it does not exist. Nevertheless, its integral over a domain containing the singular point is unity. In short, if $f(x) = 0$ for $x \neq 0$ and $\int_{-\infty}^{\infty} f(x)dx = 1$, the 'function' $f(x)$ is called the delta function and is written as $\delta(x)$. We can cite some examples. Suppose a perfectly elastic ball, moving horizontally with a uniform velocity u, strikes a smooth vertical wall so that the ball rebounds and moves with a velocity $-u$. Obviously, the acceleration of the ball vanishes at all instants except at that of the impact. Writing $f(t)$ for the acceleration, $f(t) = 0$ if $t \neq 0$, where $t = 0$ is the instant of impact. Moreover, $2u$ being the total change in velocity, $-2u = \int_{-\infty}^{\infty} f(t)dt$. Hence, recalling the definition of the delta function (see Fig. 2.7), we may write $f(t) = -2u\delta(t)$.

As a second example, take the case of the string plucked exactly at a point so that its shape is as shown in Fig. 2.8. Calling y the displacement of any point on the string and x the distance of a point on the string from the fixed point A, if $x = a$ is

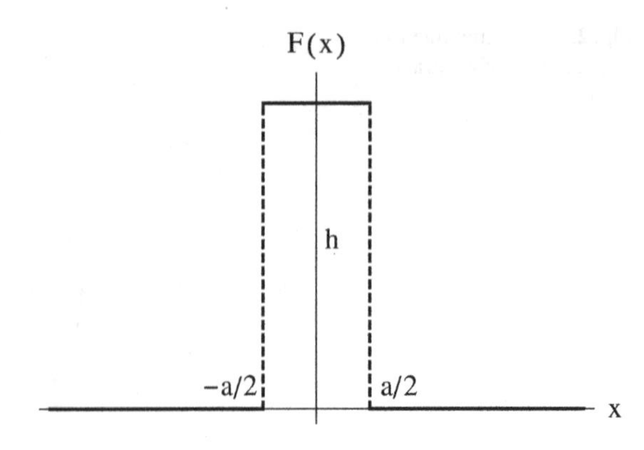

Fig. 2.7 A representation of Dirac δ-function

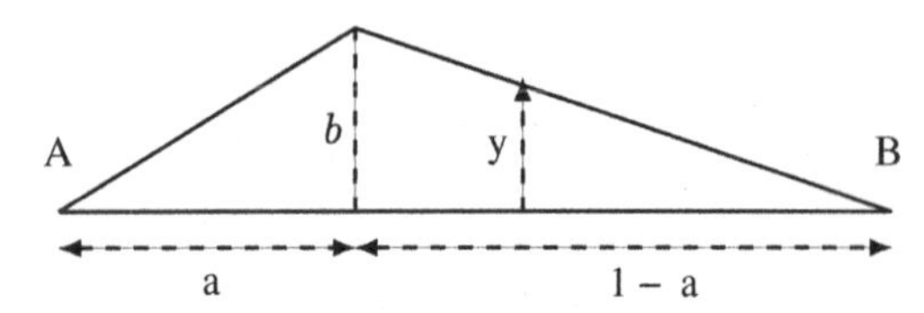

Fig. 2.8 A plucked string

the plucked point and $y = b$ is the displacement there, we have

$$\frac{dy}{dx} = \frac{b}{a} \text{ for } x < a, \qquad \text{and} \qquad \frac{dy}{dx} = -\frac{b}{\ell - a} \text{ for } x > a$$

where $\ell = AB$ is the length of the string. We note that the second derivative, d^2y/dx^2, vanishes everywhere except at the plucked point $x = a$ where it blows up. However, noting the difference in dy/dx on the two sides, we can write

$$\frac{d^2y}{dx^2} = 0 \text{ if } x \neq a, \qquad \text{and} \qquad \int_0^{\ell} \frac{d^2y}{dx^2} dx = -\frac{bl}{a(\ell - a)}$$

Again, we may give all this information by introducing the δ-function and writing

$$\frac{d^2y}{dx^2} = -\frac{bl}{a(l - a)} \delta(x - a)$$

Of course in the integral $\int_{-\infty}^{\infty} \delta(x)dx$, only the region containing $x = 0$ has a contribution. Hence, instead of taking the limits of the integral as $\pm\infty$, one can take any limit which includes the point $x = 0$. Further, consider an integral $\displaystyle\int_{-\epsilon}^{\epsilon} g(x)\delta(x)dx$ where $g(x)$ is a function, which is finite in the domain $-\epsilon \leq x \leq \epsilon$. As the contribution to the integral comes only from $x = 0$, due to the properties of the delta function, the integral will be $g(0)\int_{-\epsilon}^{\epsilon}\delta(x)dx = g(0)$, which is the value of the function $g(x)$ at $x = 0$.

It is possible to extend the idea of delta function to more than one variable. Of particular use will be the delta function $\delta(\mathbf{r})$ of the three-dimensional position vector $\mathbf{r}$. We shall define it in the following manner: $\delta(\mathbf{r})$ vanishes if $\mathbf{r} \neq 0$, i.e., unless all the three coordinates x, y and z vanish, and that the function blows up at $\mathbf{r} = 0$ (*i.e.*, at the origin) in such a manner that $\int \delta(\mathbf{r})dv = 1$, where dv is the volume differential $dxdydz$ and the domain of integration includes the origin.

For our purposes, a relation between the Laplacian operator

$$\nabla^2 \equiv \frac{\partial^2}{\partial x^2} + \frac{\partial^2}{\partial y^2} + \frac{\partial^2}{\partial z^2}$$

and the delta function will be particularly useful. Consider the expression $\nabla^2(1/r)$ where r is given by $r^2 = x^2 + y^2 + z^2$. By direct differentiation, we get

$$\nabla^2\left(\frac{1}{r}\right) = -\frac{3}{r^3} + \frac{3(x^2 + y^2 + z^2)}{r^5} \tag{2.9}$$

Thus, $\nabla^2(1/r)$ vanishes except at $\mathbf{r} = 0$. To have an idea of its behaviour at $\mathbf{r} = 0$, consider the integral

$$\lim_{a\to 0}\int \nabla^2\left(\frac{1}{\sqrt{r^2 + a^2}}\right) dv$$

A straightforward calculation gives

$$\lim_{a\to 0}\int \nabla^2\left(\frac{1}{\sqrt{r^2 + a^2}}\right) dv = -\lim_{a\to 0}\int \frac{3a^2}{(r^2 + a^2)^{5/2}} dv = -4\pi$$

More generally, define a function $f(r, \alpha)$ such that for the limiting value $\alpha \to 0$, $f(r, \alpha) \to \frac{1}{r}$ but with $\alpha \neq 0$, $f(r, \alpha)$ is regular at the origin. Then

$$\int \left(\nabla^2 f(r, \alpha)\right) dv = \int \nabla f(r, \alpha) \cdot \hat{\mathbf{n}}\, ds = 4\pi a^2 \left[\frac{df(r, \alpha)}{dr}\right]_{r=a}$$

if the surface of integration be a sphere of radius a. In the limit $\alpha \to 0$, $\frac{df(r, \alpha)}{dr} = \frac{d}{dr}\left(\frac{1}{r}\right) = -\frac{1}{r^2}$. Hence,

$$\int \nabla^2\left(\frac{1}{r}\right) dv = -4\pi \tag{2.10}$$

We may sum up these properties of $\nabla^2(1/r)$ by writing

$$\nabla^2\left(\frac{1}{r}\right) = -4\pi\delta(\mathbf{r})$$

The function $\delta(\mathbf{r})$ allows us to represent discrete distributions in a novel manner. Suppose we have a point mass (or charge) of magnitude m (respectively e) at the origin. If we introduce a density or mass (respectively charge) ρ, then obviously ρ vanishes everywhere except at the origin where it blows up. However, as the mass (or charge) is finite, we may say that the volume integral of ρ is equal to m (respectively e). Thus, we may write $\rho = m\delta(\mathbf{r})$ (respectively $\rho = e\delta(\mathbf{r})$). If instead of being at the origin, the mass (or charge) be at the position $\mathbf{r}_0$, the corresponding δ-function will be $\delta(\mathbf{r} - \mathbf{r}_0)$.

Problems

1. You are given $E_x = -\dfrac{\beta xy}{r^5}$, $E_y = \dfrac{\alpha}{r^3} - \dfrac{\beta y^2}{r^5}$ and $E_z = \dfrac{\beta yz}{r^5}$ for an electrostatic field. Find the relation between α and β. Also identify the sources of the field.
2. Two coplanar dipoles A and B have their centres fixed. Dipole B can rotate about its centre, while A is completely fixed. Show that B will be in equilibrium making an angle θ' with the line $\mathbf{r}$ joining the centres such that $\tan\theta = -2\tan\theta'$, where θ is the angle between A and $\mathbf{r}$.
3. A system is stable towards a particular displacement from equilibrium condition if the potential energy increases for such a displacement, and unstable if the potential energy decreases. Find the position of equilibrium for a charge placed in the field of two equal charges of the same sign placed at a distance. Discuss the stability for different displacements.

Chapter 3
Gauss' Theorem, Laplace and Poisson Equations

An important quantity associated with fields is the flux through a surface. The term is borrowed from fluid mechanics and appears as a measure of the amount of fluid flowing out through a surface. Thus, it is given by $\rho \mathbf{u} \cdot \hat{\mathbf{n}} ds$, where ρ is the density of the fluid, $\mathbf{u}$ its velocity and $\hat{\mathbf{n}} ds$ the oriented element of area through which the flux of fluid is being considered. The oriented area element $\hat{\mathbf{n}} ds$ is considered a vector in the direction of its outward drawn normal, thus effectively the component of the fluid velocity in the direction of the outward drawn normal matters in the flux calculation. Analogously, the flux of any vector field is defined. For the electric field, the flux is $\mathbf{E} \cdot \hat{\mathbf{n}} ds$. In case the flux over an entire closed surface is taken, it is usual to write it as $\oint \mathbf{E} \cdot \hat{\mathbf{n}} ds$. Gauss' theorem may now be enunciated thus: The flux of an electric field through any closed surface equals $1/\epsilon_0$ times the total charge contained inside the surface. Two points may be emphasized: First, it is immaterial whether we consider discrete charges or continuous distributions, and second, charges outside the region bounded by the surface make no contribution whatsoever to the flux integral. We now give a proof of Gauss' theorem.

Let e be a point charge situated at P within the closed surface, as in the left figure in Fig. 3.1. Using Coulomb's law, we get

$$\oint \mathbf{E} \cdot \hat{\mathbf{n}} ds = \frac{1}{4\pi \epsilon_0} \oint \frac{e}{r^2} ds \cos\theta = \frac{1}{4\pi \epsilon_0} \times 4\pi e = \frac{e}{\epsilon_0} \tag{3.1}$$

If, however, the point charge be outside as in the right figure in Fig. 3.1, the flux over the elements ds_1 and ds_2 cancel each other. As the whole surface would be divided into such pairs of elements, the total flux due to such a charge would vanish. Now within the surface, there may be both discrete and continuous distribution of charges. The continuous distribution may be broken up into infinitesimal elements so that it is equivalent to an infinite number of point charges. From the linear superposition

A. K. Raychaudhuri, *Classical Theory of Electricity and Magnetism*, Texts and Readings in Physical Sciences 21, https://doi.org/10.1007/978-981-16-8139-4_3

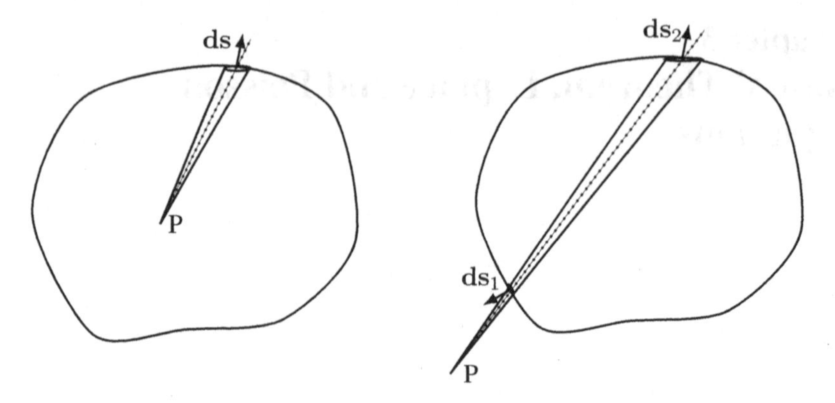

Fig. 3.1 Flux through a closed surface (L) enclosing a charge, and (R) due to a charge outside the surface

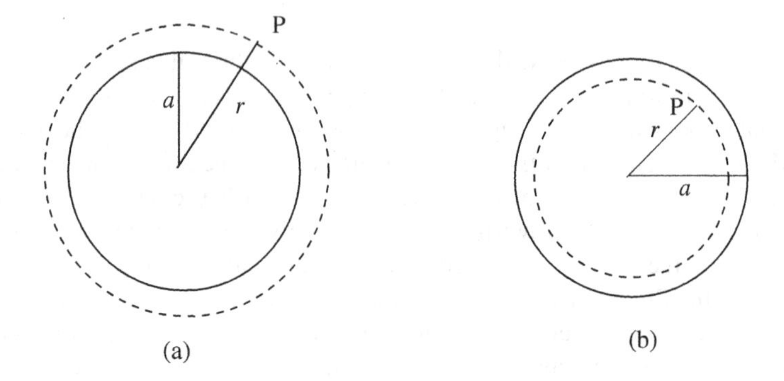

Fig. 3.2 Electric field due to a uniformly charged sphere **a** outside **b** inside

principle, we see that the flux due to this combination of charges will be simply equal to the sum of flux due to each individual charge. Hence

$$\oint \mathbf{E} \cdot \hat{\mathbf{n}} ds = \oint \mathbf{E}_1 \cdot \hat{\mathbf{n}} ds + \oint \mathbf{E}_2 \cdot \hat{\mathbf{n}} ds = \frac{1}{\epsilon_0} \sum_i e_i + \frac{1}{\epsilon_0} \int \rho dv = \frac{Q}{\epsilon_0} \tag{3.2}$$

which proves the theorem.

Gauss' theorem allows us to find the field very simply in many cases, especially where there is a high degree of symmetry. We give several illustrations.

I: *Field inside and outside a centrally symmetric distribution of charge*
If the field is to be calculated at an outside point P, we consider a concentric sphere through P, as shown in Fig. 3.2a. By symmetry, the field at P must be in the radial direction and the field has the same magnitude at all points on the sphere through P. Hence, the flux through the sphere through P is

$$\oint \mathbf{E} \cdot \hat{\mathbf{n}} ds = \mathbf{E} \cdot \oint \hat{\mathbf{n}} ds = 4\pi r^2 E \tag{3.3}$$

Using Gauss' theorem

$$4\pi r^2 E = \frac{1}{\epsilon_0} \int \rho \, dv = \frac{Q}{\epsilon_0}, \quad \text{or} \quad E = \frac{Q}{4\pi \epsilon_0 r^2} \tag{3.4}$$

In case P is inside, at a distance r from the centre of the sphere, we get similarly (see Fig. 3.2b)

$$4\pi r^2 E = \frac{1}{\epsilon_0} \int_0^r \rho dv = \frac{4\pi}{\epsilon_0} \int_0^r \rho r^2 dr \quad \text{or} \quad E = \frac{Q_r}{4\pi \epsilon_0 r^2} \tag{3.5}$$

where Q and Q_r indicate, respectively, the total charge and the charge within a radius r. In case the charge density be a constant, the field will increase linearly with r inside the charge distribution, attain a maximum value at the boundary of the distribution and then decrease, monotonically, inversely as the square of r.

II: *A uniformly charged infinite cylinder*
To find the field at a point outside the cylinder, note that symmetry requires the field to be radial. The azimuthal component E_ϕ vanishes because of irrotational character and E_z vanishes because of reflection symmetry about any plane normal to the z-axis. Now apply Gauss' theorem to a cylinder passing through the point and bounded by flat surfaces perpendicular to the axis of the cylinder (see the left part of Fig. 3.3). The flux over the flat surfaces vanishes, so that we get

$$\oint \mathbf{E} \cdot \hat{\mathbf{n}} ds = 2\pi r l E = \frac{\sigma l}{\epsilon_0} \quad \text{or} \quad E = \frac{\sigma}{2\pi \epsilon_0 r}$$

where σ is the charge per unit length of the cylinder.

III: *Infinite charged conducting plane*
Gauss' theorem is applied to a rectangular parallelepiped, whose one face (parallel to the conducting plane) passes through the point P and the opposite face is within the conductor (see the right figure in Fig. 3.3). Taking into account the vanishing of electrical intensity within the conductor and the symmetry requirement that the intensity outside is everywhere directed in the direction normal to the plane, we get for the intensity $E = \dfrac{\sigma}{\epsilon_0}$.

3.1 Equations of Laplace and Poisson

There is an important theorem in vector calculus, also known after Gauss. It states that the volume integral of the divergence of a vector is equal to the flux of the vector through the bounding surface of the volume, i.e.,

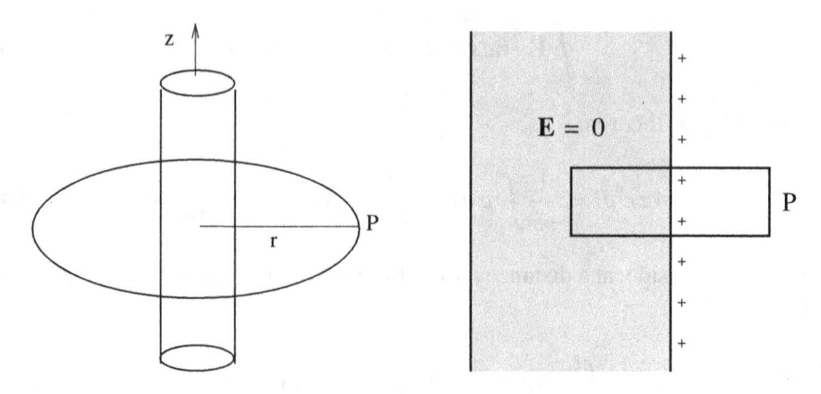

Fig. 3.3 Field due to a uniformly charged wire (left), and conducting plane (right)

$$\int \nabla \cdot \mathbf{A}\, dv = \oint \mathbf{A} \cdot \hat{\mathbf{n}} ds, \qquad \text{where} \;\; \nabla \cdot \mathbf{A} = \frac{\partial A_x}{\partial x} + \frac{\partial A_y}{\partial y} + \frac{\partial A_z}{\partial z}$$

Combining this divergence relation with Gauss' theorem in electrostatics in case of a continuous distribution of charges, we get

$$\int \nabla \cdot \mathbf{E}\, dv = \oint \mathbf{E} \cdot \hat{\mathbf{n}} ds = \frac{1}{\epsilon_0} \int \rho\, dv$$

As the above holds good for arbitrary regions of integration, we must have an equality of the integrands

$$\nabla \cdot \mathbf{E} = \frac{\rho}{\epsilon_0} \tag{3.6}$$

Equivalently, in terms of the electrostatic potential

$$\nabla^2 \phi = -\frac{\rho}{\epsilon_0} \tag{3.7}$$

The above equation is known as Poisson's equation, while in charge-free space it becomes the Laplace equation

$$\nabla^2 \phi = 0 \tag{3.8}$$

We know that the potential at a point $\mathbf{r}$ due to a point charge e at $\mathbf{r}'$ is

$$\phi(\mathbf{r}) = \frac{1}{4\pi\epsilon_0} \frac{e}{|\mathbf{r} - \mathbf{r}'|}$$

which is a solution of the equation

$$\nabla^2 \phi = -\frac{1}{\epsilon_0} e\, \delta(\mathbf{r} - \mathbf{r}')$$

Hence, Poisson's equation has the formal solution

$$\phi(\mathbf{r}) = \frac{1}{4\pi\epsilon_0}\int \frac{\rho(\mathbf{r}')}{|\mathbf{r}-\mathbf{r}'|}dv' \tag{3.9}$$

By direct differentiation and using the properties of the δ-function, one can verify that (3.9) is a solution of Poisson's equation (3.7). For a formal derivation of (3.9), see Chap. 14.

However, as we have seen earlier there may be cases where this integral diverges and thus (3.9) ceases to be meaningful. But one has still a well-defined $\mathbf{E}$ and can construct a function ϕ satisfying $\mathbf{E} = -\nabla\phi$. (Check these remarks in case of the infinite conducting plane: the electric field $E_x = \sigma/\epsilon_0$ and $\phi = -\sigma x/\epsilon_0$, but ϕ calculated from the Poisson integral diverges.)

3.1.1 *Green's Theorem in Vector Calculus*

If for a vector $\mathbf{A}$, one takes $\mathbf{A} = \psi\nabla\phi$, then Gauss' theorem gives

$$\int \left(\psi\nabla^2\phi + \nabla\psi\cdot\nabla\phi\right)dv = \oint \psi\nabla\phi\cdot\hat{\mathbf{n}}ds \tag{3.10}$$

As in the above ψ and ϕ are arbitrary scalars, we may interchange them to obtain

$$\int \left(\phi\nabla^2\psi + \nabla\phi\cdot\nabla\psi\right)dv = \oint \phi\nabla\psi\cdot\hat{\mathbf{n}}ds \tag{3.11}$$

Subtracting (3.11) from (3.10), we obtain

$$\int \left(\psi\nabla^2\phi - \phi\nabla^2\psi\right)dv = \oint (\psi\nabla\phi - \phi\nabla\psi)\cdot\hat{\mathbf{n}}ds \tag{3.12}$$

Relation (3.12) (and sometime also (3.10)) are referred to as Green's theorem.

3.1.2 *A Formula of Interest in Field Theory*

Consider a surface S enclosing the point P (i.e., a point at which we are interested in finding the value of the field). In Green's theorem, we take ϕ for the potential function and $\psi = 1/r$ (where r is the distance from P). We get

$$\int \frac{1}{r}\nabla^2\phi\, dv - \int \phi\,\nabla^2\left(\frac{1}{r}\right)dv = \oint \frac{1}{r}\nabla\phi\cdot\hat{\mathbf{n}}ds - \oint \phi\,\nabla\left(\frac{1}{r}\right)\cdot\hat{\mathbf{n}}ds$$

which gives

$$\int \phi \nabla^2\left(\frac{1}{r}\right) = \int \frac{1}{r}\nabla^2\phi\, dv - \oint \frac{1}{r}\nabla\phi\cdot\hat{\mathbf{n}}ds + \oint \phi\nabla\left(\frac{1}{r}\right)\cdot\hat{\mathbf{n}}ds$$

As we have seen $\nabla^2(\frac{1}{r}) = -4\pi\delta(\mathbf{r})$, the integral on the left gives $-4\pi\phi_P$, therefore, using Gauss' law in the first term on the right

$$\phi_P = \frac{1}{4\pi\epsilon_0}\int \frac{\rho}{r}dv + \frac{1}{4\pi}\oint \frac{1}{r}\nabla\phi\cdot\hat{\mathbf{n}}ds - \frac{1}{4\pi}\oint \phi\nabla\left(\frac{1}{r}\right)\cdot\hat{\mathbf{n}}ds \tag{3.13}$$

Equation (3.13) may be given a field-theoretic interpretation. The potential ϕ is due to the charge distribution extending over all space. However, (3.13) shows that the influence of all charges outside S may be reduced to a surface distribution of density $\epsilon_0(\boldsymbol{\nabla}\boldsymbol{\phi})_n$ (where the subscript refers to the component normal to the surface) over S plus a distribution of dipoles of moment $-\epsilon_0\phi$ per unit area directed normally to S.

3.1.3 Earnshaw's Theorem

We consider a sphere S of radius a with its centre at the origin. From Green's theorem, we get

$$\int \phi\nabla^2\left(\frac{1}{r}\right)dv - \int \frac{1}{r}\nabla^2\phi\, dv = \oint_{r=a} \phi\nabla\left(\frac{1}{r}\right)\cdot\hat{\mathbf{n}}ds - \frac{1}{a}\oint \nabla\phi\cdot\hat{\mathbf{n}}ds$$

We have already seen that the first integral on the left becomes $-4\pi\phi_P$, while the second integral vanishes, if the region be charge-free. On the right, the second integral becomes $1/a$ times the flux through the spherical surface, and this also vanishes in case the region is charge-free. Thus, we get

$$\phi_P = \frac{1}{4\pi a^2}\oint_{r=a} \phi\, dS$$

The relation shows that the average potential over the surface of the sphere S is equal to the potential at the centre, provided there is no charge within S.

Hence, it is clear that at any regular point, the electrostatic potential can have neither a maximum nor a minimum. As the potential is a measure of the potential energy, we conclude that there is no position of stable equilibrium for a charge in an electrostatic field. This statement is known as Earnshaw's theorem. (Note that ϕ gives the potential energy of a positive charge and $-\phi$ that of a negative charge, hence the condition of minimum potential energy may correspond to minimum or maximum of potential depending on the sign of the charge.)

Problems

1. Show that the total charge in the field is zero if the potential vanishes more rapidly than $1/r$, non-zero and finite if the potential varies as $1/r$, and infinite if the potential does not vanish as r goes to infinity.
2. Calculate the charge distribution for the potential $\phi = \dfrac{e^{-\alpha r}}{4\pi\epsilon_0 r}$.
3. Considering an infinite plane sheet of surface charge of density σ, show that the field is $\sigma/(2\epsilon_0)$. Explain how the field for a conducting sheet has double the value.
4. Calculate the field and the charge distribution if

$$\phi = \sin\left(\frac{\pi x}{a}\right)\sin\left(\frac{\pi y}{b}\right)\sin\left(\frac{\pi z}{c}\right)$$

for $0 \le x \le a, 0 \le y \le b$ and $0 \le z \le c$. Give a physical picture of the situation.

Problems

1. Show that the total charge in the [illegible] is zero if the potential vanishes faster [illegible] rapidly [illegible] and [illegible] of the potential [illegible] infinite [illegible] which as [illegible] to infinity.
2. [illegible] distribution for the potential [illegible]
3. [illegible] surface [illegible] [illegible] the [illegible] the charge [illegible] it.

$$[illegible]$$

[illegible]

Chapter 4
Analysis of the Electrostatic Field: Multipole Moments

We suppose that the charges are contained in a bounded region and investigate the potential at a point outside that region. We take the origin within the charge distribution. The potential is given by Eq. (3.9) in Chap. 3. We expand the function $f(\mathbf{r}') \equiv 1/|\mathbf{r}-\mathbf{r}'|$ in a Taylor series about the origin to write

$$\frac{1}{|\mathbf{r}-\mathbf{r}'|} = \frac{1}{r} + \sum_{i=1}^{3} x_i' \left[\frac{\partial}{\partial x_i'}\left(\frac{1}{|\mathbf{r}-\mathbf{r}'|}\right)\right]_0 + \frac{1}{2}\sum_i \sum_k x_i' x_k' \left[\frac{\partial^2}{\partial x_i' \partial x_k'}\left(\frac{1}{|\mathbf{r}-\mathbf{r}'|}\right)\right]_0 + \cdots$$

where x_1', x_2' and x_3' stand for x', y' and z', respectively, and the subscript zero indicates that the derivatives are evaluated at the origin. The potential can be written as

$$\phi(\mathbf{r}) = \frac{1}{4\pi\epsilon_0}\frac{1}{r}\int \rho\, dv' + \frac{1}{4\pi\epsilon_0}\sum_{i=1}^{3}\left[\left(\frac{1}{|\mathbf{r}-\mathbf{r}'|}\right)\right]_0 \int \rho\, x_i' dv' + \cdots$$

The general term in this expansion

$$\frac{1}{4\pi\epsilon_0}\sum_{i_1}\cdots\sum_{i_n}\frac{1}{n!}\left[\frac{\partial^n}{\partial x_{i_1}'\cdots\partial x_{i_n}'}\left(\frac{1}{|\mathbf{r}-\mathbf{r}'|}\right)\right]_0 \int \rho\, x_{i_1}'\cdots x_{i_n}' dv' \tag{4.1}$$

is called the 2^n-pole term. The term outside the integral is obviously a function of the coordinates of the field point only and the integral is a characteristic of the source distribution.

Let us examine the terms individually. The first term is isotropic (i.e., independent of the angular position of the field point), and varies inversely as the distance from the

A. K. Raychaudhuri, *Classical Theory of Electricity and Magnetism*, Texts and Readings in Physical Sciences 21, https://doi.org/10.1007/978-981-16-8139-4_4

origin, while the source part $\int \rho \, dv'$ is simply the total charge. This term, called the monopole term, is the same as the potential due to the total charge of the distribution concentrated at the origin. The second term can be written as

$$\phi^{(2)}(\mathbf{r}) = -\frac{1}{4\pi\epsilon_0}\left(\int \rho \, \mathbf{r}' dv'\right) \cdot \nabla \left(\frac{1}{r}\right) \tag{4.2}$$

This is the same as the potential due to a dipole of moment $\int \rho \, \mathbf{r}' dv'$ placed at the origin. The dipole moment varies inversely as the square of the distance and has a characteristic angular dependence. If the origin be shifted to $\mathbf{r}_0$, the dipole moment of the charge distribution will be

$$\int \rho \, (\mathbf{r}' - \mathbf{r}_0) dv' = \int \rho \, \mathbf{r}' dv' - \mathbf{r}_0 \int \rho \, dv'$$

If the total charge $\int \rho \, dv'$ does not vanish, the dipole moment is not independent of choice of origin and indeed may be made to vanish if we choose

$$\mathbf{r}_0 = \frac{1}{\int \rho \, dv'} \int \rho \, \mathbf{r}' dv'$$

Borrowing a term from mechanics, we may call the new origin the centroid of the charge distribution.

The next term is called the quadrupole term. It has the form

$$\phi^{(4)}(\mathbf{r}) = \frac{1}{4\pi\epsilon_0} \sum_i \sum_k \frac{1}{2} \left[\frac{\partial^2}{\partial x_i' \partial x_k'}\left(\frac{1}{|\mathbf{r} - \mathbf{r}'|}\right)\right]_0 \int \rho x_i' x_k' dv' \tag{4.3}$$

The falling off of the quadrupole term is thus even faster—varying inversely as the cube of the distance. The source term $Q_{ik} = \int \rho x_i' x_k' dv'$ has apparently six independent terms as $Q_{ik} = Q_{ki}$. It constitutes a tensor of rank two and just as any vector may be reduced to having only one non-vanishing component by a suitable choice of axes, a symmetric tensor Q_{ik} may be reduced to have only three non-vanishing components by a suitable choice of axes. This is referred to as the diagonalization of the tensor, or transformation to principal axes of the tensor. In that case, we shall have only the components Q_{11}, Q_{22} and Q_{33}. In this particular case, however, the relevant number of components is further reduced to two, since after diagonalization, the quadrupole term is

$$\frac{1}{4\pi\epsilon_0}\frac{1}{2}\left(\frac{\partial^2}{\partial x'^2}\frac{1}{|\mathbf{r} - \mathbf{r}'|}\bigg|_0 Q_{11} + \frac{\partial^2}{\partial y'^2}\frac{1}{|\mathbf{r} - \mathbf{r}'|}\bigg|_0 Q_{22} + \frac{\partial^2}{\partial z'^2}\frac{1}{|\mathbf{r} - \mathbf{r}'|}\bigg|_0 Q_{33}\right)$$

But $\nabla'^2 \left(\frac{1}{|\mathbf{r} - \mathbf{r}'|}\right) = 0$, hence the above may be rewritten as

$$\phi^{(4)}(\mathbf{r}) = \frac{1}{8\pi\epsilon_0}\left(\frac{\partial^2}{\partial y'^2}\frac{1}{|\mathbf{r}-\mathbf{r}'|}\bigg|_0 (Q_{22}-Q_{11}) + \frac{\partial^2}{\partial z'^2}\frac{1}{|\mathbf{r}-\mathbf{r}'|}\bigg|_0 (Q_{33}-Q_{11})\right) \tag{4.4}$$

Thus, only the departure from isotropy as appearing in $Q_{11} - Q_{22}$ and $Q_{33} - Q_{11}$ are effective and the two numbers suffice to describe the quadrupole effects, provided the principal axes are known. (Just as we formed the idea of dipole moment by considering two discrete charges, we may have a quadrupole field by taking four discrete charges.)

We may proceed with the discussion of further terms, however, it is important to note that each term appears as a product of a power of $1/r$ and a function of the angular coordinates of the field point, besides the coefficients which are constants so far as the coordinates of the field point are concerned. Again, as the variable parts appear as partial derivatives of $1/r$, each term separately is a solution of Laplace's equation. Taking a hint from these facts, we shall solve Laplace's equation by separation of variables in spherical polar coordinates and recover thereby the multipole expansion that we have just now studied.

In spherical polar coordinates (r, θ, ψ), Laplace's equation $\nabla^2\phi = 0$ assumes the form

$$\frac{1}{r^2}\frac{\partial}{\partial r}\left(r^2\frac{\partial\phi}{\partial r}\right) + \frac{1}{r^2\sin\theta}\frac{\partial}{\partial\theta}\left(\sin\theta\frac{\partial\phi}{\partial\theta}\right) + \frac{1}{r^2\sin^2\theta}\frac{\partial^2\phi}{\partial\psi^2} = 0 \tag{4.5}$$

We try a solution in the form $\phi = R(r)\Theta(\theta)\Psi(\psi)$, where each of R, Θ and Ψ is a function of one variable alone, r, θ and ψ, respectively. Substituting in Eq. (5.16) and dividing by $R(r)\Theta(\theta)\Psi(\psi)$, we get

$$\left[\frac{1}{R}\frac{d}{dr}\left(r^2\frac{dR}{dr}\right) + \frac{1}{\Theta\sin\theta}\frac{d}{d\theta}\left(\sin\theta\frac{d\Theta}{d\theta}\right)\right]\sin^2\theta = -\frac{1}{\Psi}\frac{d^2\Psi}{d\psi^2} \tag{4.6}$$

As the left-hand side of (4.6) is a function of r, θ (but does not involve ψ) and the right-hand side is a function of ψ alone, each side must be equal to a constant, say m^2 (such constants are called separation constants). We then have from the right-hand side

$$\Psi(\psi) = Ae^{\pm im\psi} \tag{4.7}$$

Since ψ is an angular coordinate, any change of ψ by an integral multiple of 2π does not change the position of the point. Hence, as the potential must be single-valued, we must have m restricted to integers (including zero).

Next, we split the left-hand side of (4.6) to obtain

$$\frac{1}{R}\frac{d}{dr}\left(r^2\frac{dR}{dr}\right) = -\frac{1}{\Theta}\left[\frac{1}{\sin\theta}\frac{d}{d\theta}\left(\sin\theta\frac{d\Theta}{d\theta}\right) - \frac{m^2}{\sin^2\theta}\Theta\right] \tag{4.8}$$

Again, one side is a function of r alone and the other side of θ alone, so each side must be equal to a constant. Writing the constant as $l(l+1)$, we get

$$\frac{d}{dr}\left(r^2\frac{dR}{dr}\right) - l(l+1)R = 0 \tag{4.9}$$

$$\frac{1}{\sin\theta}\frac{d}{d\theta}\left(\sin\theta\frac{d\Theta}{d\theta}\right) + \left(l(l+1) - \frac{m^2}{\sin^2\theta}\right)\Theta = 0 \tag{4.10}$$

Introducing a new variable $x = \cos\theta$, Eq. (4.10) becomes

$$\frac{d}{dx}\left((1-x^2)\frac{d\Theta}{dx}\right) + \left(l(l+1) - \frac{m^2}{1-x^2}\right)\Theta = 0 \tag{4.11}$$

in the domain $-1 \le x \le 1$. A study of Eq. (4.11) shows that Θ is finite at $x = \pm 1$ only if l is a positive integer (including zero). The solutions of Eq. (4.11) thus involve two integers l and m and are written as $P_l^m(x)$. These functions are known as associated Legendre polynomials. There is another constraint on l and m, namely $l \ge |m|$, otherwise the functions $P_l^m(x)$ vanish identically. Equation (4.9) is readily integrated to give

$$R(r) = Cr^{-(l+1)} + Dr^l$$

Collecting the expressions of R, Θ and Ψ, we have a typical solution of Laplace's equation in the form

$$Ar^{-(l+1)}P_l^m(\cos\theta)e^{\pm im\psi}$$

where we have not taken the term which diverges as r goes to infinity.

As Laplace's equation is linear, linear combination of solutions will also be a solution, therefore, the general solution which is regular everywhere except at the origin and is also single-valued may be written in the form

$$\phi(r,\theta,\psi) = \frac{1}{4\pi\epsilon_0}\sum_{l=0}^{\infty}\sum_{m=-l}^{l}\frac{Q_{lm}}{r^{l+1}}P_l^m(\cos\theta)\ (A\cos m\psi + B\sin m\psi) \tag{4.12}$$

Comparing (4.12) with the multipole expansion of potential, we see that the terms corresponding to any particular l in (4.12) give the 2^l pole term. Thus, the $l = 0$ term has necessarily $m = 0$ and gives the angle independent monopole term. The $l = 1$ condition is consistent with $m = 0, \pm 1$ and, using the values of the relevant P_l^m and choosing the zero of ψ suitably, we get three terms

$$\frac{\alpha}{r^2}\cos\theta,\quad \frac{\beta}{r^2}\sin\theta\cos\psi,\quad \frac{\gamma}{r^2}\sin\theta\sin\psi$$

Obviously, the three terms correspond to dipoles in three perpendicular directions. For $l = 2$, m will have five possible values $(0, \pm 1, \pm 2)$. Thus, apparently we would have five terms, however, we have threefold freedom in choosing the orientation of the rectangular Cartesian axes (i.e., the polar axis and the zero of ψ); these reduce the number of independent terms to two as we have seen in our previous discussion of quadrupole moment.

It is important to have some knowledge of the associated Legendre polynomials. If $m = 0$, the functions are called Legendre polynomials and satisfy the differential equation

$$\frac{d}{dx}\left[(1-x^2)\frac{dP_l(x)}{dx}\right] + l(l+1)P_l(x) = 0 \tag{4.13}$$

where $P_l(x) \equiv P_l^0(x)$. We try a power series solution of the form

$$P_l(x) = \sum_{n=0}^{\infty} a_n x^n$$

Substituting in Eq. (4.13) and demanding that the coefficient of each power of x must vanish, we get

$$(n+2)(n+1)a_{n+2} = \big(n(n+1) - l(l+1)\big)a_n$$

Thus, the series terminates at x^l if $n = l$. Normalizing the Legendre polynomials by the condition that $P_l(x) = 1$ for $x = 1$, we get the first few Legendre polynomials

$$P_0(x) = 1, \quad P_1(x) = x, \quad P_2(x) = \frac{1}{2}(3x^2 - 1), \quad P_3(x) = \frac{1}{2}(5x^3 - 3x), \ \ldots$$

The Legendre polynomials can be expressed by the general formula

$$P_l(x) = \frac{1}{2^l l!}\frac{d^l}{dx^l}(x^2-1)^l \tag{4.14}$$

The associated Legendre polynomials (for positive m) are given by

$$P_l^m(x) = (-1)^m(1-x^2)^{m/2}\frac{d^m}{dx^m}P_l(x) \tag{4.15}$$

For negative m, it is convenient to take

$$P_l^{-m}(x) = (-1)^m\frac{(l-m)!}{(l+m)!}P_l^m(x)$$

The orthonormal functions defined by

$$Y_{lm}(\theta, \psi) = \sqrt{\frac{(2l+1)}{4\pi}\frac{(l-m)!}{(l+m)!}}\,P_l^m(\cos\theta)\,e^{im\psi}$$

are called spherical harmonics so that

$$Y_{l,-m}(\theta, \psi) = (-1)^m Y^{\star}_{l,+m}(\theta, \psi)$$

and

$$\int_0^{2\pi} d\psi \int_0^{\pi} d\theta \sin\theta \, Y_{lm}(\theta, \psi) Y^{\star}_{l',m'}(\theta, \psi) = \delta_{l,l'} \, \delta_{m,m'}$$

The associated Legendre polynomials have an orthogonality property

$$\int_{-1}^{1} dx \int_0^{2\pi} d\psi \, P_l^m(x) P_{l'}^m(x) \, e^{i(m-m')\psi} = 0, \text{ if } l \neq l', m \neq m' \tag{4.16}$$

With a suitable normalization, we also have

$$\int_{-1}^{1} \left(P_l^m(x)\right)^2 dx = \frac{2}{2l+1} \frac{(l+m)!}{(l-m)!} \tag{4.17}$$

so that in particular

$$\int_{-1}^{1} \left(P_l(x)\right)^2 dx = \frac{2}{2l+1} \tag{4.18}$$

A relation which is of particular importance in potential theory is

$$\begin{aligned}\frac{1}{|\mathbf{r}-\mathbf{r}'|} &= \sum_{l=0}^{\infty} \sum_{m=-l}^{l} \frac{4\pi}{(2l+1)} \frac{r'^l}{r^{l+1}} Y^{\star}_{lm}(\theta', \psi') Y_{lm}(\theta, \psi), \qquad \text{for } r > r' \\ &= \sum_{l=0}^{\infty} \sum_{m=-l}^{l} \frac{4\pi}{(2l+1)} \frac{r'^l}{r^{l+1}} P_l^m(\cos\theta) P_l^m(\cos\theta') e^{im(\psi-\psi')}\end{aligned}$$

Using the above relation in the Poisson integral, we have

$$\begin{aligned}\phi(\mathbf{r}) &= \frac{1}{4\pi\epsilon_0} \sum_{l=0}^{\infty} \sum_{m=-l}^{l} \frac{4\pi}{(2l+1)} \frac{Y_{lm}(\theta, \psi)}{r^{l+1}} \int \rho(\mathbf{r}') r'^l Y^{\star}_{lm}(\theta', \psi') dv', \quad (r > r') \\ &= \frac{1}{\epsilon_0} \sum_{l=0}^{\infty} \sum_{m=-l}^{l} \frac{P_l^m(\cos\theta) e^{im\psi}}{(2l+1) r^{l+1}} \int \rho(\mathbf{r}') r'^l P_l^m(\cos\theta') e^{-im\psi'} dv'\end{aligned} \tag{4.19}$$

Comparing (4.19) with (4.12), we can readily obtain the Q_{lm}'s in terms of the charge distribution function $\rho(\mathbf{r}')$.

There is one more property of the $Y_{lm} = P_l^m(\theta) e^{im\psi}$ which we shall mention. This is called completeness and it signifies that any function of the angles θ and ψ may be expressed as a sum of the Y_{lm}'s. The coefficients are easily determined because of the orthogonality relation (4.16). Thus,

$$f(\theta, \psi) = \sum_l \sum_m a_{lm} Y_{lm} = \sum_l \sum_m a_m P_l^m(\theta) e^{im\psi}$$

where

$$a_{lm} = \frac{(2l+1)}{4\pi} \int_0^{2\pi} d\psi \int_{+1}^{-1} dx f(\theta, \psi) P_l^m(x) e^{-im\psi} \text{ with } x = \cos\theta$$

In particular, if we have a function of θ alone, we can expand it in terms of the Legendre polynomials

$$f(x) = \sum a_n P_n(x) \quad \text{where} \quad a_n = \frac{2n+1}{2} \int_{-1}^{1} f(x) P_n(x) dx$$

An interesting and useful result that may be obtained without much difficulty is

$$\frac{1}{\sqrt{1 - 2x\cos\theta + x^2}} = \sum_{n=0}^{\infty} P_n(\cos\theta)\, x^n \qquad \text{for } |x| < 1$$

4.1 Energy of Multipoles in an External Field

The energy of a charge distribution $\rho(\mathbf{r})$, in an external field with potential $\phi(\mathbf{r})$, is given by

$$W = \int \rho(\mathbf{r}) \phi(\mathbf{r}) dv$$

Expanding $\phi(\mathbf{r})$ in Taylor series

$$\phi(\mathbf{r}) = \phi(0) + \mathbf{r} \cdot (\nabla\phi)_0 + \sum_{i,k} \frac{1}{2} x_i x_k \left(\frac{\partial^2 \phi}{\partial x_i \partial x_k} \right)_0 + \cdots$$

we get

$$W = \phi(0) \int \rho(\mathbf{r}) dv - \mathbf{E}_0 \cdot \int \rho(\mathbf{r}) \mathbf{r} dv + \sum_{i,k} \frac{1}{2} \left(\frac{\partial^2 \phi}{\partial x_i \partial x_k} \right)_0 \int \rho(\mathbf{r}) x_i x_k dv + \cdots$$

But $\delta_{ik} \dfrac{\partial^2 \phi}{\partial x_i \partial x_k} = \nabla^2 \phi = 0$, as the external field has no source at the origin. Using this

$$W = q\phi(0) - \boldsymbol{\mu} \cdot \mathbf{E}_0 + \frac{1}{6} \sum_{i,k} \left(\frac{\partial^2 \phi}{\partial x_i \partial x_k} \right)_0 \int \rho(\mathbf{r}) \big(3x_i x_k - r^2 \delta_{ik}\big) dv + \cdots$$

The first term arises due to the total charge q (monopole energy), the second term is due to the dipole moment $\boldsymbol{\mu}$ of the distribution and the third term, due to the quadrupole moment of the distribution, is non-vanishing only if

$$\frac{\partial^2 \phi}{\partial x_i \partial x_k} = -\frac{\partial E_i}{\partial x_k} = -\frac{\partial E_k}{\partial x_i} \neq 0$$

i.e., the field is not uniform. It is clear from the above expression that the interaction energy of two monopoles varies inversely as the distance r, that between two dipoles inversely as r^3 (cf. Eq. (2.4) of Chap. 2) and in the case of quadrupoles as r^{-5}. In general, we may say that the interaction energy of two 2^l-poles varies as r^{-2l-1}. Thus, qualitatively the interaction energy of higher multipoles simulates a somewhat short-range interaction.

Problems

1. Show that the quadrupole moment remains unchanged by a shift of origin if the total charge and dipole moment of the distribution vanish. Find the corresponding proposition for the general multipole moment.
2. Calculate the energy of a quadrupole in the external potential $\phi = \dfrac{e^{-\alpha r}}{r}$.
3. Calculate the quadrupole moment of the following charge distribution: e at $(0, 0, 0)$, $-e$ at $(a, 0, 0)$, e at $(a, a, 0)$ and $-e$ at $(0, a, 0)$.
4. Calculate the dipole and quadrupole moments for

   (i) a charged disc,
   (ii) a disc with a distribution of dipoles

   as considered in Chap. 2. (Take the centres of the disc as the origin in both cases.) Do you observe anything interesting about the higher multipoles in the two cases?
5. Calculate the potential due to a uniformly charged circular ring.

Chapter 5
Dielectrics and the Uniqueness Theorem

Until now, we have been considering fields *in vacuo*, except for the existence of some charges (sources of the field). When material media are present, the field is modified. In the molecules of the medium, we have positive charges (protons in the nuclei) as well as electrons outside the nuclei. Normally, the molecules are electrically neutral but depending on their structure they may or may not have a dipole moment. For example, a carbon dioxide molecule is linear with the carbon atom at the centre, and it has no intrinsic dipole moment; on the other hand, a water molecule is an isosceles triangle with the hydrogen atoms at the vertex, and it possesses an intrinsic dipole moment. In any case, an external electric field perturbs the electronic states (i.e., the eigenfunctions in the language of quantum mechanics and the orbits of the simple Bohr-Sommerfeld theory) so that a dipole moment is generated. In case the molecules have an intrinsic dipole moment, there will be an additional effect—in the absence of any external field, the molecular dipole moments will be oriented randomly due to thermal agitations and there would be no resultant dipole moment on a macroscopic scale. However, the external field will tend to bring about an alignment of the molecular dipoles in the direction of the field. Thus, we may now expect a resultant dipole moment even on a macroscopic scale. This second effect, called the orientation effect, will obviously be temperature- dependent as it is opposed by thermal motions, while the first effect, the perturbation of eigenfunctions, will be essentially independent of temperature. This phenomenon of generation of dipole moments on a macroscopic scale is called the polarization of the dielectric, and is measured by the dipole moment per unit volume indicated by **P**. Due to this dipole distribution of moment **P** per unit volume, there will be an additional field, the potential of which is given by (cf. Eq. (2.1))

A. K. Raychaudhuri, *Classical Theory of Electricity and Magnetism*, Texts and Readings in Physical Sciences 21, https://doi.org/10.1007/978-981-16-8139-4_5

$$\begin{aligned}\phi(\mathbf{r}) &= \frac{1}{4\pi\epsilon_0}\int\frac{\mathbf{P}(\mathbf{r}')\cdot(\mathbf{r}-\mathbf{r}')dv'}{(|\mathbf{r}-\mathbf{r}'|)^3}\\ &= +\frac{1}{4\pi\epsilon_0}\int\mathbf{P}(\mathbf{r}')\cdot\nabla'\frac{1}{|\mathbf{r}-\mathbf{r}'|}dv' \\ &= -\frac{1}{4\pi\epsilon_0}\int\frac{\nabla'\cdot\mathbf{P}(\mathbf{r}')}{|\mathbf{r}-\mathbf{r}'|}dv' + \frac{1}{4\pi\epsilon_0}\oint\frac{\mathbf{P}(\mathbf{r}')\cdot\hat{\mathbf{n}}\,ds'}{|\mathbf{r}-\mathbf{r}'|}\end{aligned} \tag{5.1}$$

where ∇' represents differentiation with respect to the coordinate of the source point.

In the above, we have applied the divergence theorem to convert a volume integral over the dielectric region into a surface integral over the boundary of the dielectric. The interpretation of Eq. (5.1) is that the volume dipole distribution is equivalent to a charge density $-\nabla\cdot\mathbf{P}$ throughout the space occupied by the dielectric along with a surface distribution of charge P_n (where P_n is the component of $\mathbf{P}$ in the direction of the outward normal to the element of bounding surface). Hence, for the field within the dielectric, we may write

$$\nabla\cdot\mathbf{E} = \frac{\rho}{\epsilon_0} - \frac{\nabla\cdot\mathbf{P}}{\epsilon_0} \tag{5.2}$$

where ρ represents the original charge density, called the *true charge* or *free charge*, and the charge density $-\nabla\cdot\mathbf{P}$ arising due to polarization is called the *bound charge*. We might have introduced the surface charge density in Eq. (5.2) by using a suitable delta-function, but it is usual to take care of it by means of boundary conditions. We shall see that it results in a discontinuity of the component of electric intensity normal to the boundary. We may write Eq. (5.2) in either of the two following forms:

$$\nabla\cdot\mathbf{E} = \frac{1}{\epsilon_0}(\rho+\rho_P) = \frac{1}{\epsilon_0}\rho_{\text{total}} \tag{5.3}$$

$$\nabla\cdot\mathbf{D} = \rho \tag{5.4}$$

where $\rho_P = -\nabla\cdot\mathbf{P}$ and the vector $\mathbf{D}$, which is $\epsilon_0\mathbf{E}+\mathbf{P}$, is called the displacement vector, or sometimes, the electric induction.

The first of Eq. (5.4) shows that $\mathbf{E}$ is still a central, inverse square field and hence it will be irrotational, i.e.,

$$\nabla\times\mathbf{E} = 0 \tag{5.5}$$

Equations (5.4) and (5.5) constitute the basic equations of electrostatics. However thereby, we have only four equations involving six unknowns. In reality, the number of equations is actually one less because of the differential identity $\nabla\cdot(\nabla\times\mathbf{E}) = 0$; only two of the three equations in (5.5) are independent. An additional three equations connecting $\mathbf{D}$ and $\mathbf{E}$ are provided by phenomenological considerations

$$D_i = \sum_k \epsilon_{ik}E_k \tag{5.6}$$

where ϵ_{ik}'s are constants characteristic of the dielectric material known as the *permittivity* of the material. Such a relation is generally valid at least as long as the electric field is not too large. Further, if the material be isotropic, Eq. (5.6) reduces to

$$D_i = \epsilon E_i \tag{5.7}$$

A dielectric is called linear isotropic if Eq. (5.7) holds. The corresponding relation between the vectors **P** and **E** is then given by

$$P_i = \epsilon_0 \chi_e E_i \tag{5.8}$$

where χ_e is called the electric susceptibility. The ratio $\epsilon_r = \epsilon/\epsilon_0$ is called the dielectric constant (or the *relative permittivity*) of the medium, and we have the relation

$$\epsilon = \epsilon_0(1 + \chi_e) \tag{5.9}$$

Equation (5.4) can be converted into a flux theorem

$$\oint \mathbf{D} \cdot \hat{\mathbf{n}}\, ds = \text{(total true charge inside the surface)} \tag{5.10}$$

5.1 Boundary Conditions to be Satisfied at the Interface of Two Different Dielectrics

Before obtaining the boundary conditions by the conventional procedure, we enunciate a principle. Namely that even at the boundary, the basic Eqs. (5.4) and (5.5) remain meaningful. That is, none of the quantities occurring in these two equations will blow up unless there is some extraneous singularity in the problem (e.g., a surface density of true charge at the interface). Now suppose that the z-axis is normal to the interface at the point under consideration. Our principle tells us that in view of (5.4), D_z must be continuous. Again from Eq. (5.5), E_y and E_x must be continuous. The argument rests on the occurrence of $\partial D_z/\partial z$, $\partial E_x/\partial z$ and $\partial E_y/\partial z$ in Eqs. (5.4) and (5.5). Thus, we have the theorem: at an interface of two dielectrics, the normal component of the displacement vector and the tangential components of the electric vector are continuous. More formally, consider an infinitesimal volume whose sides, parallel to the interface, are large compared to that normal to the surface.

The volume contains a portion of the interface and parts of both media (Fig. 5.1a). Equation (5.10) gives, considering this volume,

$$(D_1)_n dA - (D_2)_n dA = \sigma dA$$
$$(D_1)_n - (D_2)_n = \sigma \tag{5.11}$$

Fig. 5.1 Illustrating boundary conditions on **E** and **D** at an interface

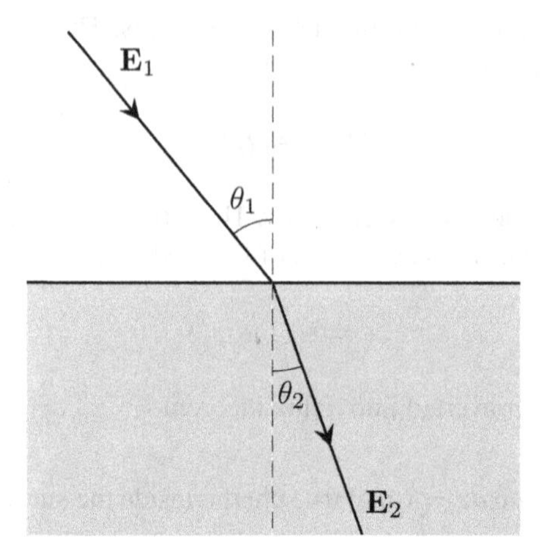

Fig. 5.2 Refraction of electric field line at the interface of two dielectrics

where the flux over the normal faces have been neglected as those dimensions are negligibly small compared to the parallel dimensions, σ is the true surface charge density and dA the element of interface area enclosed. If the surface charge density vanishes

$$(D_1)_n = (D_2)_n \tag{5.12}$$

i.e., the normal component of the displacement vector is continuous. Again, consider the line integral of **E** over the closed loop whose sides AB = CD are much larger than BC = DA; of these, AB and CD are parallel to the surface and BC, DA are normal to it. As the line integral vanishes (cf. 5.5), we get

$$(E_1)_T = (E_2)_T \tag{5.13}$$

where $(E_1)_T$ and $(E_2)_T$ are the components of **E** tangential to the interface and in the same direction. Equations (5.11)–(5.13) combine to give 'the law of refraction of lines of electric force'. We have (see Fig. 5.2)

$$\begin{aligned} \epsilon_1 E_1 \cos\theta_1 &= \epsilon_2 E_2 \cos\theta_2 \\ E_1 \sin\theta_1 &= E_2 \sin\theta_2 \\ \frac{\tan\theta_1}{\tan\theta_2} &= \frac{\epsilon_1}{\epsilon_2} \end{aligned} \tag{5.14}$$

Note that $\mathbf{D}$ may not be irrotational. From Eq. (5.7), we have

$$\nabla \times \mathbf{D} = \epsilon \nabla \times \mathbf{E} + \nabla \epsilon \times \mathbf{E} = \nabla \epsilon \times \mathbf{E}$$

Hence, if the dielectric be non-homogeneous, ϵ may not be a constant and if $\nabla \epsilon$ and $\mathbf{E}$ do not coincide in direction, $\nabla \times \mathbf{D}$ would not vanish.

5.2 The Uniqueness Theorem

The basic problem of electrostatics is the solution of Poisson's equation when the charge density is given and some boundary conditions are specified. If we can prove that under such circumstances, there is only one solution possible, then we can try to guess a solution or obtain a solution by some trick and then argue that because of the uniqueness of the solution, our solution is the actual solution of the problem. We give below an enunciation of the uniqueness theorem.

> If in a homogenous isotropic dielectric, the true charge densities are given everywhere and the electrostatic potential, or the normal component of its gradient, is specified on the boundary surface (or surfaces), and the potential vanishes sufficiently rapidly at an infinite distance, then the potential is uniquely determined at all points.

Proof: The condition of isotropic homogeneous dielectric gives

$$\mathbf{D} = \epsilon \mathbf{E}$$

$$\nabla \cdot \mathbf{E} = \frac{1}{\epsilon} \nabla \cdot \mathbf{D} = \frac{\rho}{\epsilon}$$

therefore

$$\nabla^2 \phi = -\frac{\rho}{\epsilon} \tag{5.15}$$

We are to find a solution of Eq. (5.15) in the domain shown shaded in Fig. 5.3. It extends to infinity but the regions (unshaded) bounded by the surfaces S_1, S_2, etc., are excluded. The charge density ρ is specified at all points within the shaded region. We do not know anything about the unshaded regions, except that either the potential ϕ is given at every point on the surfaces S, S_1, S_2, etc., or the component of $\nabla \phi$ in the direction of the normal to the surface element is given for every point over these surfaces. Further, we are supposing that the potential vanishes sufficiently rapidly to ensure the vanishing of some integrals over the sphere at infinity which will arise in our discussion. (This condition is relevant only if the domain under consideration is not bounded.) If possible, let there be two distinct solutions satisfying (5.15) and the conditions spelt out. Thus, if $\phi_1(r)$ and $\phi_2(r)$ indicate the two solutions,

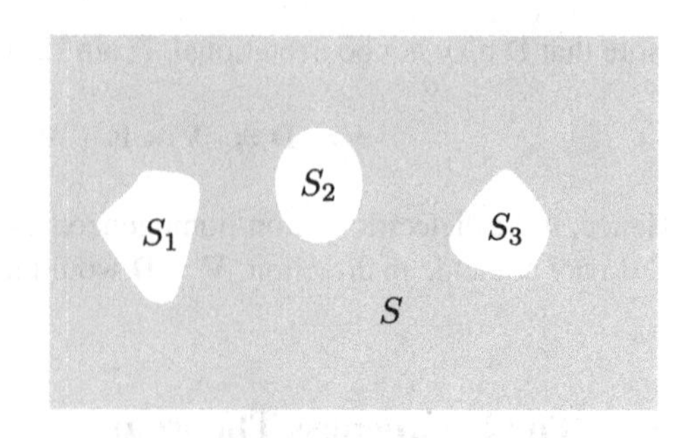

Fig. 5.3 Boundary value problem

$$\begin{aligned}\nabla^2\phi_1 = \nabla^2\phi_2 = -\frac{\rho}{\epsilon} \\ \nabla^2(\phi_1 - \phi_2) = 0\end{aligned}$$

Writing

$$\begin{aligned}\phi_1 - \phi_2 = \Psi \\ \nabla^2\Psi = 0\end{aligned} \tag{5.16}$$

and taking the region of integration our shaded domain, we use Green's theorem to write

$$\begin{aligned}\oint_S \Psi\nabla\Psi \cdot \mathbf{ds} + \oint_{S_1} \Psi\nabla\Psi \cdot \mathbf{ds} + \oint_{S_2} \Psi\nabla\Psi \cdot \mathbf{ds} + \cdots &= \int \nabla \cdot (\Psi\nabla\Psi)\, dv \\ &= \int \Psi\nabla^2\Psi dv + \int (\nabla\Psi)^2\, dv\end{aligned} \tag{5.17}$$

The first integral on the right-hand side falls off because of Eq. (5.16) and because of the specified boundary conditions, either Ψ or $(\nabla\Psi)_n$ vanishes at every point of S, S_1, S_2, etc. In case the domain under consideration is unbounded, S is the sphere at infinity, and as all other surface integrals vanish, Eq. (5.17) becomes

$$\int_{S\to\infty} \Psi\nabla\Psi \cdot \mathbf{ds} = \int (\nabla\Psi)^2 dv \tag{5.18}$$

The concept of the sphere at infinity is a sphere whose radius r is tending to infinity. Thus, any element ds of this sphere tends to infinity as r^2. If, therefore, $|\Psi\nabla\Psi|_n \to 0$ faster than $1/r^2$, the left-hand side of (5.18) would vanish and we would have

$$\int (\nabla\Psi)^2\, dv = 0$$

As the above integrand is positive definite, we have $\nabla\Psi = 0$, or $\Psi = \phi_1 - \phi_2 =$ constant. A constant difference in the potential is of no significance but even that is ruled out by the condition that the potential is specified at some points (or the condition of vanishing at infinity). Thus, the theorem is proved.

Notes

1. Instead of a dielectric, one might as well have vacuum.
2. Instead of a continuous density of charges, we may have discrete charges given by delta functions in ρ and the theorem would still hold.
3. Instead of an everywhere homogeneous dielectric, one might have a 'piece-wise' homogeneous dielectric with the boundary conditions (5.11)–(5.13) satisfied at the interfaces.
4. If some surface S that we have considered is that of a conductor, then the given potential over it must be a constant over the surface. For the conducting surface, the integral $\int_{S_i} \Psi \nabla \Psi \cdot \mathbf{ds}$ becomes

$$\Psi_i \int_{S_i} \nabla \Psi \cdot \mathbf{ds} = \Psi_i \left[\int_{S_i} \nabla \phi_1 \cdot \mathbf{ds} - \int_{S_i} \nabla \phi_2 \cdot \mathbf{ds} \right]$$

   Now $\int \nabla \phi \cdot \mathbf{ds} = -(1/\epsilon_0) \times$ (charge enclosed within S_i). Hence, for a conducting surface the boundary condition may specify the total charge on the surface instead of the potential or its normal gradient at each point.

To emphasize that the uniqueness theorem is not trivial, we shall give an illustration. Suppose there is a uniform distribution of charge throughout space extending to infinity. Admittedly, the situation appears artificial but it was not so to some nineteenth-century physicists. Poisson's equation holds good in Newtonian gravitation theory, and those physicists imagined the universe to consist of uniformly distributed matter extending up to infinity. The situation might have some relevance in electricity if the Bondi-Lyttleton speculation that proton and electron charges are unequal turned out to be correct. Now consider the potential function $\phi = (\rho/6\epsilon_0)r^2$. Substituting in Poisson's equation, we find that there is a distribution of charge with constant density ρ. But we have not as yet specified the origin, so that r and consequently the potential is non-unique. The reason for the non-applicability of the uniqueness theorem is obviously the non-vanishing of the potential at infinity.

As an application of the uniqueness theorem, we shall consider the case of a dielectric sphere placed in a uniform external field (see Fig. 5.4). Even after the introduction of the sphere, the field will be asymptotically uniform and as r tends to infinity the potential will be of the form

$$\phi = -E_0 x = -E_0 r \cos\theta = -E_0 r P_1(\cos\theta)$$

We write the solution in terms of the spherical harmonics $Y_{lm}(\theta, \phi) \equiv P_l^m(\cos\theta)e^{-im\phi}$ but note that

(a) everything is symmetric about the x-axis so we may expect the potential to be independent of the angle ϕ, hence $m = 0$, and only the Legendre polynomials occur, and

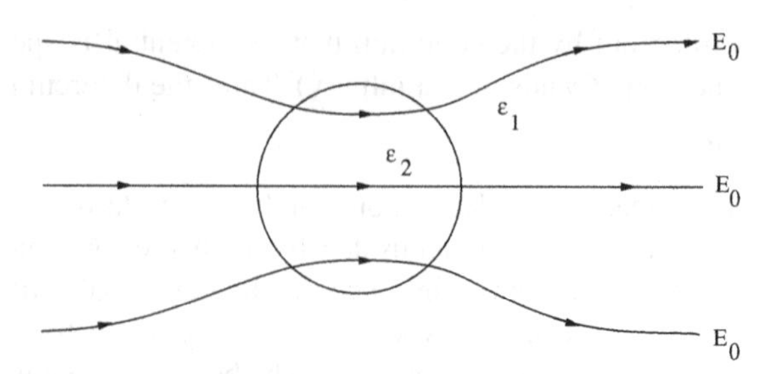

Fig. 5.4 Dielectric sphere in a uniform electric field

(b) in the region within the sphere $r \leq a$, we cannot use the functions which have a singularity at $r = 0$, moreover, in the outside region $r \geq a$, for $r \to \infty$ the potential $\phi_e \to -E_0 r P_1(\cos\theta)$ and there must not be any term in ϕ_e containing higher powers of r.

Hence, we write

$$\phi = \begin{cases} \phi_i = \sum_{l=0}^{\infty} A_l r^l P_l(\cos\theta) & \text{for } r \leq a \\ \phi_e = \sum_{l=0}^{\infty} B_l r^{-(l+1)} P_l(\cos\theta) - E_0 r P_1(\cos\theta) & \text{for } r > a \end{cases}$$

The terms under sum correspond to a charge-free field while the asymptotic field is also taken care of. Now the boundary conditions require

$$\epsilon_1 \left(\frac{\partial \phi_e}{\partial r}\right)_{r=a} = \epsilon_2 \left(\frac{\partial \phi_i}{\partial r}\right)_{r=a}$$
$$(\phi_e)_{r=a} = (\phi_i)_{r=a}$$

for all θ. These conditions must be satisfied for each l separately because the P_ℓ's are orthogonal to one another. We thus have

$$A_l = B_l = 0 \quad \text{if } l \neq 1$$
$$A_1 a = \frac{B_1}{a^2} - E_0 a$$
$$\epsilon_2 A_1 = -\epsilon_1 \left(E_0 + \frac{2B_1}{a^3}\right)$$

so that

$$A_1 = -\frac{3E_0\epsilon_1}{2\epsilon_1 + \epsilon_2} \quad \text{and} \quad B_1 = E_0 a^3 \frac{(\epsilon_2 - \epsilon_1)}{2\epsilon_1 + \epsilon_2}$$

and finally

$$\phi = \begin{cases} -\dfrac{3\epsilon_1}{2\epsilon_1 + \epsilon_2} E_0 r \cos\theta & \text{for } r \leq a \\ \dfrac{(\epsilon_2 - \epsilon_1)}{2\epsilon_1 + \epsilon_2} E_0 \dfrac{a^3}{r^2} \cos\theta - E_0 r \cos\theta & \text{for } r > a \end{cases}$$

The field within the sphere is uniform and is in the x-direction but is modified by the factor $3\epsilon_1/(2\epsilon_1 + \epsilon_2)$ which is less than unity if $\epsilon_2 > \epsilon_1$ (case of a material sphere in vacuum). Outside, the sphere brings in an additional field corresponding to a dipole of moment

$$\mu = 4\pi\epsilon_0 \left(\frac{\epsilon_2 - \epsilon_1}{\epsilon_2 + 2\epsilon_1} \right) a^3 \mathbf{E}_0 \tag{5.19}$$

so that the polarization is

$$\mathbf{P} = 3\epsilon_0 \left(\frac{\epsilon_2 - \epsilon_1}{\epsilon_2 + 2\epsilon_1} \right) \mathbf{E}_0 \tag{5.20}$$

The polarization is thus independent of the dimension of the sphere.

Note
If the sphere be conducting, the field inside vanishes so that $A_l = 0$ for all l. The condition that the tangential field must vanish at the boundary in the exterior as well, gives $B_1 = E_0 a^3$, so for a conducting sphere

$$\phi = \begin{cases} \phi_i = 0 & \text{for } r \leq a \\ \phi_e = \frac{E_0 a^3}{r^2} \cos\theta - E_0 r \cos\theta & \text{for } r > a \end{cases} \tag{5.21}$$

Note that formula (5.21) follows from the general case if $\epsilon_2/\epsilon_1 \to \infty$.

The uniqueness theorem allows us to conclude that the field in the hollow within a closed conductor of arbitrary shape vanishes, if there be no charge in the hollow itself. For example, consider the region of the hollow—the charge density is given and also the potential over the boundary surface is specified (say equal to ϕ_0). Then obviously $\phi = \phi_0$ everywhere in the hollow satisfies the Poisson (in reality the Laplace, in this case) equation as well as the boundary condition. Because of the uniqueness theorem, this is the actual solution to the problem.

5.3 Field in a Cavity in a Dielectric

Suppose we are interested in determining the intensity in a dielectric by measuring the force on a charge. It turns out that the force depends on the form of the cavity in the dielectric we make to introduce our test charge.

(a) We consider a cavity bounded by two planes parallel to $\mathbf{E}$ and a very small distance apart; see Fig. 5.5. From the result that the tangential components of

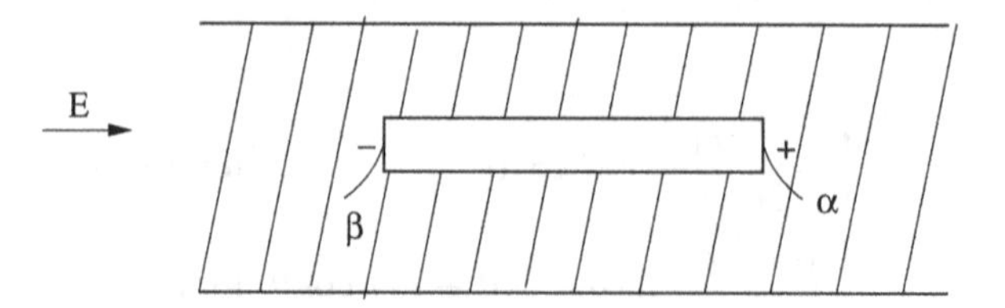

Fig. 5.5 Cavity in a dielectric bounded by planes parallel to **E**

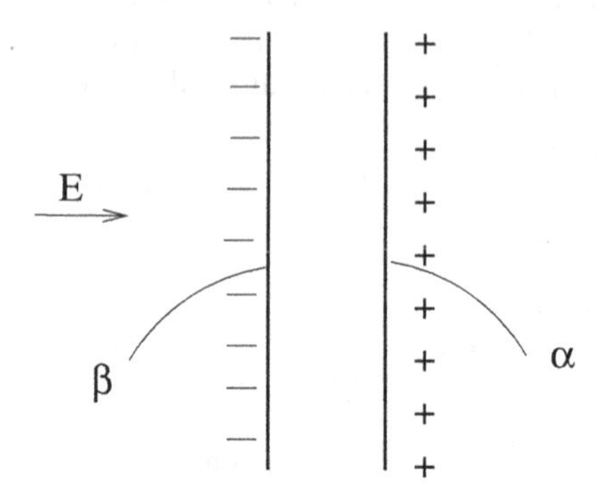

Fig. 5.6 Cavity in a dielectric bounded by planes perpendicular to **E**

E are continuous, we get that the field in the cavity will be also **E**. The charges that appear on the faces α and β are too small and too far to cause any effective change of the field in the cavity.

(b) The cavity is now bounded by planes normal to **E** and only a small distance apart; see Fig. 5.6. The continuity of the normal component of the displacement now shows that the field in the cavity is $\epsilon\mathbf{E}$. The increase of field is due to the polarization charges on planes α and β which are 'infinite planes' for small enough separations.

(c) In the case of a spherical cavity, the field can be obtained from our formulae for a dielectric sphere in an asymptotically uniform field by taking $\epsilon_1 = \epsilon, \epsilon_2 = \epsilon_0$.

$$\mathbf{E} = \frac{3\epsilon\mathbf{E}_0}{2\epsilon + \epsilon_0}$$

5.4 Molecular Polarizability and Clausius-Mossotti Relation

To build up a molecular theory of dielectrics, we must consider the actual field which acts on a molecule of the dielectric. So far the discussions have been on the assumption of a continuous distribution of dipoles. This may be a permissible approximation so long as we take a macroscopic view. But on a microscopic scale, the

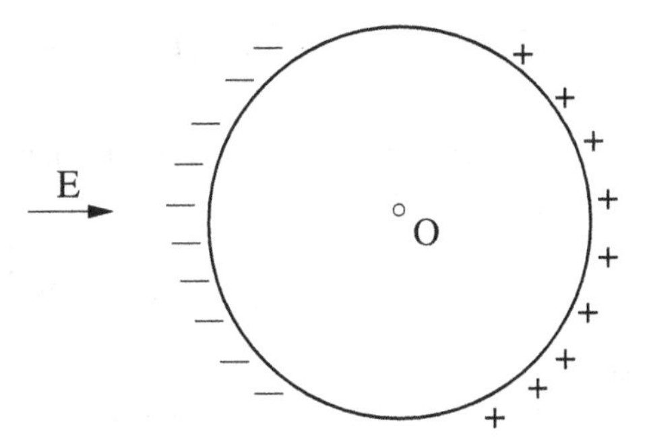

Fig. 5.7 Microscopic spherical cavity inside a dielectric

individual molecule can 'see' the discrete molecular distribution and the continuous assumption must be given up. Hence, to calculate the effective field on a particular molecule, we proceed as follows.
(a) Consider a sphere of radius r about the molecule; r is large compared to molecular dimensions but small macroscopically; see Fig. 5.7. In the field **E**, the contribution of this region is contained on the basis of a continuous distribution of dipole moments of volume density **P**. As this is wrong, we first deduct this contribution from **E**. The contribution is easy to calculate. As **P** is constant, $\nabla \cdot \mathbf{P} = 0$, hence there is no volume distribution of charge due to polarization but only a surface density of magnitude P_n. Obviously at O, this gives a field in the direction opposite to **E**. It is

$$\frac{1}{4\pi\epsilon_0} \oint \frac{P_n ds \cos\theta}{r^2} = -\frac{\pi}{2\epsilon_0} \int_0^{\pi} P \sin\theta \cos^2\theta d\theta = -\frac{\mathbf{P}}{3\epsilon_0}$$

Hence on deducting this, the field at 0 due to distant regions is

$$\mathbf{E} + \frac{\mathbf{P}}{3\epsilon_0}$$

(b) We must now add to the above the actual field due to the distribution of discrete molecules within the sphere of radius r. If the arrangement of the molecules be random (as in a dilute gas) or isotropic or highly symmetric as in a cubic crystal, then it is fairly easy to see that the contributions from different molecules completely cancel out and the net contribution vanishes. (We omit the formal proof of this statement.) For more general cases, one must know the actual distribution of molecules for a calculation. Restricting ourselves to the simple cases where this contribution vanishes, we get

$$\mathbf{E}_{\text{effective}} = \mathbf{E} + \frac{\mathbf{P}}{3\epsilon_0}$$

We define the polarizability of a molecule as the dipole moment developed per molecule due to unit field strength. Thus, if α be the polarizability of a molecule, the dipole moment per molecule will be

$$\boldsymbol{\mu} = \alpha \mathbf{E}_{\text{effective}} = \alpha \left(\mathbf{E} + \frac{\mathbf{P}}{3\epsilon_0} \right)$$

The polarization $\mathbf{P} = n\boldsymbol{\mu}$, where n is the number density of molecules so that

$$\mathbf{P} = n\alpha \left(\mathbf{E} + \frac{\mathbf{P}}{3\epsilon_0} \right) = \frac{n\alpha}{1 - \frac{n\alpha}{3\epsilon_0}} \mathbf{E}$$

Hence

$$\epsilon - \epsilon_0 = \frac{n\alpha}{1 - \frac{n\alpha}{3\epsilon_0}}$$

which gives

$$\frac{\epsilon_r - 1}{\epsilon_r + 2} = \frac{n\alpha}{3\epsilon_0} \tag{5.22}$$

This last equation is known as the Clausius-Mossotti relation. As we have already remarked, the polarizability may arise from two distinct causes—a perturbation of the electronic states and an orientation of the molecules if they have an intrinsic dipole moment. The second effect is temperature-dependent. If $\boldsymbol{\mu}$ be the permanent dipole moment of a molecule, the probability of this dipole moment making an angle between θ and $\theta + d\theta$ with the direction of the field is, according to Boltzmann statistics,

$$e^{\mu E \cos\theta / k_B T} 2\pi \sin\theta d\theta / 4\pi$$

where $-\mu E \cos\theta$ is the energy of a dipole in the field $\mathbf{E}$, T is the Kelvin temperature, k_B the Boltzmann constant and $2\pi \sin\theta d\theta$ the solid angle lying between θ and $\theta + d\theta$. Each of these dipoles gives a component dipole moment $\mu \cos\theta$ in the direction of the field and the perpendicular components cancel out. Hence, if n be the number density of molecules, the polarization due to the orientation effect is

$$P = \frac{n\mu \displaystyle\int_0^{\pi} \cos\theta e^{\mu E \cos\theta / k_B T} \sin\theta d\theta}{\displaystyle\int_0^{\pi} e^{\mu E \cos\theta / k_B T} \sin\theta d\theta}$$

This gives, after integration,

$$P = n\mu \left(\coth\left(\frac{\mu E}{k_B T} \right) - \frac{k_B T}{\mu E} \right)$$

Under usual laboratory conditions, $\mu E / k_B T$ is small compared to unity and then we can approximate

$$P \approx \frac{n\mu^2 E}{3k_B T}$$

Thus due to this effect, the polarizability is increased from α to $\alpha + \frac{\mu^2}{3k_BT}$ and so the Clausius-Mossotti relation will assume the form

$$\frac{\epsilon_r - 1}{\epsilon_r + 2} = \frac{n}{3\epsilon_0}\left(\alpha + \frac{\mu^2}{3k_BT}\right) \tag{5.23}$$

Thus, if intrinsic molecular dipole moment be present, a plot of $(\epsilon_r - 1)/(\epsilon_r + 2)$ against $1/T$ will be a straight line from the slope of which we can determine μ and from the value of $(\epsilon_r - 1)/(\epsilon_r + 2)$ as $1/T \to 0$, we get the value of α. Thus, the measurement of dielectric constant gives very useful information about molecular characteristics. In Eq. (5.23), the molecular number density is often replaced by $N\rho/M$, where N is the Avogadro number, ρ the density and M the molecular weight, so that

$$\frac{\epsilon_r - 1}{\epsilon_r + 2}\frac{M}{\rho} = \frac{N}{3\epsilon_0}\left(\alpha + \frac{\mu^2}{3k_BT}\right)$$

In some cases, the electromagnetic theory result connecting the refracting index with the dielectric constant $\epsilon_r = n^2$ is used giving

$$\frac{n^2 - 1}{n^2 + 2}\frac{M}{\rho} = \frac{N}{3\epsilon_0}\left(\alpha + \frac{\mu^2}{3k_BT}\right)$$

It is of some interest to consider the observational data. For air, the following table gives the dielectric constant at different pressures at a temperature ~300 K.

Pressure (in atmospheres)	Dielectric constant
20	1.0108
40	1.0218
60	1.0333
80	1.0439
100	1.0548

The ratio $(\epsilon_r - 1)/p$ is thus seen to be constant with the value 54×10^{-5} per atmosphere and α may be calculated to have the value $\sim 1.6 \times 10^{-24}$ cm^3. A complete theory must be able to deduce this from theoretical considerations but that is possible only on the basis of quantum mechanics. A water molecule has an intrinsic dipole moment, and we give below a table showing the variation of the dielectric constant of water vapour with temperature.

Temperature (in degrees Kelvin)	Pressure (in cm of mercury)	Dielectric constant
393	56.5	1.00400
423	60.9	1.00372
453	65.3	1.00349
483	69.8	1.00329

The calculation of dipole moment of a water molecule from these data is left to the student as an exercise.

Problems

1. A small dielectric sphere of radius a is placed at the origin in an electric field with potential $\phi = \alpha r^2$. Calculate the force on the sphere.
2. Show that if for a given arrangement of conductors (ϕ, ρ, q_i) and (ϕ', ρ', q_i') be two possible distributions of potential, charge density and charge on the ith conductor, then

$$\int \phi\rho' dv + \sum_i \phi_i q_i' = \int \phi' \rho dv' + \sum_i \phi_i' q_i$$

   Hence, find the potential of an insulated uncharged spherical conductor when a point charge Q is placed at a distance r from the centre.
3. The dielectric constant for gaseous SO_2 is 1.00993 at 273 K and 1.00569 at 373 K at a pressure of 1 atmosphere. Calculate the dipole moment of the SO_2 molecule. Consider ideal gas behaviour.
4. The dipole moment of a HBr molecule is approximately 0.8 D (1D $= 3.33 \times 10^{-30}$ C-m). If the dielectric constant be 1.0031 at STP, calculate the molecular polarizability.
5. Calculate the polarization charge density at different points on the surface of a dielectric sphere placed in an asymptotical uniform field. Show that the fields can be directly calculated from this charge distribution.
6. Calculate the field at different points for a spherical shell of dielectric constant ϵ_r placed in vacuum with an asymptotically uniform field.

Chapter 6
Solution of the Laplace Equation

We have already seen how Laplace's equation can be solved in spherical polar coordinates. The method used there is known as separation of variables and can be applied with other coordinates as well. We shall now consider the cases of rectangular Cartesian and cylindrical polar coordinates.

6.1 Rectangular Cartesian Coordinates

We take $\phi = X(x)Y(y)Z(z)$ where X is a function of x alone, etc.

$$\nabla^2\phi = 0$$

yields

$$\frac{1}{X}\frac{d^2X}{dx^2} + \frac{1}{Y}\frac{d^2Y}{dy^2} + \frac{1}{Z}\frac{d^2Z}{dz^2} = 0$$

Obviously, each term on the left must be equal to a constant

$$\frac{1}{X}\frac{d^2X}{dx^2} = K_1$$
$$\frac{1}{Y}\frac{d^2Y}{dy^2} = K_2$$
$$\frac{1}{Z}\frac{d^2Z}{dz^2} = K_3$$

with $K_1 + K_2 + K_3 = 0$, so that constants cannot all be of the same sign. (In case all the K's vanish, we get a uniform field.) Thus, in X, Y and Z both trigonometric

A. K. Raychaudhuri, *Classical Theory of Electricity and Magnetism*, Texts and Readings in Physical Sciences 21, https://doi.org/10.1007/978-981-16-8139-4_6

and exponential functions will occur. Besides this, there is no restriction so far on the K's. Imposition of boundary conditions, however, generally restricts the K's to discrete values. For example, suppose it is given that a hollow cube of edge length a is bounded by conducting walls except one wall whose equation is $x = a$ and the potential distribution over this side is given as $V(y, z)$ while all other walls are at zero potential. The solution satisfying the boundary conditions is of the form

$$\phi = \sum_{n=1}^{\infty}\sum_{m=1}^{\infty} A_{mn} \sin \frac{m\pi y}{a} \sin \frac{n\pi z}{a} \sinh px$$

where n, m and p are related to K_2, K_3 and K_1, respectively, and

$$p^2 = \frac{\pi^2}{a^2}(m^2 + n^2) \tag{6.1}$$

The coefficients A_{mn} may be evaluated using the orthogonality of Fourier series terms. Thus,

$$A_{mn} = \frac{4}{a^2 \sinh(pa)} \int_0^a \int_0^a V \sin \frac{m\pi y}{a} \sin \frac{n\pi z}{a} \, dy\, dz$$

In particular, if $V = V_0 \sin \frac{\pi y}{a} \sin \frac{\pi z}{a}$

$$A_{mn} = \frac{V_0}{\sinh(pa)} \delta_{m1}\, \delta_{n1} = \frac{V_0}{\sinh(\pi\sqrt{2})} \delta_{m1}\, \delta_{n1}$$

where we have used Eq. (6.1).

6.2 Cylindrical Polar Coordinates

Laplace's equation now reads

$$\frac{1}{r}\frac{\partial}{\partial r}\left(r\frac{\partial \phi}{\partial r}\right) + \frac{1}{r^2}\frac{\partial^2 \phi}{\partial \psi^2} + \frac{\partial^2 \phi}{\partial z^2} = 0$$

Assuming a solution $\phi = R(r)\Psi(\psi)Z(z)$, we get

$$r\frac{d}{dr}\left(r\frac{dR}{dr}\right) + (k^2r^2 - n^2)R = 0$$

$$\frac{d^2\Psi}{d\psi^2} + n^2\Psi = 0$$

$$\frac{d^2Z}{dz^2} - k^2Z = 0$$

where k and n are constants arising out of the separation of variables. The particular choice of sign is dictated by the requirement that ψ being an angular coordinate, Ψ must be periodic in ψ with period 2π, while a periodicity in z would make the potential vanish periodically over planes orthogonal to the axis of symmetry which is an unlikely behaviour. Also, the potential would not vanish as $z \to \infty$. The equation in r is known as Bessel's equation. Writing $kr = x$, the equation becomes

$$x\frac{d}{dx}\left(x\frac{dR}{dx}\right) + (x^2 - n^2)R = 0$$

We attempt a solution of this equation in the form

$$R = \sum a_s x^s$$

and obtain the recurrence relation

$$a_{s-2} = a_s(n^2 - s^2)$$

Thus if n be a positive integer, the series will begin with $s = n$, and have only terms involving powers of x as n, $n + 2$, $n + 4$, etc. A little manipulation allows one to write R in the following form:

$$R = \sum_{p=0}^{\infty} \frac{(-1)^p}{p!\,(n+p)!}\left(\frac{x}{2}\right)^{n+2p}$$

The series converges for all finite values of x, further for integral n, in which we are interested here; the solutions for n and $-n$ are related by $R_{-n}(x) = (-1)^n R_n(x)$. These functions are called Bessel functions and have been normalized by the condition that the coefficient of x^n is $1/(2^n n!)$. There is another linearly independent solution of Bessel's equation for integral n but that diverges at x =0. We write the Bessel function as $J_n(kr)$ and if k and k' be two distinct roots of the equation $J_n(ka)$ = 0, then it can be shown

$$\int_0^a J_n(kr)\,J_n(k'r)\,r\,dr = 0$$

which is a sort of orthogonality of these functions. We have also the relation $\int_0^\infty J_n^2(kr)r\,dr = \frac{1}{2}a^2 J_n'^2(ka)$, where $J_n' = \frac{d}{dx}J_n(x)$.

These allow us to express any function of x in the interval $x = 0$ to $x = a$, provided $f(a) = f(0) = 0$, as a sum of Bessel functions in the form

$$f(x) = \sum C_s J_r(k_s x)$$

where the coefficients C_s's are unique and k_s's are the roots of the equation $J_n(ka)$ = 0. Thus, the solution of Laplace's equation in terms of Bessel functions may be made to satisfy the given boundary conditions.

6.3 Method of Electrical Images

This method of solution of Laplace's equation depends heavily on the uniqueness theorem. We replace an apparently 'awkward' distribution of charges by a system of charges whose field can be easily calculated and which are also consistent with the given boundary conditions. We then argue that the uniqueness theorem ensures that the field calculated in this manner is the actual solution to our problem.

6.3.1 A Point Charge in Front of an Infinite Conducting Plane

The charge q is at a distance r from the infinite plane AOA′ which is at zero potential, Fig. 6.1. The potential function ϕ will satisfy the equation

$$\nabla^2\phi = -\frac{1}{\epsilon_0} q\,\delta(\mathbf{r} - \mathbf{OP})$$

(we are taking the origin at O and the x-axis along OP) and the boundary condition is that it vanishes everywhere on the plane AA′ and the sphere at infinity. Now consider a charge $-q$ at P′ such that OP′ = OP, i.e., P′ is the position of the optical image of a source at P by a reflecting plane AA′. The potential everywhere on the plane due to these two charges will be (R is any point on the plane)

$$\frac{q}{\mathrm{RP}} - \frac{q}{\mathrm{RP'}} = 0$$

i.e., the requisite boundary condition is satisfied and also the combined potential due to these two charges will satisfy the equation $\nabla^2\phi = -(1/\epsilon_0)\,q\,\delta(\mathbf{r} - \mathbf{OP})$ everywhere on the right-hand side of the plane AA′. In the left-half space, the hypothetical

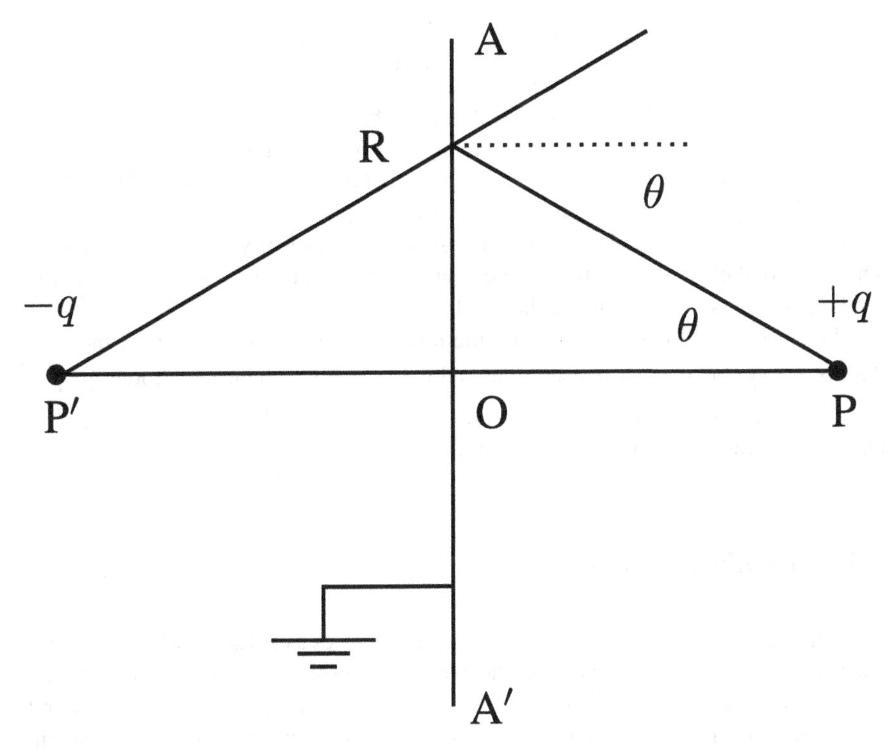

Fig. 6.1 A charge in presence of a grounded conducting plane

charge is not to be considered and we assert that on the left side the potential is simply zero everywhere. Thus, the proposed solution is

$$\phi(\mathbf{r}) = \begin{cases} \dfrac{q}{4\pi\epsilon_0|\mathbf{r}-\mathbf{OP}|} - \dfrac{q}{4\pi\epsilon_0|\mathbf{r}-\mathbf{OP}'|} = 0 & \text{for } x \geq 0 \\ 0 & \text{for } x \leq 0 \end{cases}$$

and because of the uniqueness theorem is the actual solution. We observe that the complicated induced charge distribution is effectively equivalent to the image charge $-q$ at P' so far as the right side of the plane is considered.

It is easy to calculate the charge on the plane. The field at a point on the plane is directed normally, i.e., along the x-axis and is given by

$$E_x = -\frac{qa}{2\pi\epsilon_0(a^2+r^2)^{3/2}}$$

where we have written a for the distance OP, r is the distance of the point on the plane from O and the negative sign indicates that the intensity is directed towards the negative direction of the x-axis for positive q. Hence, using Gauss' theorem, the charge density at the point on the plane is

$$\sigma = \epsilon_0 E_x \big|_{x=0} = -\frac{qa}{2\pi(a^2+r^2)^{3/2}}$$

One can see by direct integration that the total charge on the plane is equal to $-q$. However, that result can be inferred directly by considering the flux over a closed surface consisting of the hemisphere on the right at infinity and surface either within the conducting plane or entirely on the left of the plane. The field vanishes faster than $1/r^2$ at points on the hemisphere at right and is identically zero on the left side. Thus, the total charge consisting of the charge q at P and the entire induced charge on the plane must vanish. Hence the result.

The field at P blows up. However, the force on P is due to the part of the field due to charges other than q itself. This part is $-\frac{q^2}{4\pi\epsilon_0(2a)^2}$, hence the force on q is $-\frac{q^2}{4\pi\epsilon_0(2a)^2}$. Because of Newton's third law, we may say that the plane experiences an attraction towards q of the same magnitude.

6.3.2 Conducting Sphere

A spherical conductor of radius a is insulated and carries a charge Q. There is a point charge q placed at a distance r from the centre of the sphere (see Fig. 6.2). Consider the domain bounded by the sphere at infinity (where the potential vanishes) and the conducting sphere, the total charge over which is specified. Within this domain, the source distribution is also specified, namely $\rho = q\delta(\mathbf{r} - \mathbf{OP})$. Hence, we know that the solution of Poisson's equation will be unique. We now introduce an image charge $q' = -qa/r$ at P' where $OP' = \frac{a^2}{r} = r'$ (say). This is outside our domain of interest and hence does not disturb the relevant source distribution. However, it serves to make the potential everywhere on the surface of the sphere zero, since the potential at any point on the sphere is

$$\begin{aligned}\phi &= \frac{1}{4\pi\epsilon_0}\left[\frac{q}{(a^2+r^2-2ar\cos\theta)^{1/2}} + \frac{q'}{(a^2+r'^2-2ar'\cos\theta)^{1/2}}\right] \\ &= \frac{1}{4\pi\epsilon_0}\left[\frac{q}{(a^2+r^2-2ar\cos\theta)^{1/2}} - \frac{a}{r}\frac{q}{(a^2+\frac{a^4}{r^2}-\frac{2a^3}{r}\cos\theta)^{1/2}}\right] \\ &= 0\end{aligned}$$

However, the sphere is not earthed but insulated and carries a charge Q. To satisfy this boundary condition, we introduce another charge $Q - q'$ at the centre. This will not disturb the equipotential nature of the spherical surface, which is demanded because the sphere is conducting. Also, if we apply Gauss' theorem to a surface surrounding the sphere but not containing P, the total charge contained will be Q in agreement with the specified charge on the sphere. (Note that the potential on the sphere is

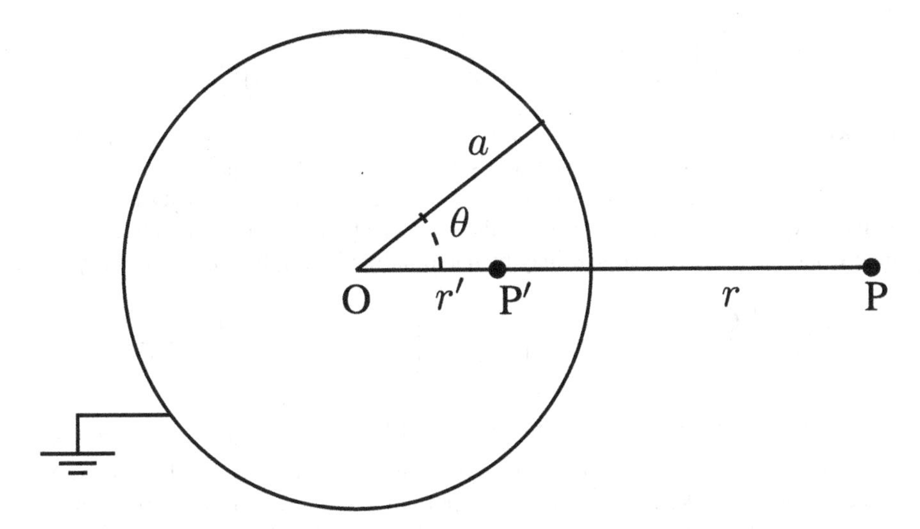

Fig. 6.2 Method of images: a charge in presence of a conducting sphere

$(Q - q')/4\pi\epsilon_0 a$ and thus depends on both the position and magnitude of the charge at P.)

At any point (R, θ) outside the sphere (i.e., $R \geq a$), the potential is

$$\phi(R,\theta) = \frac{1}{4\pi\epsilon_0}\left[\frac{Q-q'}{R} + \frac{q}{(R^2+r^2-2Rr\cos\theta)^{1/2}} - \frac{qa}{r(R^2+r'^2-2Rr'\cos\theta)^{1/2}}\right]$$

The radial component of the field is therefore

$$\begin{aligned} E_R &= -\frac{\partial\phi}{\partial R} \\ &= \frac{1}{4\pi\epsilon_0}\frac{Q-q'}{R^2} + \frac{q}{4\pi\epsilon_0}\left[\frac{(R-r\cos\theta)}{(R^2+r^2-2Rr\cos\theta)^{3/2}} - \frac{a}{r}\frac{(R-r'\cos\theta)}{(R^2+r'^2-2Rr'\cos\theta)^{3/2}}\right] \end{aligned}$$

For $R = a$, the field is purely radial and its magnitude is

$$\begin{aligned} E_R &= \frac{1}{4\pi\epsilon_0}\frac{Q-q'}{a^2} + \frac{q}{4\pi\epsilon_0}\left[\frac{(a-r\cos\theta)}{(a^2+r^2-2ar\cos\theta)^{3/2}} - \frac{a}{r}\frac{(a-r'\cos\theta)}{(a^2+r'^2-2ar'\cos\theta)^{3/2}}\right] \\ &= \frac{1}{4\pi\epsilon_0}\left[\frac{Q-q'}{a^2} + \frac{q}{a}\frac{(a^2-r^2)}{(a^2+r^2-2ar\cos\theta)^{3/2}}\right] \end{aligned}$$

The charge density on the sphere at an angle θ to OP is therefore

$$\frac{1}{4\pi}\left[\frac{Q-q'}{a^2}+\frac{q}{a^2}\frac{1-(r^2/a^2)}{[1+(r^2/a^2)-2(r/a)\cos\theta]^{3/2}}\right]$$

The second term in the above expression is everywhere of a sign opposite to q while the first term is everywhere of the same sign as $Q-q'$. If the charge q be close to the sphere, i.e., $r=a(1+\delta)$, where $\delta<1$, the negative charge density for $\theta=0$ would be quite large and consequently the net force between the positively charged sphere and the positive charge q may be one of attraction. In fact, the force on q is

$$\frac{q}{4\pi\epsilon_0}\left[\frac{Q-q'}{r^2}+\frac{q'}{(r-r')^2}\right]=\frac{q}{4\pi\epsilon_0}\left[\frac{Q-q'}{r^2}-\frac{qa}{r^3(1-a^2/r^2)^2}\right]$$

So the force will be one of attraction if $\left(1-\frac{a^2}{r^2}\right)<\frac{qa^2}{Qr^2}\left(2-\frac{a^2}{r^2}\right)$. Putting $r=a(1+\delta)$ and neglecting higher powers of δ, this gives $\delta<\frac{q}{4Q}$. This force of attraction between like charges plays an important role in the work function of electrons being extracted from an already negatively charged body.

Somewhat different boundary conditions may be easily dealt with. Thus, if the sphere be earthed, there would be no image charge at the centre while if the sphere be maintained at a constant potential the image charge at the centre will be $4\pi\epsilon_0 Va$ where V is the given potential of the sphere.

6.3.3 *Two Spheres Intersecting at Right Angle*

Figure 6.3 shows two spheres with centres at A and B and intersecting at right angle $\angle ACB=90°$ and from geometry, Cd is perpendicular to AB. We assume the spheres to be conducting so that they are at the same potential. We have by simple geometry

$$Cd=ab/\sqrt{a^2+b^2}$$
$$Ad=a^2/\sqrt{a^2+b^2}$$
$$Bd=b^2/\sqrt{a^2+b^2}$$

where a and b are the radii of the two spheres, i.e., AC and BC, respectively.

Consider now the effect of the following three charges:

(i) a charge of magnitude $+4\pi\epsilon_0 a$ placed at A;
(ii) a charge of magnitude $+4\pi\epsilon_0 b$ placed at B;
(iii) a charge of magnitude $-4\pi\epsilon_0\frac{ab}{\sqrt{a^2+b^2}}$ at d.

From our previous discussion of image charges in the case of a spherical conductor, we see that the net effect of all the three charges is to produce unit potential over the entire system of the two spheres. Hence, if we are given the potential of the system as V, we may multiply all the three charges by the factor V and obtain the field

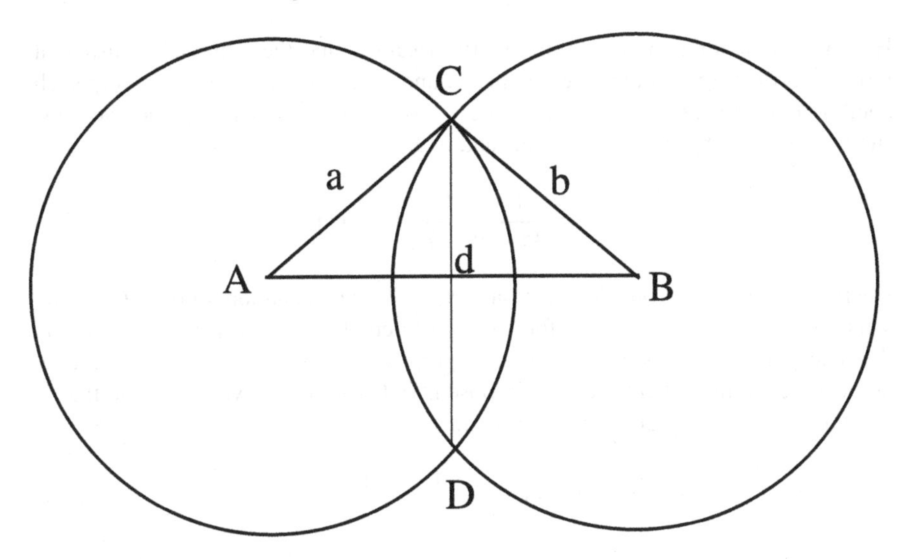

Fig. 6.3 Intersecting spheres

at any outside point due to the three charges. We can also solve the case when the total charge on the system is given, for obviously the capacity of the system (i.e., the charge required to raise the potential by unity) is

$$4\pi\epsilon_0\left[a+b-\frac{ab}{\sqrt{a^2+b^2}}\right]$$

If $a \gg b$, the system reduces to a large sphere with a hemispherical boss on it. The capacity can then be approximated by $4\pi\epsilon_0 a\left(1+\frac{1}{2}\frac{b^3}{a^3}\right)$. The increase of capacity thus depends on the relative volumes of the boss and the sphere.

6.4 Green's Function Method

Considering Poisson's equation for a unit point charge at $\mathbf{r}'$

$$\nabla^2\phi = -\frac{1}{\epsilon_0}\delta(\mathbf{r}-\mathbf{r}') \tag{6.2}$$

One usually writes the solution as $\phi(\mathbf{r})$, where

$$\phi(\mathbf{r}) = \frac{1}{4\pi\epsilon_0}\frac{1}{|\mathbf{r}-\mathbf{r}'|} \tag{6.3}$$

However, in writing (6.3), it is implicitly assumed that the potential vanishes at infinity and there is no other restricting boundary condition. If there be any such condition, as in the case of a point charge in front of a conducting plane, we must supplement (6.3) by some extra terms and in general we may write

$$\phi(\mathbf{r}) = \frac{1}{4\pi\epsilon_0}\frac{1}{|\mathbf{r}-\mathbf{r}'|} + \chi(\mathbf{r},\mathbf{r}') \tag{6.4}$$

where in the domain considered, χ satisfies Laplace's equation so that (6.4) may satisfy (6.2). We write χ in the form shown to emphasize that it depends both on the field point and the source point. The purpose of χ is to satisfy the additional boundary conditions that have been imposed. To fix our ideas, we may recall that in the point charge-conducting plane problem

$$\chi(\mathbf{r},\mathbf{r}') = -\frac{1}{4\pi\epsilon_0}\frac{1}{|\mathbf{r}-\mathbf{r}'|} \tag{6.5}$$

the potential due to the image charge, where $\mathbf{r}''$ is the position vector of the image point of $\mathbf{r}'$.

Now in a general electrostatic problem, owing to the linear superposition principle, we may consider the potential due to a charge distribution to be the sum (or rather the integral) of the potential due to point charges. In that case, we call the right-hand side of (6.4) Green's function of the problem and denote it by $G(\mathbf{r},\mathbf{r}')$

$$G(\mathbf{r},\mathbf{r}') = \frac{1}{|\mathbf{r}-\mathbf{r}'|} + \chi(\mathbf{r},\mathbf{r}') \tag{6.6}$$

where $\nabla'^2\chi(\mathbf{r},\mathbf{r}') = 0$.

Thus, $G(\mathbf{r},\mathbf{r}')$ is, essentially, the potential at $\mathbf{r}$ due to a charge $4\pi\epsilon_0$ (or, equivalently, the potential due to a unit charge in Gaussian units) at $\mathbf{r}'$ with some specific boundary condition. (Note that in Green's function, $\mathbf{r}$ and $\mathbf{r}'$ have a reciprocal role. It gives as well the potential at $\mathbf{r}'$ due to a unit charge at $\mathbf{r}$ subject to the same boundary condition.)

Now consider Green's formula

$$\int \left(\phi(\mathbf{r}')\nabla'^2 G - G\,\nabla'^2\phi(\mathbf{r}')\right) dv' = \int \left(\phi(\mathbf{r}')\nabla' G - G\,\nabla'\phi(\mathbf{r}')\right)\cdot \mathbf{ds} \tag{6.7}$$

where $\phi(\mathbf{r})$ is the potential function to be determined and Green's function $G(\mathbf{r},\mathbf{r}')$ as just defined satisfies the equation

$$\nabla'^2 G(\mathbf{r},\mathbf{r}') = \nabla^2 G(\mathbf{r},\mathbf{r}') = -4\pi\delta(\mathbf{r}-\mathbf{r}') \tag{6.8}$$

Using (6.8) in (6.7), we get

$$-4\pi\phi(\mathbf{r}) - \int G\nabla'^2\phi(\mathbf{r}')dv' = \int [\phi(\mathbf{r}')\nabla' G - G\nabla'\phi(\mathbf{r}')] \cdot \mathbf{ds} \tag{6.9}$$

We are now to specify $G(\mathbf{r}, \mathbf{r}')$ further keeping in mind the conditions (i.e., we are to specify χ). The boundary conditions are usually given in either of two ways.

(a) The value of the potential may be specified over the boundary surface. We call this the Dirichlet boundary condition.
(b) The value of the gradient of the potential (i.e., the electrostatic field intensity) normal to the boundary surface may be specified. In the case of a conducting surface boundary, this means the specification of charge density on the surface. This is known as Neumann's boundary condition.

Obviously, the two types of boundary conditions are such as to ensure the uniqueness of potential function. In case (a), the Dirichlet boundary condition, we choose $\chi(\mathbf{r}, \mathbf{r}')$ such that on the boundary surface $G(\mathbf{r}, \mathbf{r}') = 0$.

Recall that this has been attained previously by choosing an image charge. Thus, in this case, the success of Green's function method depends on our ability to find out image charges. We have already seen that image charges are found quite easily for plane and spherical surfaces. Once $G(\mathbf{r}, \mathbf{r}')$ is chosen in that way, Eq. (6.9) becomes

$$\phi(\mathbf{r}) = \frac{1}{4\pi\epsilon} \int \rho(\mathbf{r}')G(\mathbf{r}, \mathbf{r}')dv' - \frac{1}{4\pi} \oint \phi(\mathbf{r}')\nabla' G \cdot \mathbf{ds}' \tag{6.10}$$

As everything on the right-hand side is known, the problem is solved. It is not difficult to read Eq. (6.10) physically. As $G(\mathbf{r}, \mathbf{r}')$ gives the potential at $\mathbf{r}$ due to a positive charge of magnitude $4\pi\epsilon_0$ at $\mathbf{r}'$, and the first integral on the right-hand side gives the potential due to the given charge distribution. The second integral which apparently ensures that the boundary condition is satisfied arises from charges which lie either outside the domain considered, or as surface charges on the boundary.

In case (b), $G(\mathbf{r}, \mathbf{r}')$ is so chosen that its gradient normal to the boundary surface has a constant value over the entire surface, being equal to $-4\pi/S$, where S is the total area of the boundary surface. With this choice, Eq. (6.9) assumes the form

$$\phi(\mathbf{r}) = \frac{1}{4\pi\epsilon_0} \int \rho(\mathbf{r}')G(\mathbf{r}, \mathbf{r}')dv' + \langle\phi_{s'}\rangle + \frac{1}{4\pi} \oint \nabla'\phi(\mathbf{r}')G(\mathbf{r}, \mathbf{r}') \cdot \mathbf{ds}' \tag{6.11}$$

where

$$\langle\phi_{s'}\rangle = \frac{1}{S} \oint \phi \, dS$$

Everything on the right except $\langle\phi_{s'}\rangle$ is known. However, $\langle\phi_{s'}\rangle$ is not a function of r, so that $\phi(\mathbf{r})$ is determined up to an additive constant which is taken to vanish if the domain extends to infinity.

Note that we have not proved that the potentials given by (6.10) and (6.11) are really the solutions to the problem with the specified boundary conditions. The proof is difficult and will be left out.

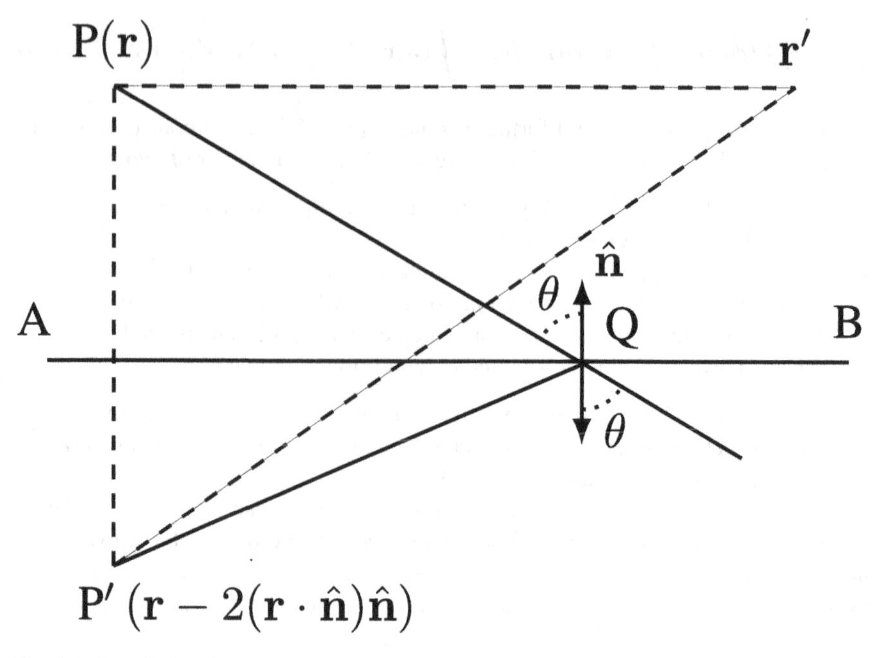

Fig. 6.4 Green's function for an image charge for a conducting plane

We have already noted the relation between $\chi(\mathbf{r}, \mathbf{r}')$ and the image charge potential. We discuss two examples where this is utilized.

(I) The potential is given over an infinite plane, the plane is of course not a simple conducting surface. Corresponding to the position $\mathbf{r}'$, the image position is $\mathbf{r}'' = \mathbf{r}' - 2(\mathbf{r}' \cdot \mathbf{n})\mathbf{n}$. Hence,

$$G(\mathbf{r}, \mathbf{r}') = \frac{1}{|\mathbf{r} - \mathbf{r}'|} - \frac{1}{|\mathbf{r} - \mathbf{r}' + 2(\mathbf{r}' \cdot \mathbf{n})\mathbf{n}|}$$

where $\mathbf{n}$ is the unit vector normal to the plane; see Fig. 6.4. Note that $G(\mathbf{r}, \mathbf{r}') = G(\mathbf{r}', \mathbf{r})$, hence the image point may be considered at $\mathbf{r} - 2(\mathbf{r} \cdot \mathbf{n})\mathbf{n}$. Let AB be the plane over which the potential is specified, P the field point $\mathbf{r}$. Then the image position is P′. As the charge density vanishes everywhere, the first integral in (6.10) falls off and hence we have merely to find the normal gradient of $G(\mathbf{r}, \mathbf{r}')$ for points on the plane AB. This is because our choice of $G(\mathbf{r}, \mathbf{r}')$ is merely the resultant field due to unit positive charge at P and unit negative charge at P′. Hence, it is

$$(\nabla G)_Q = -\frac{2\cos\theta}{\mathrm{PQ}^2} = -2\frac{d\omega}{ds}$$

where $d\omega$ is the solid angle subtended by the element of the plane ds at P. We have therefore, from (6.10), that the potential at P is

$$\phi(\mathbf{r}) = \frac{1}{2\pi} \int_{\text{plane AB}} \phi \, d\omega \tag{6.12}$$

Note that if we apply (6.12) to a point infinitesimally close to the plane AB we get the specified value of the potential at the plane, as indeed we should since the solution satisfies the given boundary conditions.

(II) Next, we consider the boundary to be a sphere of radius a over which the potential is specified, see Fig. 6.5. Again $G(\mathbf{r}, \mathbf{r}')$ is just the combined potential due to a unit positive charge at P and its image charge. This is, as seen before

$$G(\mathbf{r}, \mathbf{r}') = \frac{1}{(r^2 + r'^2 - 2rr' \cos \psi)} - \frac{a}{r\left(r'^2 + \frac{a^4}{r^2} - \frac{2a^2 r' \cos \psi}{r}\right)}$$

$$(\nabla G)_{(r'=a)} = \frac{(r^2 - a^2)}{a(a^2 + r^2 - 2ar \cos \psi)^{3/2}}$$

Hence from (6.10), the potential at any point r (for $r > a$) is given by

$$\phi(\mathbf{r}) = \frac{1}{4\pi} \int \phi(a, \theta', \varphi') \frac{(r^2 - a^2)}{a(a^2 + r^2 - 2ar \cos \psi)^{3/2}} ds'$$

where the ϕ within the integral is the specified potential at the point (a, θ', φ'), ds' is the element of surface area of the sphere at that point and ψ is the angle between directions (θ, φ) and (θ', φ')) so that $\cos \psi = \cos \theta \cos \theta' + \sin \theta \sin \theta' \cos(\varphi - \varphi')$. For $r < a$, the outward normal changes sign and hence in the above expression $(r^2 - a^2)$ is replaced by $(a^2 - r^2)$. Thus, we may write (Fig. 6.5)

$$\phi(\mathbf{r}) = \begin{cases} \dfrac{1}{4\pi} \displaystyle\int\int \phi(a, \theta', \varphi') \frac{(r^2 - a^2) a \sin \theta'}{(a^2 + r^2 - 2ar \cos \psi)^{3/2}} d\theta' d\varphi' & (r > a) \\ \dfrac{1}{4\pi} \displaystyle\int\int \phi(a, \theta', \varphi') \frac{(a^2 - r^2) a \sin \theta'}{(a^2 + r^2 - 2ar \cos \psi)^{3/2}} d\theta' d\varphi' & (r < a) \end{cases}$$

6.5 Method Using Complex Variables

Not infrequently we come across two-dimensional (planar) problems. In potential theory, Laplace's equation then becomes

$$\nabla^2 \phi = \frac{\partial^2 \phi}{\partial x^2} + \frac{\partial^2 \phi}{\partial y^2} = 0 \tag{6.13}$$

so that both the real and the imaginary parts of any analytic function of the complex variable $(x + iy)$ would separately satisfy Eq. (6.13). One can proceed in the reverse:

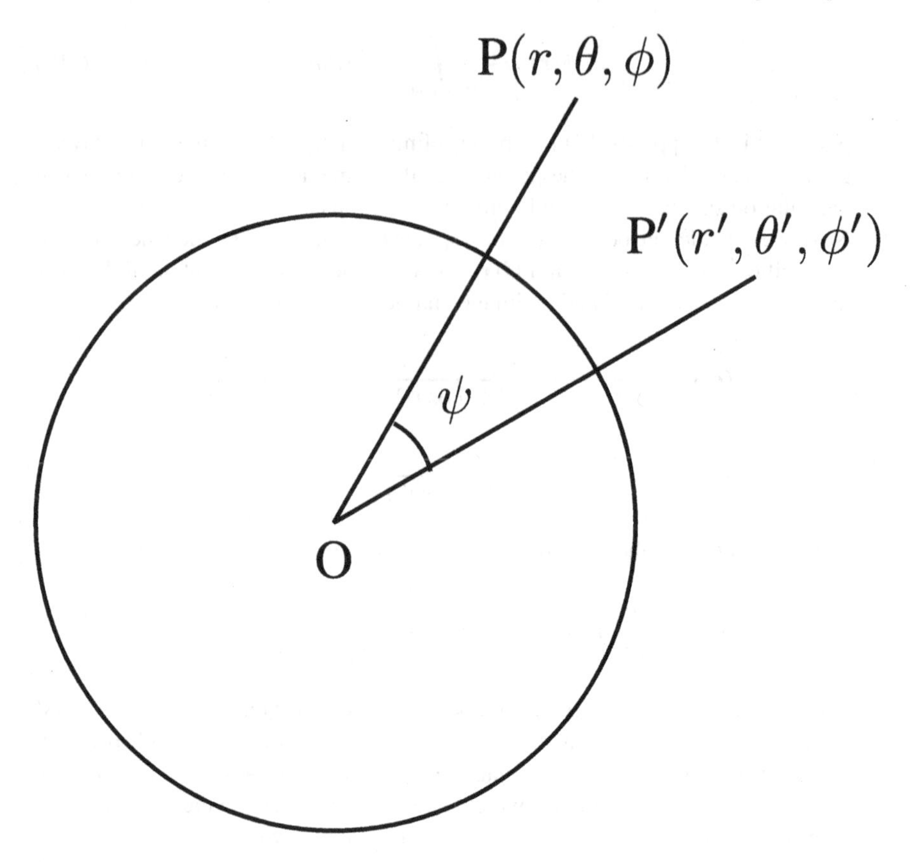

Fig. 6.5 Green's function for a charge and a conducting sphere

take any analytic function of $(x + iy)$, identify the real (imaginary) part with the potential and see what type of boundary condition it may satisfy. If this boundary condition agrees with some physical situation, you have a solution to a physically relevant problem. If ψ be the imaginary (real) part corresponding to ϕ, then the ψ-constant lines are everywhere orthogonal to the ϕ-constant lines. Thus, while ϕ-constant lines are equipotentials, ψ-constant lines give the direction of the electrical intensity at any point and are called the lines of force. We have thus

$$E_x = -\frac{\partial \phi}{\partial x} = \frac{\partial \psi}{\partial y}$$
$$E_y = -\frac{\partial \phi}{\partial y} = \frac{\partial \psi}{\partial x}$$

We may consider a few examples.

(a) Let $\phi + i\psi \equiv f(x+iy) = a(x+iy)$, where a is a real constant. Then, $\phi = ax$ and $\psi = ay$, hence the only non-zero component of the electric field is $E_x = -a$, i.e., it corresponds to a charged infinite plane (the yz-plane).

(b) Let

$$f(x+iy) = A\ln(x+iy) = A\ln(re^{i\theta})$$
$$= iA\theta + A\ln r$$

where $r^2 = x^2 + y^2$ and $\tan\theta = y/x$. Then $\phi = A\log r$ and $\psi = A\theta$. The electric field is in the radial direction and has the magnitude A/r corresponding to a charged infinite cylinder.

(c) Let $f(x+iy) = ae^{ik(x+iy)} = ae^{-ky}(\cos kx + i\sin kx)$. Then $\phi = ae^{-ky}\cos kx$. This type of potential function may be attained in the rectangular box shown Fig. 6.6. The two ends AA′ and BB′ are at zero potential, the coordinates of A and B being $x = -\pi/2k$ and $\pi/2k$, respectively. The side AB is at potential $\phi = a\cos kx$, harmonically varying from A to B. At a very large distance $y \to \infty$, $\phi \to 0$. Note that by taking a sum of such terms, one may obtain any desired potential in AB.

(d) Let $\phi + i\psi = f(x+iy) = \frac{1}{a}\cosh^{-1}(x+iy)$, or

$$x + iy = \frac{1}{2}\left(e^{a(\phi+i\psi)} + e^{-a(\phi+i\psi)}\right)$$

which gives

$$x = \cos a\psi \cosh a\phi, \qquad y = \sin a\psi \sinh a\phi$$

The ϕ-constant lines (for $\phi = \phi_0$) are given by

$$\frac{x^2}{\cosh^2 a\phi_0} + \frac{y^2}{\sinh^2 a\phi_0} = 1$$

i.e., they are confocal ellipses, while the ψ-constant lines (for $\psi = \psi_0$) are

$$\frac{x^2}{\cos^2 a\psi_0} - \frac{y^2}{\sin^2 a\psi_0} = 1$$

which are hyperbolae. Hence, by choosing ϕ as the potential function we have the field due to an infinite cylinder of elliptic cross-section.

(e) As a last example, let $f(x+iy) = (x+iy)^n$, for n a real constant which need not be integral. Writing $x + iy = re^{i\theta}$, we get $\phi + i\psi = r^n[(\cos n\theta + i\sin n\theta)]$, so that $\phi = r^n\cos n\theta$ and $\psi = r^n\sin n\theta$. We may as well take $\phi = r^n\cos n\theta$ and $\psi = \psi_0 + r^n\sin n\theta$.
Taking ψ as the potential, we see that at $\theta = 0, \pi/n$ and $\psi = \psi_0$ for all r. Hence, we may take this to be the potential due to a conducting edge with sides making an angle $\alpha = \pi/n$ (Fig. 6.7a). The field components are

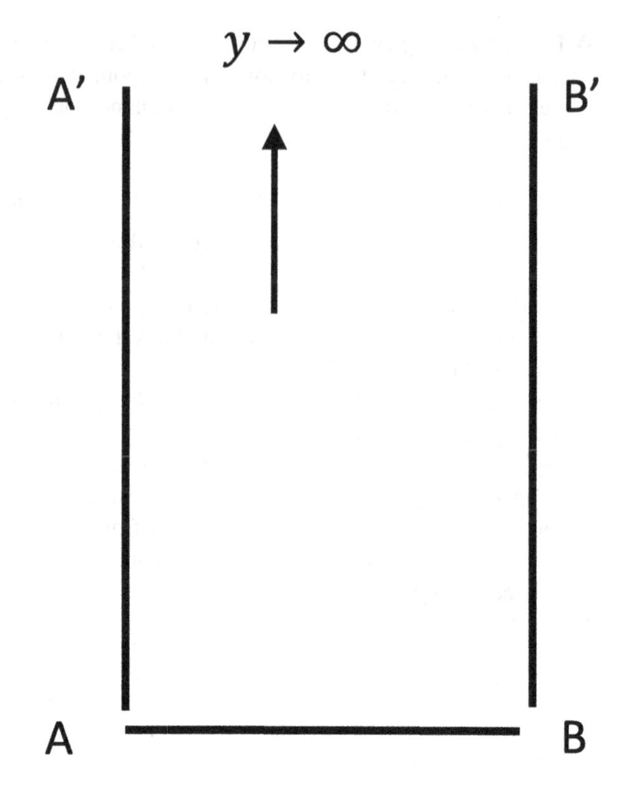

Fig. 6.6 An infinite rectangular box with specified boundary condition

$E_r = \dfrac{\partial \psi}{\partial r} = -nr^{n-1} \sin n\theta$, which vanishes over the conducting edges (as it must for $\mathbf{E}$ perpendicular to the conducting surfaces) and $E_\theta = -\dfrac{1}{r}\dfrac{\partial \psi}{\partial \theta} = -nr \cos n\theta$. Thus, $|E|$ has the value nr^{n-1} over the edges. The charge density σ is $n\epsilon_0 r^{n-1}$. If $n > 1$, the field as well as the charge density vanish at the corner. If, however, $n < 1$ and $\alpha > \pi$, the wedge (see Fig. 6.7b) has an obtuse angle and E_0 as well as σ diverge. In the limit $\alpha \to 2\pi$, $n \to 1/2$, the charge density $\sigma \sim r^{-1/2}$. This divergence of $\mathbf{E}$ at a sharp edge accounts for the breakdown of the air near the lightning conductor and consequent passage of the discharge through these conductors.

Problems

1. Calculate the field due to a point charge Q placed at $(a, 0, 0)$ in a dielectric of constant ϵ_1, the dielectric constant in the region $x < 0$ being ϵ_2 (i.e., the plane $x = 0$ separates the two dielectrics). Use the boundary conditions to find the image charges at $(-a, 0, 0)$ and $(a, 0, 0)$.
2. An octant of a sphere is bounded by conducting surfaces and earthed. Find the system of image charges for a point charge placed inside it.

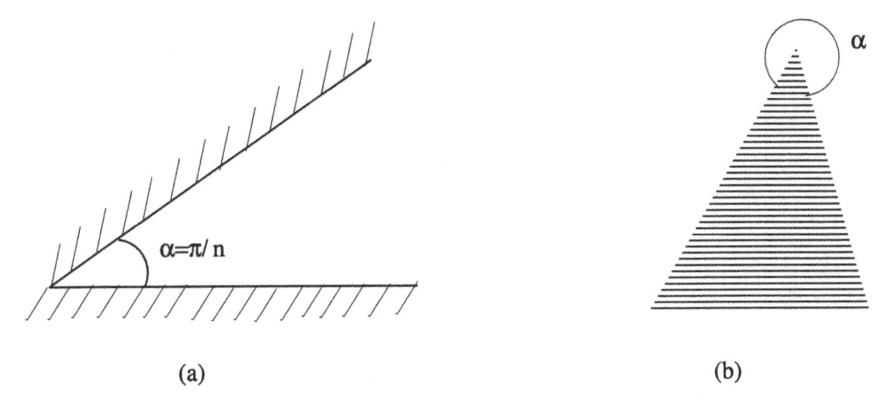

Fig. 6.7 A conducting wedge with **a** acute angle, and **b** obtuse angle

3. A dipole is placed in front of an earthed infinite conducting plane. Show that the net charge induced on the plane vanishes and find the equation of the curve where the surface density of induced charge vanishes. Find also the force and torque on the dipole.
4. An earthed conducting sphere of radius a is surrounded by a dielectric shell of constant ϵ up to radius b, and a point charge q is placed beyond that in air at a distance r from the centre. Calculate the total charge induced on the sphere.
5. Find the field everywhere if a point charge is placed within an earthed conducting spherical shell.
6. An infinite conducting plane has a conducting hemisphere in contact with it. A point charge is placed on the axis of the hemisphere at a distance of $4a$ from its centre and on the same side of the plane as the hemisphere. If the radius of the hemisphere be $3a$, show that the induced charges on the hemisphere and the plane are in the ratio 13 : 7.

Chapter 7
Field Energy and Forces in Electrostatics

Electrostatic forces being derivable from a potential, one can associate potential energy with charges moving under electrostatic forces. Field theory, however, makes it possible to consider that this energy is associated with the field. Thus, we shall express the energy solely in terms of the field variables **E** and **D** (as distinct from the charge distributions or the magnitude and positions of discrete charges). Indeed, the ideas go further: in the case of electromagnetic radiation (which does not however belong to electrostatics), we need hardly consider the sources when we consider the absorption of radiation energy. Also in the general theory of relativity, where electromagnetic energies have gravitational influence, one considers the energy to be spread out in the field region rather than localized in the charges and currents.

In order to bring the charge $\delta q = \delta\rho\, dv$ to the volume element dv, so as to increase the charge density there from ρ to $\rho + \delta\rho$ the work done is $\phi\delta\rho\, dv$, where ϕ is the potential at the volume element. Hence, the total work done to establish a charge density distribution ρ from a state of vanishing charge density is

$$\int_v \int_\rho \phi\delta\rho\, dv$$

where the integral extends over all space and the limits of ρ are from 0 to ρ. We have

$$\nabla \cdot \mathbf{D} = \rho \quad \text{or} \quad \nabla \cdot (\delta\mathbf{D}) = \delta\rho$$

Hence, the work done is

$$\begin{aligned}\int\int \phi\nabla \cdot (\delta\mathbf{D})\, dv &= \int\int \nabla \cdot (\phi\delta\mathbf{D})\, dv - \int\int \nabla\phi \cdot \delta\mathbf{D}\, dv \\ &= \oint \phi\delta\mathbf{D} \cdot \mathbf{ds} + \int \mathbf{E} \cdot \delta\mathbf{D}\, dv\end{aligned}$$

A. K. Raychaudhuri, *Classical Theory of Electricity and Magnetism*, Texts and Readings in Physical Sciences 21, https://doi.org/10.1007/978-981-16-8139-4_7

The surface integral is over the sphere at infinity and vanishes in case the potential vanishes sufficiently rapidly at infinity—this would be the case if the charges lie within a bounded region. Thus, the work done, which we are considering to be the field energy, is

$$\int\int \mathbf{E}\cdot\delta\mathbf{D}\,dv$$

If the dielectric be linear isotropic, $\mathbf{D} = \epsilon\mathbf{E}$ with ϵ independent of $\mathbf{E}$, then we can integrate the above expression to obtain

$$W = \frac{1}{2}\int \epsilon E^2\,dv = \frac{1}{2}\int \mathbf{E}\cdot\mathbf{D}\,dv$$

Hence, we have the result that the field energy density is $\frac{1}{2}\mathbf{E}\cdot\mathbf{D}$ or $\frac{1}{2}\epsilon E^2$.

Notes:

1. In the process of bringing the charges from infinity, it is assumed that electrostatic conditions are not affected. Thus, the process is quasi-static or virtual.
2. If the dielectric be linear but not isotropic, $D_i = \sum_k \epsilon_{ik} E_k$ and the expression for work done becomes

$$\int\int \left(\sum_k\sum_i \epsilon_{ik} E_i \delta E_k\right) dv$$

   This expression is completely integrable (which means the field energy will depend only on the state of the field and not on the path by which that state has been attained) if the integrand be an exact differential, i.e., if

$$\frac{\partial}{\partial E_l}\left(\sum \epsilon_{ik} E_i\right) = \frac{\partial}{\partial E_k}\left(\sum \epsilon_{il} E_i\right)$$

   or

$$\epsilon_{lk} = \epsilon_{kl}$$

   Thus, the symmetry of the dielectric tensor is established.
3. Considering two point charges, one has the force between them as $e_1 e_2/(4\pi\epsilon_0 r^2)$ and hence the work in bringing them to a separation r from an initial infinite separation is

$$-\frac{1}{4\pi\epsilon_0}\int_\infty \frac{e_1 e_2}{r^2}\,dr = \frac{1}{4\pi\epsilon_0}\frac{e_1 e_2}{r^2}$$

   which may be considered to be the mutual potential energy of the two charges. The result may be generalized to any number of charges:

$$W = \frac{1}{2}\frac{1}{4\pi\epsilon_0}\sum_i \sum_{\substack{k \\ k\neq i}} \frac{e_i e_k}{r_{ik}} = \frac{1}{2}\sum_i e_i \phi_i$$

where ϕ_i is the potential at the position of the ith charge due to all charges other than e_i. There are apparently some differences between the above expression and the field-theoretic expression, e.g., the field-theoretic expression is positive definite and blows up if there be any discrete charge where **E** blows up. The interaction-at-a-distance expressions, however, may have either sign and never blow up unless two charges are coincident. The following analysis clarifies the situation.

Consider for simplicity two point charges. Starting with the field-theoretic expression for energy, we get

$$W = \frac{\epsilon_0}{2}\int E^2\,dv = \frac{\epsilon_0}{2}\int E_1^2\,dv + \frac{\epsilon_0}{2}\int E_2^2\,dv + \epsilon_0\int \mathbf{E}_1\cdot\mathbf{E}_2\,dv$$

where $\mathbf{E}_1$ and $\mathbf{E}_2$ are the fields due to the charges e_1 and e_2, respectively. The last integral may be reduced using Green's theorem.

$$\begin{aligned}
\epsilon_0\int \mathbf{E}_1\cdot\mathbf{E}_2\,dv &= \epsilon_0\int \nabla\phi_1\cdot\nabla\phi_2\,dv \\
&= \epsilon_0\int \nabla\cdot(\phi_1\nabla\phi_2)\,dv - \epsilon_0\int \phi_1\nabla^2\phi_2\,dv \\
&= \epsilon_0\oint(\phi_1\nabla\phi_2)\cdot\mathbf{ds} - \epsilon_0\int\phi_1\nabla^2\left(\frac{1}{4\pi\epsilon_0}\frac{e_2}{\mid\mathbf{r}-\mathbf{r}_2\mid}\right)dv \\
&= \frac{1}{4\pi\epsilon_0}\frac{e_1e_2}{\mid\mathbf{r}_1-\mathbf{r}_2\mid} \\
W &= \frac{\epsilon_0}{2}\int E^2\,dv \\
&= \frac{\epsilon_0}{2}\int E_1^2\,dv + \frac{\epsilon_0}{2}\int E_2^2\,dv + \frac{1}{4\pi\epsilon_0}\frac{e_1e_2}{r_{12}}
\end{aligned}$$

Thus, the field-theoretic expression includes, besides the mutual potential energy, the energy due to the separate fields of the individual charges irrespective of the presence of other charges. These are called self-energy terms and arise from the work required to be done to build up the discrete charges from a state where the charge is spread out in space in infinite dilution. In the mutual potential energy expression, the discrete charges are assumed to be already there. (In the above, we have considered the charges to be in vacuo, the extension to the case of dielectrics is purely formal and is not of any particular significance so far as the basic idea is considered.)

The Stresses in the Dielectric

Field theory demands that forces should be represented in a way communicated through the medium rather than acting at a distance. Hence, if we enclose a charge by a surface in the dielectric, then the forces on this charge due to charges lying outside the surface must be representable as forces acting on this surface. These surface forces are called tensors, a term which we now proceed to explain.

The position of a point in space is given by three numbers which we call the coordinates of the point. These coordinates may be of different types, but we shall restrict ourselves to orthogonal Cartesian coordinates, i.e., we choose three axes which are mutually perpendicular and meet at a point O. The coordinates of any point P are then simply the projections of OP on the three axes. Call these x_1, x_2, x_3, or in general x_i. If we now change over to a new system of three mutually perpendicular axes also meeting at O, the coordinates of P will change to x_i' given by

$$x_i' = \sum \alpha_{ik} x_k \tag{7.1}$$

where the nine coefficients α_{ik} give the cosines of the angles between the two sets of axes. They are, of course, not independent. We express OP^2 in either of the two coordinate systems to obtain

$$\begin{aligned} \mathrm{OP}^2 &= \sum_i x_i x_i = \sum_i x_i' x_i' == \sum_i \sum_k \sum_l (\alpha_{ik}\alpha_{il}) x_k x_l \\ \text{or,} \ \sum_k \sum_l x_k x_l \delta_{kl} &= \sum_i \sum_l \sum_k (\alpha_{ik}\alpha_{il}) x_k x_l \end{aligned}$$

As the above equality holds for all values of the coordinates, we must have

$$\sum_i \alpha_{ik}\alpha_{il} = \delta_{kl} \tag{7.2}$$

Again, we can write the inverse transformation

$$x_i = \sum_k \beta_{ik} x_k'$$

so that $x_i' = \sum_k \sum_l \alpha_{ik}\beta_{kl} x_l'$. Hence, we have

$$\sum_k \alpha_{ik}\beta_{kl} = \delta_{il} \tag{7.3}$$

Using Eqs. (7.2) in (7.3), we get

$$\beta_{kl} = \alpha_{lk}$$
$$\sum_i \beta_{ik}\beta_{il} = \sum_i \alpha_{ki}\alpha_{li} = \delta_{kl}$$

Equation (7.2) is a matrix multiplication relation and shows that if J indicates the determinant $|\alpha_{ik}|$, then $J^2 = 1$ or $J = \pm 1$. For any change of coordinates which does not involve a reflection (i.e., a change from left-handed to right-handed coordinate system or vice versa) $J = +1$, while for changes from left-handed to right-handed (or vice versa) $J = -1$. The transformations that we have considered are called orthogonal transformations and Eq. (7.2) may be taken as the definition of an orthogonal transformation.

A vector is said to be represented by the coordinates of a point so that it will obey the same transformation formulæ as the coordinates. Thus, if A_k be the components of a vector in the unprimed coordinate system and A'_i after the transformation (7.1), we shall have

$$A'_i = \sum_k \alpha_{ik} A_k$$

In fact, this transformation property may be taken as the definition of a vector: a vector is a set of three numbers (called the components of the vector) which obeys the same law of transformation as the coordinates of a point. Now consider the nine numbers formed by multiplying three components of a vector A_i by three components of another vector B_i. Their transformation will obviously be of the form

$$A'_i B'_k = \left(\sum_l \alpha_{il} A_l\right)\left(\sum_m \alpha_{km} B_m\right) = \sum_l \sum_m \alpha_{il}\alpha_{km} A_l B_m$$

Now writing $C_{ik} = A_i B_k$

$$C'_{ik} = \sum_l \sum_m \alpha_{il}\alpha_{km} C_{lm} \tag{7.4}$$

The transformation law (7.4) is said to define a tensor of rank two.

Notes

1. We can in this way go on forming tensors of rank, three, four, etc.
2. A scalar and a vector may be called, respectively, tensors of rank zero and one.
3. One can also get tensors of lower rank by a process called contraction. Put two indices identical and then add up over all values of this index, e.g., $A_1 B_1 + A_2 B_2 + A_3 B_3 = \sum A_i B_i$ is a scalar since

$$\sum A'_i B'_i = \sum_i \sum_k \sum_l \alpha_{ik}\alpha_{il} A_k B_l = \sum_k \sum_l \delta_{kl} A_k B_l = \sum_k A_k B_k$$

where we have used Eq. (7.2). Similarly, $\sum_i C_{ii}$ is a scalar, $\sum_k C_{ik} B_k$ is a vector and so on.

4. The definitions of vectors, tensors, etc., need not be restricted to three dimensions. Indeed in the special theory of relativity, one has a four-dimensional formalism. There, a tensor of rank two has sixteen components, etc.
5. One can also generalize to non-orthogonal or even non-linear transformations. Thus, if the transformation be given as $x_i' = f_i(x_k)$, we consider the coordinate differentials

$$dx_i' = \sum_k \frac{\partial x_i'}{\partial x_k} dx_k = \sum_k \alpha_{ik}(x)\, dx_k, \qquad \left(\alpha_{ik}(x) \equiv \frac{\partial x_i'}{\partial x_k}\right)$$

and these α_{ik}'s then determine the transformation of vectors and tensors. However, in these cases some complexities come in, into which we need not enter here.

6. The relation

$$A_l = lA_x + mA_y + nA_z \tag{7.5}$$

giving the component of the vector A_l in the direction with direction cosines (l, m, n) may be looked upon as an explicit form of the vector transformation formula.

Now let us consider the term stress. If there be forces acting on a surface, they need in general two indices for their specification. First, the index gives the direction of the component. Secondly, the force also depends in general on the orientation of the surface element considered—to fix ideas consider a fluid in equilibrium. If we consider an element in the yz-plane, the force will be in the x-direction alone. But for an element in xy- or xz-plane, the x-component of the force vanishes. Thus, to give complete information about the surface forces we write them as T_{ik} where the index i refers to the component of the force considered and k to the direction of the normal to the surface element. Thus, T_{xy}, T_{yz} and T_{zx} represent tangential forces and T_{ii}'s represent normal forces. In the case of fluids, we thus have $T_{ik} = -p\delta_{ik}$. As defined, the stresses are forces per unit area, so the total x-component of the stress forces on a finite surface would be written as $\int T_{xk} ds_k$, the index k corresponding to the normal to the surface element. Now consider an infinitesimal volume of pyramidal shape bounded by infinitesimal elements of three coordinate planes and a slant face whose normal is indicated by v (see Fig. 7.1).

As the volume of such an element is of a higher order of smallness compared to surface areas, in the limit, forces other than surface forces may be neglected. We thus obtain for the equilibrium in x-direction

$$\begin{aligned} T_{xx} ds_x + T_{xy} ds_y + T_{xz} ds_z + T_{xv} ds_v &= 0 \\ \text{or,}\quad T_{xv} &= lT_{xx} + mT_{xy} + nT_{xz} \end{aligned} \tag{7.6}$$

Fig. 7.1 Stress in a pyramidal volume element

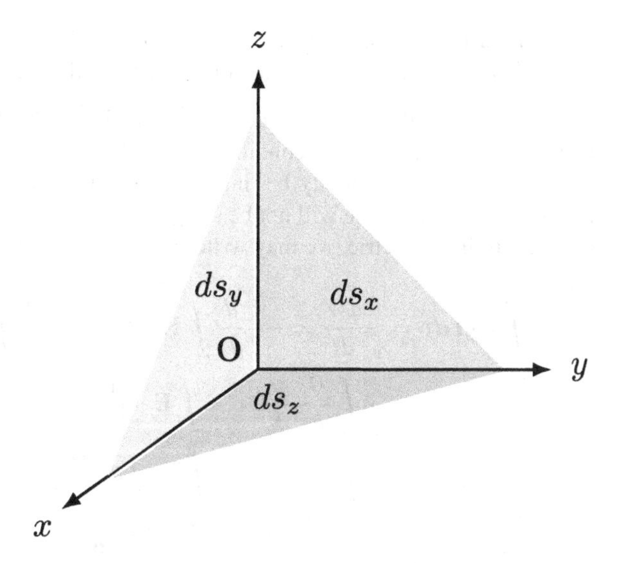

where (l, m, n) are the direction cosines of v, the normal ton face XYZ. Equation (7.6) shows that the second index in T_{ik} is a vector index (cf. Eq. (7.5)). The first index indicating as it does a force component is obviously a vector index. Hence, the stress components form a tensor of rank two. The surface forces acting on a closed surface may be mathematically reduced to a volume distribution of forces over the enclosed volume. Thus, the force in the ith direction is

$$\begin{aligned} F_i &= \oint T_{iv} ds_v = \oint \left(l T_{ix} + m T_{iy} + n T_{iz}\right) ds_v \\ &= \int \left(\frac{\partial T_{ix}}{\partial x} + \frac{\partial T_{iy}}{\partial y} + \frac{\partial T_{iz}}{\partial z}\right) dv \\ &= \int \left(\sum_k \frac{\partial T_{ik}}{\partial x_k}\right) dv = \int f_i \, dv \end{aligned} \tag{7.7}$$

Thus, the volume force density is $f_i = \sum_k \frac{\partial T_{ik}}{\partial x_k}$. Reversing the argument, we may say that if we can represent the force density in the form $\sum_k \frac{\partial T_{ik}}{\partial x_k}$ (called the divergence of the tensor T_{ik}), then we can consider the force to be spread over the boundary surface as stress. The standpoint of the field theory will then be vindicated.

Notes

1. In deducing (7.7), we have used Gauss's theorem. Consider a vector **P** whose x-component alone is non-vanishing, so that $\nabla \cdot \mathbf{P} = \frac{\partial P_x}{\partial x} = \frac{\partial \phi}{\partial x}$ (say). Gauss' theorem now gives

$$\int \nabla \cdot \mathbf{P} \, dv = \oint \mathbf{P} \cdot \mathbf{ds} \rightarrow \int \frac{\partial \phi}{\partial x} dv = \oint l \phi \, ds$$

Similar formulæ have been used in deducing (7.7).

2. In $\sum_k \frac{\partial T_{ik}}{\partial x_k}$, there is a 'contraction' over the index k, hence it is a vector.

Consider now the matter in an electrostatic field to be moving with velocity $\mathbf{u}$. In fact, our idea is of a limiting condition $\mathbf{u} \to 0$ for we are regarding the field to be still electrostatic. (If the velocity be finite, the moving charges will generate magnetic fields and the interaction will not be simply electrostatic.) If $\mathbf{f}$ be the volume density of force in the dielectric, we may write from the conservation of energy principle

$$\begin{aligned}
\int \mathbf{u} \cdot \mathbf{f}\, dv &= -\frac{dW}{dt} = -\frac{1}{2}\frac{d}{dt}\int \mathbf{E} \cdot \mathbf{D}\, dv = -\frac{1}{2}\int \frac{\partial}{\partial t}\left(\frac{D^2}{\epsilon}\right) dv \\
&= \frac{1}{2}\int \frac{D^2}{\epsilon^2}\frac{\partial \epsilon}{\partial t} dv - \int \mathbf{E} \cdot \frac{\partial \mathbf{D}}{\partial t} dv \\
&= \frac{1}{2}\int E^2 \frac{\partial \epsilon}{\partial t} dv + \int \nabla \cdot \left(\phi \frac{\partial \mathbf{D}}{\partial t}\right) dv - \int \phi \frac{\partial}{\partial t}(\nabla \cdot \mathbf{D}) dv \\
&= \frac{1}{2}\int E^2 \frac{\partial \epsilon}{\partial t} dv + \int \phi \frac{\partial \mathbf{D}}{\partial t} \cdot \mathbf{ds} - \int \phi \frac{\partial}{\partial t}(\nabla \cdot \mathbf{D}) dv \\
&= \frac{1}{2}\int E^2 \frac{\partial \epsilon}{\partial t} dv - \int \phi \frac{\partial \rho_e}{\partial t} dv
\end{aligned}$$

where we have assumed the surface integral over the sphere at infinity to vanish. The two integrals on the right show the influence of the change of dielectric constant and the charge distribution separately. We now reduce these by considering the conservation of charge and mass. Considering a volume V with boundary surface S, we have (ρ_m being the mass density)

$$\int \frac{\partial \rho_m}{\partial t} dv + \int \rho_m \mathbf{u} \cdot \mathbf{ds} = 0$$

$$\text{or} \quad \frac{\partial \rho_m}{\partial t} + \nabla \cdot (\rho_m \mathbf{u}) = 0 \tag{7.8}$$

Similarly, if ρ_e be the charge density, conservation of charge gives

$$\frac{\partial \rho_e}{\partial t} + \nabla \cdot (\rho_e \mathbf{u}) = 0 \tag{7.9}$$

Note that we are considering the charges to be tied down with matter, so that their velocities are identical. We have also the operator relation

$$\frac{d}{dt} = \frac{\partial}{\partial t} + \mathbf{u} \cdot \nabla$$

where d/dt signifies differentiation considering a specified element of fluid and $\partial/\partial t$ differentiation considering a fixed point in space. Thus, from (7.8)

$$\frac{d\rho_m}{dt} = -\rho_m \nabla \cdot \mathbf{u}$$

Also, assuming a functional relation between ϵ and ρ_m

$$\begin{aligned}\frac{\partial \epsilon}{\partial t} &= \frac{d\epsilon}{dt} - (\mathbf{u} \cdot \nabla)\epsilon = \frac{d\epsilon}{d\rho_m}\frac{d\rho_m}{dt} - \mathbf{u} \cdot \nabla\epsilon \\ &= -\frac{d\epsilon}{d\rho_m}\rho_m \nabla \cdot \mathbf{u} - \mathbf{u} \cdot \nabla\epsilon \end{aligned} \tag{7.10}$$

Using (7.8) and (7.10) in the energy conservation equation, we get

$$\begin{aligned}\int \mathbf{u} \cdot \mathbf{f}\, dv &= -\frac{1}{2}\int E^2 \rho_m \frac{d\epsilon}{d\rho_m} \nabla \cdot \mathbf{u}\, dv - \frac{1}{2}\int E^2 \mathbf{u} \cdot \nabla\epsilon + \int \phi \nabla \cdot (\rho_e \mathbf{u}) dv \\ &= \frac{1}{2}\int \mathbf{u} \cdot \left[\nabla\left(E^2 \rho_m \frac{d\epsilon}{d\rho_m}\right) - E^2 \nabla\epsilon\right] dv - \int \rho_e \mathbf{u} \cdot \nabla\phi\, dv + \\ &\quad + \text{surface integrals over the sphere at infinity}\end{aligned}$$

Assuming the surface integrals to vanish, we get for the volume density of force

$$\mathbf{f} = \rho_e \mathbf{E} + \frac{1}{2}\nabla\left(E^2 \rho_m \frac{d\epsilon}{d\rho_m}\right) - \frac{1}{2}E^2 \nabla\epsilon \tag{7.11}$$

Now $E^2 \nabla\epsilon = \nabla(E^2\epsilon) - \epsilon\nabla(E^2)$ and $\frac{\partial}{\partial x_i}(E^2) = 2\sum_k E_k \frac{\partial E_k}{\partial x_i} = 2\sum_k E_k \frac{\partial E_i}{\partial x_k}$ (because $\nabla \times \mathbf{E} = 0$) so that the first expression may be written as $E^2 \frac{\partial \epsilon}{\partial x_i} = \frac{\partial}{\partial x_i}\sum(E_k D_k) - 2\sum D_k \frac{\partial E_i}{\partial x_k}$. Hence, the force density assumes the form

$$f_i = \rho_e E_i - \frac{1}{2}\frac{\partial}{\partial x_i}(\sum_k E_k D_k) + \sum_k D_k \frac{\partial E_i}{\partial x_k} + \frac{1}{2}\frac{\partial}{\partial x_i}\left(E^2 \rho_m \frac{\partial \epsilon}{\partial \rho_m}\right)$$

Replacing the charge density ρ by $\sum_k \frac{\partial D_k}{\partial x_k}$, we obtain

$$\begin{aligned}2f_i &= 2E_i \sum_k \frac{\partial D_k}{\partial x_k} - \frac{\partial}{\partial x_i}(\sum_k E_k D_k) + 2\sum_k D_k \frac{\partial E_i}{\partial x_k} + \frac{\partial}{\partial x_i}\left(E^2 \rho_m \frac{\partial \epsilon}{\partial \rho_m}\right) \\ &= 2\sum_k \frac{\partial}{\partial x_k}(E_i D_k) - \frac{\partial}{\partial x_i}\sum_k (E_k D_k) + \frac{\partial}{\partial x_i}\left(E^2 \rho_m \frac{\partial \epsilon}{\partial \rho_m}\right)\end{aligned}$$

If we now write

$$T_{ik} = \frac{1}{2}\left(-\sum_l E_l D_l + E^2 \rho_m \frac{d\epsilon}{d\rho_m}\right) + E_i D_k \tag{7.12}$$

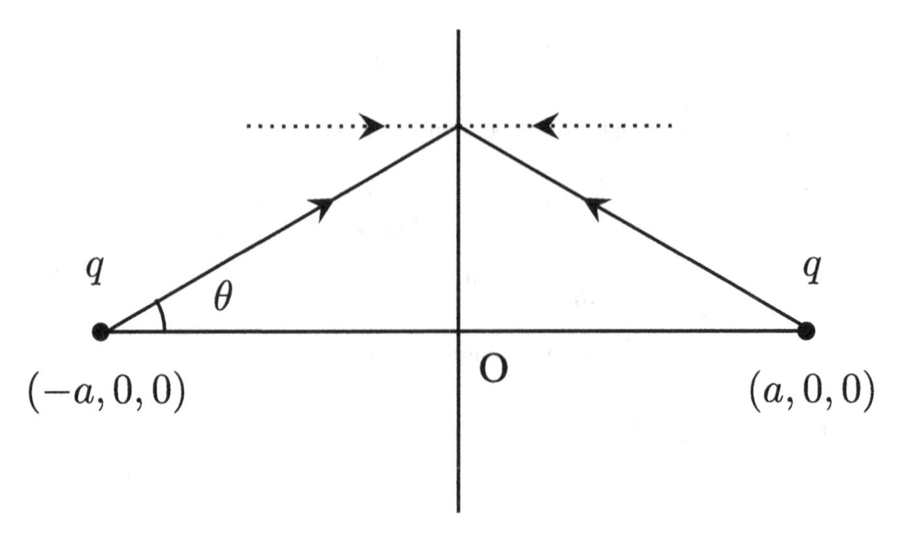

Fig. 7.2 Force between two similar charges using Maxwell stress

the force density assumes the desired form

$$f_i = \sum_k \frac{\partial T_{ik}}{\partial x_k}$$

The expression (7.12) may therefore be considered to be the stress tensor in the dielectric medium. The stress tensor is symmetric, hence, by a well-known theorem, may be diagonalized by an orthogonal transformation.

Note that the force expression (7.11) has some additional terms besides the term $\rho_e \mathbf{E}$ due to true charges. These additional terms involve the gradient of the dielectric constant and may be interpreted as due to the polarization charges. In many cases, the term involving $d\epsilon/d\rho_m$ may be disregarded and the resulting stresses are known as Maxwell stresses. Thus, from (7.12) the Maxwell stresses for an electric field in the x-direction are

$$T_{xx} = \frac{\epsilon}{2}E^2, \quad T_{yy} = T_{zz} = -\frac{\epsilon}{2}E^2$$

with all other components vanishing. (These explain Faraday's ideas of pressure and tension of lines of force.)

One can use the stress tensor to calculate the force between two charges and thus check the correctness of our calculations. As an Example, we consider (see Fig. 7.2) two equal and similar charges at $(-a, 0, 0)$ and $(a, 0, 0)$. Take the surface to consist of the infinite yz-plane through the origin and the hemisphere at right infinity bounded by this plane. (Actually, we might have chosen any surface which encloses one charge but not both. However, for convenience of calculation we choose the surface as described.) As the stress tensor is of second degree in $\mathbf{E}$, its components at points on the hemisphere at infinity will vanish as $1/r^4$ while the surface elements will behave as r^2, therefore, the surface integral over the hemisphere will make no contribution. At any point on the yz-plane, the resultant intensity will be perpendicular to the x-axis and hence T_{xx} (which is the only tensor component coming in the calculation of the surface integral) is given by

$$T_{xx} = -\frac{\epsilon}{2}E^2 = -\frac{\epsilon}{2}\left(\frac{2q\sin\theta}{4\pi\epsilon a^2\sec^2\theta}\right)^2$$

Remembering that the outward drawn normal to the yz-plane is along the negative x-axis, the surface integral $\int T_{ik}ds_k$ becomes

$$\int T_{xx}ds_x = \frac{\epsilon}{2}\int_0^{\pi/2}\left(\frac{2q\sin\theta}{4\pi\epsilon a^2\sec^2\theta}\right)^2 2\pi a^2\tan\theta\sec^2\theta\,d\theta = \frac{q^2}{16\pi\epsilon a^2}$$

which is the correct expression for the force between the two charges.

7.1 Electrostriction

The force f_i must be taken into account in finding the conditions of equilibrium of matter in an electrostatic field. For simplicity, we consider a fluid in which the only stress (non-electrical) is an isotropic pressure p and the consequent volume density of force is $-\nabla p$. Hence for equilibrium (dropping the subscript e in ρ_e),

$$\rho E_i - \frac{1}{2}E^2\frac{\partial\epsilon}{\partial x_i} + \frac{1}{2}\frac{\partial}{\partial x_i}\left(E^2\rho_m\frac{d\epsilon}{d\rho_m}\right) - \frac{\partial p}{\partial x_i} = 0 \tag{7.13}$$

Suppose the fluid is uncharged so that the first term falls off. As ϵ is assumed to be a function of ρ_m alone, the second term is

$$-\frac{1}{2}E^2\frac{d\epsilon}{d\rho_m}\frac{\partial\rho_m}{\partial x_i}$$

therefore (7.13) becomes

$$\frac{1}{2}\rho_m\frac{\partial}{\partial x_i}\left(E^2\frac{\partial\epsilon}{\partial\rho_m}\right) - \frac{\partial p}{\partial x_i} = 0 \tag{7.14}$$

Equation (7.14) shows that p, E^2, $d\epsilon/d\rho_m$ are all functionally related. We shall consider two cases: (a) an incompressible fluid and (b) and ideal gas. In case (a), Eq. (7.14) gives

$$p_1 - p_0 = \frac{1}{2}\rho_m E^2\frac{d\epsilon}{d\rho_m} \tag{7.15}$$

where p_0 is the pressure in a region outside the field. The derivative $d\epsilon/d\rho_m$ may be determined from the Clausius-Mossotti relation Eq. (5.22), which may be re-expressed as (see discussion following derivation of Eq. (5.22))

$$\frac{d\epsilon}{d\rho_m} = \frac{(\epsilon_r - 1)(\epsilon_r + 2)}{3\rho_m}\epsilon_0$$

Hence Eq. (7.15) becomes

$$p_1 - p_0 = \frac{1}{6}(\epsilon_r - 1)(\epsilon_r + 2)\,\epsilon_0 E^2 \tag{7.16}$$

In case of an ideal gas $p = \frac{\rho_m RT}{M}$ (M being the molecular weight of the gas) and assuming isothermal condition, Eq. (7.14) gives

$$\frac{RT}{M}\ln\frac{p_1}{p_0} = \frac{1}{2}E^2\frac{d\epsilon}{d\rho_m} \tag{7.17}$$

For a gas $\epsilon \approx 1$, and so the Clausius-Mossotti formula may be written as

$$\epsilon_r - 1 = n\alpha = \frac{n\alpha}{\epsilon_0} = \frac{N\rho_m\alpha}{M\epsilon_0} \tag{7.18}$$

which leads to $\frac{d\epsilon}{d\rho_m} = \frac{N\alpha}{M}$. Hence, we get

$$kT\ln\frac{p_1}{p_0} = \frac{1}{2}\alpha E^2, \quad \text{or} \quad p_1 = p_0 e^{\alpha e^2/2kT}$$

Thus, for a gas in equilibrium in an electrostatic field, the pressure will be non-uniform, exponentially increasing with the square of the magnitude of the field. The result may be physically understood in terms of the energy of the dipoles that develop due to polarization.

7.2 Pressure Discontinuity at the Boundary of a Dielectric

For simplicity, we shall consider a dielectric in contact with vacuum; the generalization to the case of two dielectrics offers little difficulty. In spite of the discontinuities in the dielectric constant and the field variables, we shall assume that the relevant integrals exist and can be evaluated in a manner disregarding the discontinuities. We have from Eq. (7.13) with $\rho = 0$

$$p_a - p_b = \frac{1}{8\pi}\int_a^b E^2 \frac{\partial \epsilon}{\partial x} dx - \frac{1}{2}\int_a^b \frac{\partial}{\partial x}\left(E^2 \frac{d\epsilon}{d\rho_m}\rho_m\right) dx$$
$$= \frac{1}{2}\int_a^b E^2 \frac{\partial \epsilon}{\partial x} dx - \frac{1}{2}\left[E^2 \frac{d\epsilon}{d\rho_m}\rho_m\right]_b \qquad (7.19)$$

where a and b refer to points on the vacuum side and the matter side, respectively, hence ρ_m vanishes at a. To evaluate the first integral, we split up $\mathbf{E}$ into its components E_n, normal to the boundary and E_t, tangential to the boundary. Hence, keeping in mind the boundary conditions the integral in (7.19) is

$$\frac{1}{2}\int_a^b \left(E_t^2 + \frac{(\epsilon E_n)^2}{\epsilon^2}\right)\frac{\partial \epsilon}{\partial x} dx = \frac{1}{2}(\epsilon - \epsilon_0)E_t^2 + \frac{1}{2}\left(\frac{1}{\epsilon_0} - \frac{1}{\epsilon}\right)(\epsilon E_n)^2$$
$$= \frac{1}{2}(\epsilon - \epsilon_0)\left(E^2 + (\epsilon_r - 1)E_n^2\right)$$

Therefore, Eq. (7.19) gives

$$p_a - p_b = \frac{1}{2}E^2\left((\epsilon - \epsilon_0) - \frac{d\epsilon}{d\rho_m}\rho_m\right) + \frac{1}{2}\epsilon_0 E_n^2(\epsilon - 1)^2$$

Using now the Clausius-Mossotti formula to evaluate $d\epsilon/d\rho_m$, we get

$$p_a - p_b = \frac{1}{2}(\epsilon - \epsilon_0)^2\left(\frac{2}{3}E_n^2 - \frac{1}{3}E_t^2\right)$$

Thus, the influence on pressure discontinuity of E_n and E_t is in opposite directions. (Note that we have put $\rho_m = 0$ and $\epsilon = 1$ at a but retained p_a—the underlying idea is that the medium at a may be a gas instead of vacuum, in which case also the above formulae apply.)

7.3 Force Sucking in a Dielectric into a Capacitor

A capacitor is an arrangement of conductors to facilitate the storage of charge at a relatively low potential. Thus, consider an isolated Sphere; if it is given a charge Q, its potential ϕ_a is $Q/(4\pi\epsilon_0 a)$ where a is the radius of the sphere. The charge required to raise the potential by unity is called the capacity, thus the capacity of an isolated spherical conductor is its radius $4\pi\epsilon_0 a$. If now it is enclosed by a larger conducting sphere of radius b, the potential difference between the two spheres will be $\phi_a - \phi_b = \frac{Q}{4\pi\epsilon_0}\left(\frac{1}{a} - \frac{1}{b}\right)$. If the outer sphere be earthed, the potential of the inner sphere will be $\frac{Q}{4\pi\epsilon_0}\frac{b-a}{ab}$ and consequently the capacity of the system is $4\pi\epsilon_0 ab/(b-a)$. Thus the capacity is increased. Another common form of the capacitor is an arrangement of two parallel plates, separated by a distance small compared to the dimensions of the plates. In this case, the field is uniform, at least in the central region. Hence, if σ be the surface density of charge on the plates (the two plates will be oppositely charged), the electric intensity will be σ/ϵ_0 and the potential difference between the plates will be $\sigma t/\epsilon_0$ where t is the separation between the plates. Thus, the capacity per unit area is ϵ_0/t. In case a dielectric is interposed, the field within the dielectric will be reduced by the factor ϵ and the capacity increased by the same factor.

It is easy to calculate the energy associated with a capacitor. In general, if C be the capacity, the charge $Q = C\phi$, with ϕ being the potential difference between the plates of the capacitor. So the work required to increase the charge by dQ will be $\phi\, dQ = QdQ/C$ and the total work required to charge up the capacitor is

$$\int_0^Q \frac{Q}{C} dQ = \frac{Q^2}{2C} = \frac{1}{2} Q\phi$$

which is the expression of the energy stored in the capacitor. We may express the energy also in the field theoretic form $\frac{1}{2}\int \epsilon_0 E^2\, dv$. Thus, for the spherical capacitor, the energy is

$$\begin{aligned}\frac{1}{2}\int_a^b \epsilon_0 E^2\, dv &= \frac{\epsilon_0}{2}\int_a^b \frac{Q^2}{(4\pi\epsilon_0 r^2)^2}\, dv \\ &= \frac{Q^2}{8\pi\epsilon_0}\left(\frac{1}{a} - \frac{1}{b}\right) \\ &= \frac{Q^2}{2C} = \frac{1}{2} Q\phi\end{aligned}$$

in agreement with our previous expression. The energy is decreased if vacuum be replaced by a dielectric. Thus, considering the parallel plate capacitor the energy per unit area of the capacitor plates, with air or vacuum in the intervening space, is

$$\frac{1}{2}\int_0^t \epsilon_0 E^2\, dx = \frac{\sigma^2 t}{2\epsilon_0} = \frac{\sigma^2}{2C}$$

while with a dielectric of constant ϵ_r, the energy is

$$\frac{1}{2}\int_0^t \mathbf{E}\cdot\mathbf{D}\, dx = \frac{1}{2}\int_0^t \epsilon\left(\frac{\sigma}{\epsilon_0}\right)^2 dx = \frac{\epsilon_r \sigma^2 t}{2\epsilon_0}$$

Consider now an arrangement as shown in Fig. 7.3 in which AA′ and BB′ are the two plates of a parallel plate capacitor distance t apart, and CDEF is a dielectric slab of thickness d which is partially in the space between the plates. If now we consider a virtual displacement of the dielectric slab inwards, there will be a decrease of potential energy as we have just now seen. As a system tends towards the configuration of minimum potential energy, there will be a force trying to suck in the dielectric slab. If $\mathbf{E}$ be the electric intensity in air, that in the dielectric slab will be $\mathbf{E}/\epsilon$, hence the

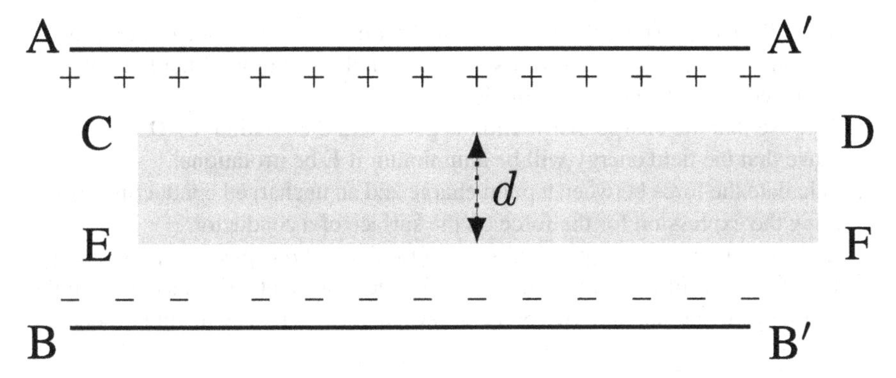

Fig. 7.3 A dielectric slab partly inside a parallel plate capacitor

energy decrease for a virtual displacement dx of the dielectric slab is

$$\frac{1}{2}\epsilon_0\left(E^2 A\,dx - \epsilon\left(\frac{E}{\epsilon_r}\right)^2 A\,dx\right) = \frac{1}{2}\epsilon_0 E^2\frac{\epsilon_r - 1}{\epsilon_r}A\,dx$$

This must be equal to $FA\,dx$ where F is the force per unit area of cross-section and A is the area of cross-section of the dielectric slab. Thus

$$F = \frac{1}{2}\frac{\epsilon_r - 1}{\epsilon_r}\epsilon_0 E^2$$

7.4 Force on the Surface of a Charged Conductor

Suppose we have a charged conducting surface. If we consider a virtual displacement **dl** of a surface element **ds** of the conductor, then the volume of space swallowed by the conductor is **ds** · **dl**. As there is no electric field in the body of the conductor, the field energy has been decreased by an amount $\frac{1}{2}\epsilon_0 E^2 \mathbf{ds}\cdot\mathbf{dl}$ by this virtual displacement. Hence, the force per unit area of the surface of the conductor is $\epsilon_0 E^2/2$. As at the surface the intensity is normal to the surface and of magnitude σ/ϵ_0, we say that there is a normal outward force of magnitude $\sigma^2/2\epsilon_0$ per unit area, where σ is the surface density of the charge at the point.

Problems

1. Calculate the energy of a spherically symmetric distribution of charge of uniform density (a) by considering that the distribution is built up by piling up charges and (b) from the field-theoretic expression.
2. Calculate the force between two equal and opposite charges by considering the surface integral of Maxwell stresses.

3. A small metallic sphere is in a non-uniform electric field. The dimensions of the sphere are so small that over such a region the variation of the field may be neglected. Calculate the force on the sphere.
4. Suppose that the charge distribution is given and the relation $\nabla \cdot \mathbf{D} = \rho$ holds. Prove that the field energy will be a minimum if **E** be irrotational.
5. Calculate the force between a point charge and an uncharged conducting sphere using the expression for the force on the surface of a conductor.
6. At the centre of a thick spherical shell of radii a and b is placed a point charge. The shell is insulated and uncharged. Imagine a plane at a distance p from the centre to cut the shell into two parts. Show that the two parts will tend to separate if $p > \frac{ab}{\sqrt{a^2 + b^2}}$.
7. The field energy due to a monopole blows up as $1/r$, while that due to a dipole as $1/r^3$. Explain this result by considering the model of a dipole as two equal and opposite charges. Extend your consideration to higher multipoles.

Chapter 8
Stationary Currents and Magnetic Fields

The study of magnetism began with permanent magnets. Although, unlike in electrostatics, the elementary sources were dipoles rather than monopoles, the basic law of interaction could be considered by introducing hypothetical poles. The fundamental laws could be written as follows:

1. The torque on an elementary magnet in a field **B** is given by $\boldsymbol{\mu} \times \mathbf{B}$—a relation which could also be used as a definition of **B**, the magnetic flux density or induction. With the given background field of the earth, the relation also enabled a comparison of the magnetic moments of different magnets, denoted by $\boldsymbol{\mu}$.
2. The field due to a magnet of moment $\boldsymbol{\mu}$ is given by $\mathbf{B} = -\nabla\phi = -C\nabla\left(\frac{\boldsymbol{\mu}\cdot\mathbf{r}}{r^3}\right)$, a result identical to that of electrical dipoles, with the constant $\frac{1}{4\pi\epsilon}$ being replaced by a constant C which is appropriate to the system of units being used. This indicates an inverse square central field for the hypothetical poles. (The magnet must of course be of infinitesimal dimensions.)

With the discovery of the interaction between electrical currents and permanent magnets, as also between electrical currents themselves, the idea slowly gained ground that the basic sources of magnetic fields are electrical currents and formally permanent magnets were also represented as distributions of hypothetical microscopic currents. To study this formalism, we start with the laws of interaction as followed from the investigations of Oersted, Ampere and Biot and Savart. As in the case of electrostatics, we describe the interactions by introducing the concept of an intermediary field. The elementary field due to an infinitesimal element of length $d\mathbf{l}_1$ carrying a current I_1 is given by

$$d\mathbf{B} = \frac{\mu_0}{4\pi} I_1 \frac{d\mathbf{l}_1 \times \mathbf{r}}{r^3} \tag{8.1}$$

A. K. Raychaudhuri, *Classical Theory of Electricity and Magnetism*, Texts and Readings in Physical Sciences 21, https://doi.org/10.1007/978-981-16-8139-4_8

where the constant μ_0 on the right-hand side is the magnetic permeability of free space given by

$$\frac{\mu_0}{4\pi} = 10^{-7}\,\text{N/A}^2$$

which is analogous to the permittivity of free space in electrostatics.[1] The force on the second element of conductor of length $d\mathbf{l}_2$ carrying a current I_2 in a field $\mathbf{B}$ is given as

$$d\mathbf{F} = I_2 d\mathbf{l}_2 \times \mathbf{B} \tag{8.2}$$

Equations (8.1) and (8.2) provide a definition of the unit of current, Ampere, in terms of which the unit of charge Coulomb is defined.

The current at a point in a wire is defined as the charge crossing that point per unit time. We shall find it convenient to introduce the idea of a current density vector; it will be defined as the amount of charge crossing unit area of a surface element held normal to the direction of charge flow, the direction of the vector identified with the direction of flow. Thus, if $\mathbf{j}$ be the current density vector, we have

$$I d\mathbf{l} = \mathbf{j} dv$$

Thus, Eqs. (8.1) and (8.2) can be rewritten in terms of the current vectors and considering a finite distribution we write

$$\mathbf{B}(\mathbf{r}) = \frac{\mu_0}{4\pi} \int \frac{\mathbf{j}(\mathbf{r}') \times (\mathbf{r} - \mathbf{r}')}{|\mathbf{r} - \mathbf{r}'|^3} dv' \tag{8.3}$$

$$\mathbf{F} = \int \mathbf{j} \times \mathbf{B}\, dv \tag{8.4}$$

We consider the condition to be stationary—by which we mean that the observables $\mathbf{B}, \mathbf{j}$, the charge density ρ, etc., are not changing with time although the very existence of currents signifies that things are not altogether static. We shall assume the charge conservation principle, so that considering a volume V bounded by a surface S, the change of total charge in V must be completely accounted for by the flux of charge through the surface S, i.e.,

$$\int \frac{\partial \rho}{\partial t} dV + \oint \mathbf{j} \cdot \mathbf{ds} = 0$$

$$\text{or,} \quad \frac{\partial \rho}{\partial t} + \nabla \cdot \mathbf{j} = 0$$

so that in the steady state

$$\nabla \cdot \mathbf{j} = 0 \tag{8.5}$$

[1] In Gaussian units, in which the current is measured in esu/s and the magnetic induction in emu, the constant of proportionality is taken to be unity.

With discrete wires carrying currents, the above relation means that the algebraic sum of currents at any point must vanish—a result known in elementary physics as Kirchhoff's law.

It is interesting to note that as the sources—the current densities—are now vectors unlike the case of electrostatics, our presumption of a central field in that case no longer holds good here. Indeed, a look at Eq. (8.3) shows that the field is not central. Consequently, **B** is not in general an irrotational vector. However, taking the divergence of Eq. (8.3) and using the vector identities

$$\nabla \cdot (\mathbf{A} \times \mathbf{B}) = \mathbf{B} \cdot (\nabla \times \mathbf{A}) - \mathbf{A} \cdot (\nabla \times \mathbf{B})$$
$$\frac{\mathbf{r} - \mathbf{r}'}{|\mathbf{r} - \mathbf{r}'|^3} = -\nabla \left(\frac{1}{|\mathbf{r} - \mathbf{r}'|} \right)$$

we get

$$\nabla \cdot \mathbf{B} = 0 \tag{8.6}$$

for $\mathbf{j}(\mathbf{r}')$ is not a function of the coordinates of the field point. Hence, **B** is a solenoidal vector and there exists a vector **A** such that

$$\mathbf{B} = \nabla \times \mathbf{A} \tag{8.7}$$

Equation (8.3) gives

$$\begin{aligned} \mathbf{B}(\mathbf{r}) &= -\frac{\mu_0}{4\pi} \int \mathbf{j}(\mathbf{r}') \times \nabla \left(\frac{1}{|\mathbf{r} - \mathbf{r}'|} \right) dv' \\ &= \nabla \times \left(\frac{\mu_0}{4\pi} \int \frac{\mathbf{j}(\mathbf{r}')}{|\mathbf{r} - \mathbf{r}'|} dv' \right) \end{aligned} \tag{8.8}$$

Comparing (8.7) and (8.8), we get

$$\mathbf{A}(\mathbf{r}) = \frac{\mu_0}{4\pi} \int \frac{\mathbf{j}(\mathbf{r}')}{|\mathbf{r} - \mathbf{r}'|} dv' + \nabla \psi \tag{8.9}$$

where ψ is an arbitrary scalar. The above is called a vector potential and the fact that **A** can be altered by changing the arbitrary scalar ψ without bringing any change in **B** is referred to as a gauge invariance. Such a change of **A** is called a gauge transformation.

We next calculate $\nabla \cdot \mathbf{B}$ using (8.8) and the vector identity

$$\nabla \times (\nabla \times \mathbf{A}) = \nabla(\nabla \cdot \mathbf{A}) - \nabla^2 \mathbf{A}$$

to get

$$\begin{aligned}\nabla \times \mathbf{B} &= \nabla\left(\nabla \cdot \frac{\mu_0}{4\pi}\int \frac{\mathbf{j}(\mathbf{r}')}{|\mathbf{r}-\mathbf{r}'|}dv'\right) - \nabla^2 \frac{\mu_0}{4\pi}\int \frac{\mathbf{j}(\mathbf{r}')}{|\mathbf{r}-\mathbf{r}'|}dv' \\ &= \frac{\mu_0}{4\pi}\nabla \int \mathbf{j}(\mathbf{r}')\cdot\nabla\left(\frac{1}{|\mathbf{r}-\mathbf{r}'|}\right)dv' - \frac{\mu_0}{4\pi}\int \mathbf{j}(\mathbf{r}')\nabla^2\left(\frac{1}{|\mathbf{r}-\mathbf{r}'|}\right)dv'\end{aligned}$$

Now using the relations,

$$\begin{aligned}\nabla^2\left(\frac{1}{r}\right) &= -4\pi\delta(\mathbf{r}) \\ \nabla f(|\mathbf{r}-\mathbf{r}'|) &= -\nabla' f(|\mathbf{r}-\mathbf{r}'|) \\ \nabla \times \mathbf{B} &= -\frac{\mu_0}{4\pi}\nabla \int \mathbf{j}(\mathbf{r}')\cdot\nabla'\left(\frac{1}{|\mathbf{r}-\mathbf{r}'|}\right)dv' + \mu_0\mathbf{j}(\mathbf{r})\end{aligned} \tag{8.10}$$

the integrand in the first term of (8.10) is

$$\mathbf{j}(\mathbf{r}')\cdot\nabla'\left(\frac{1}{|\mathbf{r}-\mathbf{r}'|}\right) = \nabla'\cdot\left(\frac{\mathbf{j}(\mathbf{r}')}{|\mathbf{r}-\mathbf{r}'|}\right) - \left(\nabla'\cdot\mathbf{j}(\mathbf{r}')\right)\frac{1}{|\mathbf{r}-\mathbf{r}'|}$$

The first part being a divergence can be converted to a surface integral over a surface where $\mathbf{j}$ vanishes (we are considering a current distribution within a bounded region). The second part vanishes because of the stationary assumption (Eq. (8.5)). Hence, finally

$$\nabla \times \mathbf{B} = \mu_0\mathbf{j} \qquad \boxed{\nabla \times \mathbf{B} = \frac{4\pi}{c}\mathbf{j} \text{ in Gaussian units}} \tag{8.11}$$

Equations (8.6) and (8.11) are the fundamental equations of stationary magnetic fields so long as there is no material medium and especially no permanent magnet. It may be noted that Eq. (8.11), on taking the divergence, gives Eq. (8.5), emphasizing that (8.11) is valid only under stationary conditions.

Applying the operator ∇^2 on Eq. (8.9), we get

$$\nabla^2\mathbf{A} = -\mu_0\mathbf{j} + \nabla(\nabla^2\psi)$$

If we now take ψ satisfying the Laplace equation $\nabla^2\psi = 0$ or $\psi = \text{constant}$, we have the Poisson equation for the vector potential

$$\nabla^2\mathbf{A} = -\mu_0\mathbf{j} \tag{8.12}$$

Also from (8.9) the condition $\nabla^2\psi = 0$ is equivalent to the condition

$$\nabla\cdot\mathbf{A} = 0 \tag{8.13}$$

which is known as the Coulomb gauge condition. For stationary fields, there is no difference of this with the Lorenz gauge condition which we shall meet later. With this gauge condition, (8.9) reduces to

$$\mathbf{A}(\mathbf{r}) = \frac{\mu_0}{4\pi} \int \frac{\mathbf{j}(\mathbf{r}')}{|\mathbf{r} - \mathbf{r}'|} dv' \qquad \boxed{\mathbf{A}(\mathbf{r}) = \frac{1}{c} \int \frac{\mathbf{j}(\mathbf{r}')}{|\mathbf{r} - \mathbf{r}'|} dv' \quad \text{in Gaussian units}} \tag{8.14}$$

As we have the Poisson equation for the components of **A**, Eq. (8.12), we have the uniqueness theorem as in electrostatics, provided the source **j** and suitable boundary conditions are specified.

8.1 Magnetic Moment or a Current Distribution

As in electrostatics, we shall break up the field due to a distribution of currents in a bounded region into multipole fields having characteristic dependence on distance and angle. We utilize the Taylor series expansion of $1/|\mathbf{r} - \mathbf{r}'|$

$$\frac{1}{|\mathbf{r} - \mathbf{r}'|} = \frac{1}{r} + \sum_i \left[\frac{\partial}{\partial x_i'} \left(\frac{1}{|\mathbf{r} - \mathbf{r}'|} \right) \right]_0 x_i' + \cdots$$

Substituting in expression (8.14) for **A**, we get the first terms as

$$\frac{\mu_0}{4\pi r} \int \mathbf{j}(\mathbf{r}') dv'$$

Now

$$\int j_x dx = \int \nabla \cdot (x\mathbf{j})\, dv - \int x \nabla \cdot \mathbf{j}\, dv$$

The first integral on the right may be converted into a surface integral over a surface where the integrand $x\mathbf{j}$ will vanish (because the current distribution is in a bounded region) while the second integral vanishes because of Eq. (8.5). Thus, we have no monopole term in the magnetic field. This corresponds to the fact that apparently isolated magnetic poles do not exist. The leading term is thus the second term

$$\frac{\mu_0}{4\pi r^3} \sum_i x_i \int x_i' \mathbf{j}(\mathbf{r}') dv' = \frac{\mu_0}{4\pi r^3} \int (\mathbf{r} \cdot \mathbf{r}') \mathbf{j}(\mathbf{r}') dv' \tag{8.15}$$

However

$$\begin{aligned} 0 = \int (\nabla \cdot \mathbf{j}) x_i x_k\, dv &= \int \nabla \cdot (x_i x_k \mathbf{j})\, dv' - \int \mathbf{j} \cdot \nabla (x_i x_k)\, dv' \\ &= \text{surface integral which vanishes} - \int (j_i x_k + j_k x_i)\, dv' \end{aligned}$$

If in the above equation we put $i = k$ and sum up, we get

$$\int \mathbf{r} \cdot \mathbf{j}\, dv = 0$$

Note $\int j_i x_k\, dv$ is an antisymmetric tensor and hence its trace vanishes.

Using this result in (8.15), we get finally for the leading term in the vector potential

$$\mathbf{A}_1 = \frac{\mu_0}{4\pi}\frac{\boldsymbol{\mu} \times \mathbf{r}}{r^3} \tag{8.16}$$

$$\text{where } \boldsymbol{\mu} = \frac{1}{2}\int (\mathbf{r} \times \mathbf{j}) dv \tag{8.17}$$

is called the magnetic moment of the current distribution.

1. In the above, we have omitted the prime over source coordinates where no confusion is likely to arise.
2. In going from (8.15) to (8.16), we have used the antisymmetry of the tensor $\int x_i j_k\, dv'$ in

$$\frac{x_i}{r^3}\int x_i' j_k\, dv' = \frac{x_i}{2r^3}\int (x_i' j_k - x_k' j_i)\, dv'$$
$$= \frac{1}{r^3}\int \left[(\mathbf{r}' \times \mathbf{j}) \times \mathbf{r}\right]_k dv'$$

3. At sufficiently large distances from the source region, the higher power terms in $1/r$ may be neglected and thus we very often consider the vector potential to be identical with this dipole term.
4. In the region outside the source distribution, (8.11) shows that $\mathbf{B}$ is irrotational and hence can be expressed as the gradient of a potential function. Thus, there should be formal agreement between the formulae of magnetism and electrostatics except at the origin where the sources are concentrated. Consequently, the potential functions (both vector and scalar) blow up. If we take $\phi = (\boldsymbol{\mu} \cdot \mathbf{r})/r^3$, we get keeping in mind the constancy of $\boldsymbol{\mu}$

$$\mathbf{B} = \nabla \times \mathbf{A} = -\nabla \times \left[\boldsymbol{\mu} \times \nabla\left(\frac{1}{r}\right)\right] = -\boldsymbol{\mu}\,\nabla^2\left(\frac{1}{r}\right) - \nabla\phi$$

showing that the two differ only at $r = 0$ by $-\boldsymbol{\mu}\,\nabla^2\,(1/r)$. The name magnetic moment for $\boldsymbol{\mu}$ is further justified by a calculation of the torque on a current distribution due to an external magnetic field. This, according to (8.4) will be given by

$$\mathbf{L} = \int \mathbf{r}' \times (\mathbf{j} \times \mathbf{B})\, dv' = \int \mathbf{j}(\mathbf{r}' \cdot \mathbf{B})\, dv' - \mathbf{B}\int (\mathbf{r}' \cdot \mathbf{j})\, dv' = \boldsymbol{\mu} \times \mathbf{B}$$

where in the last step we have used the antisymmetry of $\int j_i x_k \, dv$.

We can proceed to analyse the quadrupole and higher multipole terms, however, they do not offer anything of special interest and we shall not go into that exercise here.

8.2 Relation Between Magnetic Moment and Angular Momentum for a Classical System

The angular momentum is defined as

$$\mathbf{L} = \int \mathbf{r} \times \mathbf{p} \, dv \tag{8.18}$$

where $\mathbf{p}$ is the linear momentum density vector at the point $\mathbf{r}$. If we suppose that both the electric current and the linear mechanical momentum are due to the motion of particles of charge e, mass m and their number density is n, then

$$\mathbf{j} = ne\mathbf{u} \tag{8.19}$$

$$\mathbf{p} = nm\mathbf{u} \tag{8.20}$$

where $\mathbf{u}$ is the velocity vector of the particles. Substituting from (8.19) and (8.20) in Eqs. (8.17) and (8.18), we get

$$\boldsymbol{\mu} = \frac{e}{2m}\mathbf{L} = g\mathbf{L} \tag{8.21}$$

where g is called the gyromagnetic ratio. In case of orbital angular momentum of electrons, (8.21) apparently holds good; however, for electron spin the ratio between the magnetic moment and spin angular momentum has very nearly double the value given by (8.21). This is evidence, if indeed such evidence is at all necessary, that electron spin is not a classically describable property.

8.2.1 *Larmor Precession*

If a magnetic field acts on a system such as we have just now considered, the torque due to the magnetic field $\mathbf{B}$ will be $\boldsymbol{\mu} \times \mathbf{B}$ and we shall have the equation

$$\frac{d\mathbf{L}}{dt} = \boldsymbol{\mu} \times \mathbf{B} = \frac{e}{2m}(\mathbf{L} \times \mathbf{B}) \tag{8.22}$$

Therefore, forming the scalar product of the above equation with $\mathbf{L}$ and $\mathbf{B}$, respectively, the right-hand side vanishes and one gets

$$\mathbf{L}^2 = \text{constant}$$
$$\mathbf{L} \cdot \mathbf{B} = \text{constant}$$

which mean that the angular momentum vector will have a constant magnitude and maintain a constant inclination with the magnetic field. The only 'motion' that $\mathbf{L}$ can have is thus a precession about the magnetic field direction. The precessional angular velocity can be easily obtained. We have from Eq. (8.22)

$$\frac{d}{dt}(\mathbf{L} \times \mathbf{B}) = \frac{e}{2m}(\mathbf{L} \times \mathbf{B}) \times \mathbf{B}$$

Comparing this with the familiar relation between the linear velocity and angular velocity $\frac{d\mathbf{r}}{dt} = \omega \times \mathbf{r}$, we conclude that the precessional angular velocity is equal to $Be/(2m)$. This phenomenon is known as the Larmor precession, and the corresponding frequency $Be/(4\pi m)$ is called the Larmor frequency.

8.3 Permanent Magnets and the Vector H

We shall attempt to include permanent magnets in our formalism by considering them to be a volume distribution of magnetic moments. The situation has some similarity with our treatment of dielectrics which were represented as a volume distribution of electrical dipoles. However, there is an important difference—in case of dielectrics, the dipole moments owed their origin to an externally applied electric field, whereas for permanent magnets the dipoles are all the time there independent of the external field, if any. If $\mathbf{M}$ be the magnetic moment per unit volume, from (8.16) the contribution to the vector potential will be

$$\mathbf{A}(\mathbf{r}) = \frac{\mu_0}{4\pi}\int \frac{\mathbf{M} \times (\mathbf{r} - \mathbf{r}')}{|\mathbf{r} - \mathbf{r}'|^3} dv' = \frac{\mu_0}{4\pi}\int \mathbf{M} \times \nabla'\left(\frac{1}{|\mathbf{r} - \mathbf{r}'|}\right) dv'$$
$$= \frac{\mu_0}{4\pi}\left(\int \frac{\nabla' \times \mathbf{M}}{|\mathbf{r} - \mathbf{r}'|} dv' - \int \nabla' \times \left(\frac{\mathbf{M}}{|\mathbf{r} - \mathbf{r}'|}\right) dv'\right)$$

But

$$\int \nabla' \times \left(\frac{\mathbf{M}}{|\mathbf{r} - \mathbf{r}'|}\right) dv' = -\oint \frac{\mathbf{M} \times d\mathbf{s}}{|\mathbf{r} - \mathbf{r}'|} \tag{8.23}$$

The proof of the above relation is as follows. Consider a vector $\mathbf{B} = (\phi, 0, 0)$, then from Gauss's theorem

$$\int \nabla \cdot \mathbf{B}\, dv = \int \frac{\partial \phi}{\partial x} dv = \oint l\phi\, ds$$

Similarly, $\int \frac{\partial \phi}{\partial y} dv = \oint m\phi \, ds$ and $\int \frac{\partial \phi}{\partial z} dv = \oint n\phi \, ds$ where l, m, n are the direction cosines of the normal to ds. Hence

$$\int (\nabla \times \mathbf{A})_x dv = \oint (mA_z - nA_y)\, ds$$
$$= -\int (\mathbf{A} \times \mathbf{ds})_x$$

The surface integral in (8.23) vanishes for a bounded region of permanent magnetization and so

$$\mathbf{A(r)} = \frac{\mu_0}{4\pi} \int \frac{\nabla' \times \mathbf{M}}{|\mathbf{r} - \mathbf{r}'|} dv' \tag{8.24}$$

Comparing (8.24) with Eq. (8.9), we see that the magnetic moment distribution is equivalent to a current distribution of volume density $\mathbf{j}_m = \nabla \times \mathbf{M}$ so far as the magnetic field is concerned. Hence, in the presence of permanent magnets Eq. (8.11) will be modified to

$$\nabla \times \mathbf{B} = \mu_0(\mathbf{j} + \nabla \times \mathbf{M})$$
$$\text{or,} \quad \nabla \times (\mathbf{B} - \mu_0 \mathbf{M}) = \mu_0 \mathbf{j}$$

Defining a new vector

$$\mathbf{H} = \frac{\mathbf{B}}{\mu_0} - \mathbf{M} \tag{8.25}$$

we have

$$\nabla \times \mathbf{H} = \mathbf{j} \tag{8.26}$$

Thus, in the presence of permanent magnets, the fundamental equations of the magnetic field are (8.6) and (8.26). These constitute four equations for six unknowns—the current vector **j** is assumed given and the unknowns are the components of **B** and **H**. For dia- and para-magnetics, the response to any applied magnetic field is weak, but still there is some induction effect giving rise to a difference between **B** and **H** but as in dielectrics one can use the relation $\mathbf{B} = \mu \mathbf{H}$, where μ is called the permeability of the medium and is independent of **B** (or **H**, if you like). For ferromagnetics, μ has a large value and is not only not a constant but its value (defined as $\mathbf{B}/\mathbf{H}$) depends on the history of the specimen as well.

In case there are permanent magnets, one must of course know the dipole moment distribution (called the intensity of magnetization) before going to solve the problem. Equation (8.26) may be given an integral form by using Stokes' theorem

$$\int \mathbf{H} \cdot d\mathbf{l} = \int (\nabla \times \mathbf{H}) \cdot \mathbf{ds} = \int \mathbf{j} \cdot \mathbf{ds} = I \tag{8.27}$$

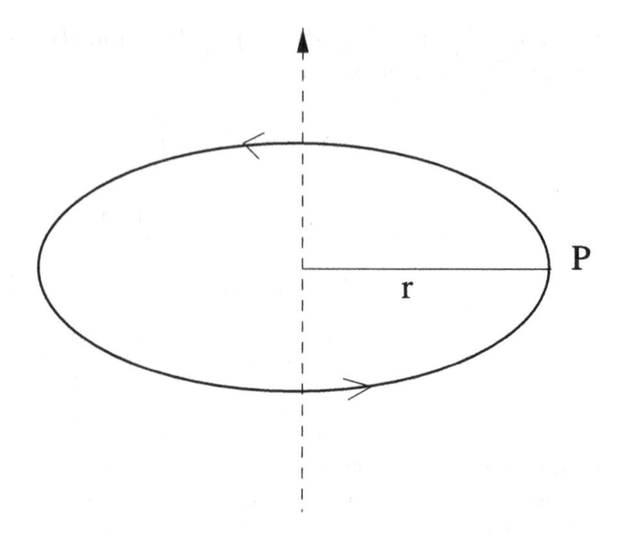

Fig. 8.1 Amperian loop for a long straight wire carrying current

This relation is usually referred to as Ampere's circuital theorem—stated in words—the line integral of **H** over a closed path is equal to the current which crosses any surface bounded by the line of integration. (Mark the words *any surface*—this is the case because as the divergence of the current density vector vanishes, the surface integral of the current over any closed surface also vanishes.) Ampere's theorem may be used to calculate the field due to coils carrying currents in cases of high symmetry.

Consider a long straight wire carrying current I. To have $\nabla \cdot \mathbf{j} = 0$, we have to consider the wire endless. To find the field at a point P distance r from the wire, consider the line integral over a circle through P whose centre is on the wire (Fig. 8.1). We have

$$\oint \mathbf{B} \cdot d\mathbf{l} = 2\pi r B_t = \mu_0 I$$

$$\text{or} \quad B_t = \frac{\mu_0 I}{2\pi r}$$

This gives the component of **B** tangential to the circle. That there is no component in perpendicular directions follows from symmetry—because of the translational symmetry along the z-axis $\frac{\partial B}{\partial z} = 0$ and hence from $\nabla \cdot \mathbf{B} = 0$ ($\partial B_r/\partial r = 0$). Constant values of B_z and B_r are eliminated by the usual boundary condition that the field vanishes at infinity. (With no permanent magnets or material medium present, **B** and **H** are related by $\mathbf{B} = \mu_0 \mathbf{H}$.)

The theorem may also be applied for an infinitely long solenoid. The field outside may be seen to vanish (this satisfies the field equations and by uniqueness theorem which exists in the case of magnetic fields as well, we may conclude that this is actually the situation). We find the field inside from the integral over the line shown (Fig. 8.2). Thus,

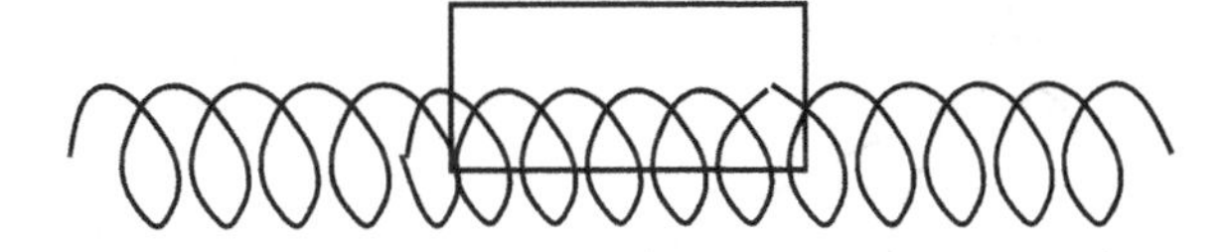

Fig. 8.2 Field inside a solenoid using Amperian loop

$$\oint B_z dl = \mu_0 nlI$$

$$\oint B_z\, dl = l B_z = \mu_0 n I \tag{8.28}$$

where n is the number of turns per unit length. That B_r and B vanish may be seen by combining symmetry consideration with the field equations.

There are other methods of calculating the field. In regions where $\mathbf{j}$ vanishes, $\nabla \times \mathbf{H} = 0$. Further in the absence of magnets and material media, this can be written as $\nabla \times \mathbf{B} = 0$, and we can regard $\mathbf{B}$ as the gradient of a scalar potential function ϕ_m, i.e., $\mathbf{B} = -\mu_0 \nabla \phi_m$. Hence, for a closed loop of wire carrying current I (the thickness of the wire being negligibly small), we have

$$\begin{aligned} -d\phi_m &= \phi_m(\mathbf{r}) - \phi_m(\mathbf{r} + d\mathbf{r}) = -\nabla \phi_m \cdot d\mathbf{r} = \frac{1}{\mu_0}\mathbf{B} \cdot d\mathbf{r} \\ &= \frac{I}{4\pi} \int \frac{d\mathbf{l} \times (\mathbf{r} - \mathbf{r}')}{\mid \mathbf{r} - \mathbf{r}' \mid^3} \cdot d\mathbf{r} = \frac{I}{4\pi} \oint (d\mathbf{r} \times d\mathbf{l}) \cdot \frac{(\mathbf{r} - \mathbf{r}')}{\mid \mathbf{r} - \mathbf{r}' \mid^3} \end{aligned} \tag{8.29}$$

As the expression for $\mathbf{B}$ shows, the field is dependent on $(\mathbf{r} - \mathbf{r}')$ rather than $\mathbf{r}$ or $\mathbf{r}'$ separately. Hence, the change $-d\phi_m$ may be thought to be brought about by a translation of the coil through $-d\mathbf{r}$, the field point remaining fixed. The surface element traced out by $d\mathbf{l}$ would then be $-(d\mathbf{r} \times d\mathbf{l}]$ and

$$-(d\mathbf{r} \times d\mathbf{l}) \cdot \frac{(\mathbf{r} - \mathbf{r}')}{\mid \mathbf{r} - \mathbf{r}' \mid^3} = \mathbf{ds} \cdot \frac{(\mathbf{r} - \mathbf{r}')}{\mid \mathbf{r} - \mathbf{r}' \mid^3}$$

is the solid angle subtended by the surface element at the field point. The integral in (8.29) will be $-d\Omega$, with $d\Omega$ being the change in the solid angle subtended by the coil at the field point. Finally

$$\phi_m = \frac{I}{4\pi} d\Omega$$

The potential function is thus, apart from a possible constant, identical with that due to a double layer of moment per unit area $\frac{I}{4\pi}$. Hence the theorem:

The magnetic field due to a closed coil carrying current I is the same as due to a surface distribution of dipoles, the boundary of the surface coinciding with the coil and the moment per unit area is equal to that of the current strength.

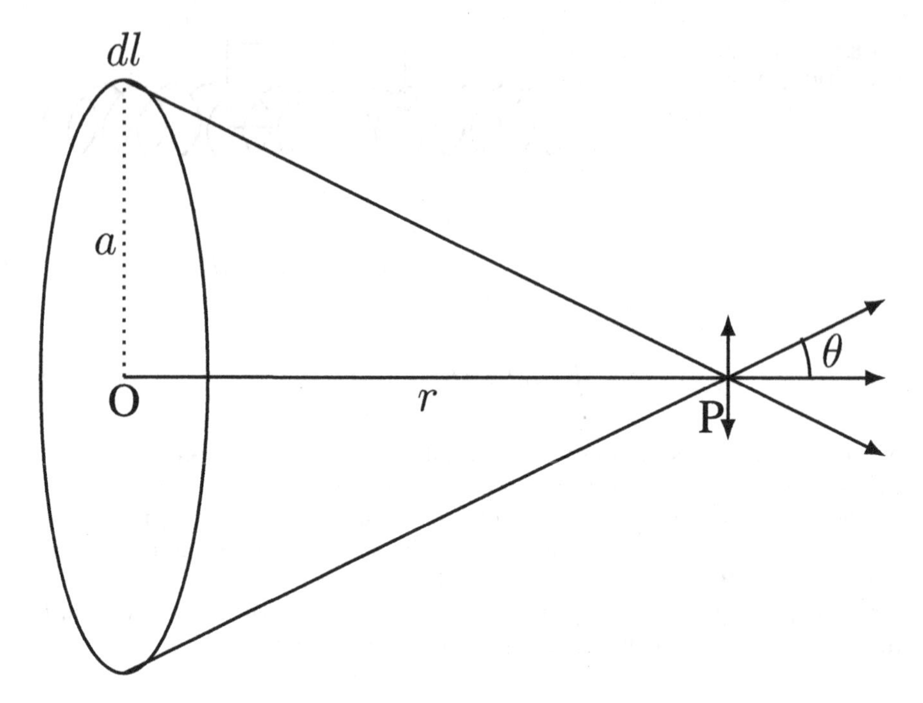

Fig. 8.3 Field on the axis of a circular coil

This theorem allows us to reduce the calculation of magnetic field due to closed coils to electrostatic problems. As an example, consider a circular coil carrying current I. The field may be written down from the known result for a circular disc of dipoles. Thus, the field at a point on the axis normal to the plane of the coil at the centre is

$$\frac{\mu_0 I}{2}\frac{a^2}{(a^2+r^2)^{3/2}} \tag{8.30}$$

where r is the distance of the field point and a the radius of the coil. However, the magnetic scalar potential, thus introduced, is in general, not single-valued. If we consider a line integral where the surface bounded by the line of integration is pierced by a current I, we would have this integral equal to $\mu_0 I$. Thus, going round this closed path the potential apparently increases by $\mu_0 I$.

In some cases, a direct calculation of the field from Eq. (8.3) is not difficult. Thus, take again the case of the circular coil (Fig. 8.3). At a point on the axis, only the component along the axis will survive and is given by

$$\frac{\mu_0 I}{4\pi}\oint\frac{\cos\theta\, dl}{(a^2+r^2)} = \frac{\mu_0 I}{4\pi}\oint\frac{a\, dl}{(a^2+r^2)^{3/2}} = \frac{\mu_0 I}{2}\frac{a^2}{(a^2+r^2)^{3/2}}$$

We may obtain the field due to an infinite solenoid by considering the solenoid as a collection of closely packed coils and obtaining the field at an arbitrary point on the axis by integrating over the length of the solenoid. Considering field at the position $x = r$ on the axis of the solenoid due to a stack of coils of width dl located at $x = l$ (with $l - r = a \tan \theta$)

$$\frac{\mu_0 n I}{2} \int_{-\infty}^{\infty} \frac{a^2 dl}{(a^2 + (l - r)^2)^{3/2}} = \frac{\mu_0 n I}{2} \int_{-\frac{\pi}{2}}^{\frac{\pi}{2}} \frac{d\theta}{\sec \theta} = \mu_0 n I$$

in agreement with the result obtained from Ampere's theorem.

8.4 Boundary Conditions at the Interface of Two Media

As in electrostatics, one can obtain the boundary conditions by application of the principle that any object that appears in the basic field equation does not blow up at the boundary unless there are singular sources. Hence, the normal component of **B** and the tangential component of **H** must be continuous. The result may also be obtained by (a somewhat uncritical) application of Eqs. (8.6) and (8.26). If, however, there be surface currents (which constitute singular sources with **j** arbitrarily large), $\nabla \times \mathbf{H}$ would blow up and one would have

$$\mathbf{H}'_T = \mathbf{H}_T + K$$

where K is the surface current density.

8.5 Scalar and Vector Potentials for a Static Magnetic Field

The equations to be integrated are (8.6) and (8.26). Let us write $\mathbf{B} = \mu \mathbf{H}$ where μ the permeability can be regarded as independent of **H** only in the absence of ferromagnets (whether permanently magnetized or not). The equation for the vector potential becomes, if μ be a constant,

$$\nabla(\nabla \cdot \mathbf{A}) - \nabla^2 \mathbf{A} = \mu \mathbf{j}$$

We utilize a gauge transformation to make $\nabla \cdot \mathbf{A} = 0$ and then we have a Poisson equation for **A**

$$\nabla^2 \mathbf{A} = -\mu \mathbf{j}$$

The uniqueness of the solution may then be deduced as we have already considered in electrostatics.

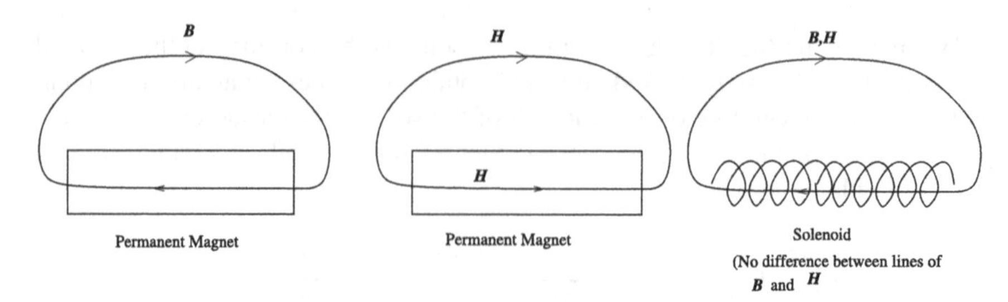

Fig. 8.4 Lines of **B** and **H** for a bar magnet and for a solenoid

When the sole sources of the field are permanent magnets, historical convention is to use a scalar potential function, $\mathbf{H} = -\nabla\phi$. Substituting in (8.6), we get Laplace's equation. Of course, as sources we have the distribution of dipoles in the body of the permanent magnet. We know from electrostatics that such a volume distribution of dipoles is equivalent to a volume distribution of 'charge' of density $-\nabla \cdot \mathbf{M}$ (where **M** is the magnetic moment per unit volume, in this case corresponding to the polarization, in electrostatics) plus a surface distribution of 'charge' $\mathbf{M}_n$. We thus come to the idea of 'charge' or monopoles in magnetism although no such thing apparently exists (at least in the picture presented here). Thus, we have

$$\phi(\mathbf{r}) = -\frac{1}{4\pi}\int \frac{\nabla' \cdot \mathbf{M}(\mathbf{r}')}{|\mathbf{r}-\mathbf{r}'|}dv' + \frac{1}{4\pi}\oint \frac{\mathbf{M}(\mathbf{r}')}{|\mathbf{r}-\mathbf{r}'|} \cdot \mathbf{ds}' \tag{8.31}$$

Of course, permanent magnets may also be dealt with in terms of the vector potential. We have already seen that from Eq. (8.23), we then have

$$\mathbf{A}(\mathbf{r}) = \frac{\mu}{4\pi}\int \frac{\nabla' \times \mathbf{M}(\mathbf{r}')}{|\mathbf{r}-\mathbf{r}'|}dv' + \frac{\mu}{4\pi}\oint \frac{\mathbf{M}(\mathbf{r}') \times \mathbf{ds}'}{|\mathbf{r}-\mathbf{r}'|} \tag{8.32}$$

However, we would now consider the volume integral to be over the region occupied by the magnets (instead of over the entire space) and the surface integral to be over the boundary of magnets (instead of the sphere at infinity). This is only to have a conceptually simple distribution of sources.

The case of a uniformly magnetized bar magnet is simple and interesting so far as the difference between the sources in the two cases (i.e., the cases of scalar and vector potential). In both (8.31) and (8.32), the volume integrals fall off as **M** is a constant vector. Thus, the sources of the scalar potential are a distribution of poles over the boundary faces normal to the length of the bar (which is the direction of **M**) and we have the picture of a magnet with opposite poles at the two ends.

For the vector potential **A**, on the other hand, the sources are only over the plane surfaces parallel to the length and simulate a solenoid carrying current. However, the fields reveal a difference between the solenoid and the permanent magnet (Fig. 8.4). For the permanent magnet, the lines of **B** and **H** are identical outside the magnet but are oppositely directed inside. This can be understood from the vanishing of the line integral of **H** and the continuity of the normal component of **B** at a boundary. In the case of the current-carrying solenoid, however, **B** and **H** are everywhere identical—the line integral of **H** no longer vanishes—as in (8.25) $\mathbf{j} \neq 0$.

8.6 Uniformly Magnetized Sphere

In this case also, $\nabla \times \mathbf{M}$ as well as $\nabla \cdot \mathbf{M}$ vanish everywhere so that we have only a surface distribution of sources. The scalar potential is

$$\phi = \frac{1}{4\pi}\oint \frac{\mathbf{M}(\mathbf{r}')}{|\mathbf{r}-\mathbf{r}'|}\mathbf{ds}' = \frac{1}{4\pi}\oint \frac{M_0 \cos\theta}{|\mathbf{r}-\mathbf{r}'|}ds$$

where θ is the angle between **M** (taken to be in the z-direction) and the normal to the sphere (i.e., the radius at the point) and M_0 is the magnitude of **M**. The integral can be easily evaluated by the series expansion of $1/|\mathbf{r}-\mathbf{r}'|$ in spherical harmonics. One gets

$$\phi(r,\theta) = \begin{cases} \dfrac{M_0}{3}\dfrac{a^3}{r^2}\cos\theta & \text{for } r > a \\[2ex] \dfrac{M_0}{3} r\cos\theta & \text{for } r < a \end{cases}$$

The interior potential is thus $M_0 z/3$, so that the field **H** is also in the z-direction and can be written as

$$\mathbf{H} = -\frac{1}{3}\mathbf{M}_0 \qquad \text{for } r < a$$

Consequently

$$\mathbf{B} = \mu(\mathbf{H}+\mathbf{M}) = \frac{2}{3}\mu\mathbf{M} \qquad \text{for } r < a$$

which is in the direction opposite to **H** (Figs. 8.5 and 8.6). Outside the potential is a dipole potential due to a dipole of moment $4\pi a^3\mathbf{M}/3$ and $\mathbf{B} = \mu_0\mathbf{H}$. It is of some interest to compare these results with those in electrostatics for a dielectric sphere in an asymptotically uniform field—of course here there is no such asymptotic field and there the dielectric sphere had only induced dipoles.

The vector potential method may also be used for solution of the above problem. One gets, as the product $\mathbf{M} \times \mathbf{ds}$ has only a ϕ component,

$$A_r = A_\theta = 0$$

$$A_\phi(r,\theta) = \begin{cases} \dfrac{\mu M_0}{3}\dfrac{a^3}{r^2}\sin\theta & \text{for } r > a \\[2ex] \dfrac{\mu_0 M_0}{3} r\sin\theta & \text{for } r < a \end{cases}$$

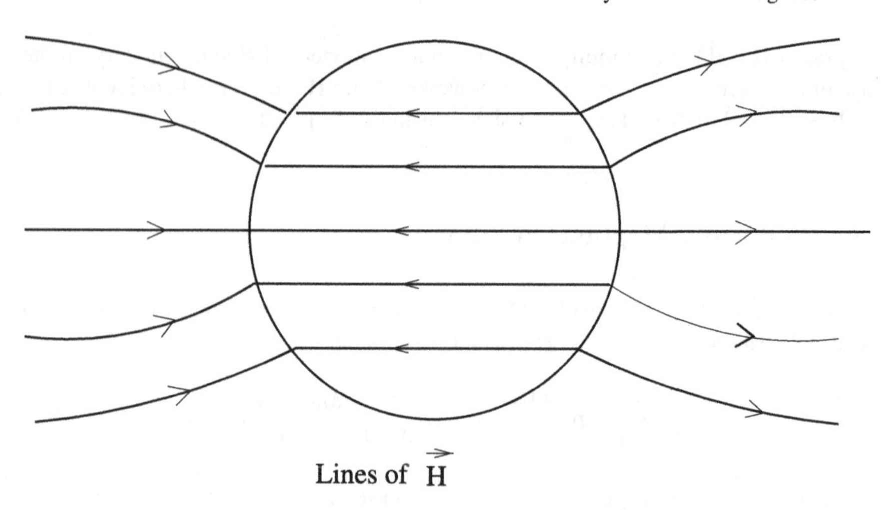

Fig. 8.5 Lines of **H** for a uniformly magnetized sphere

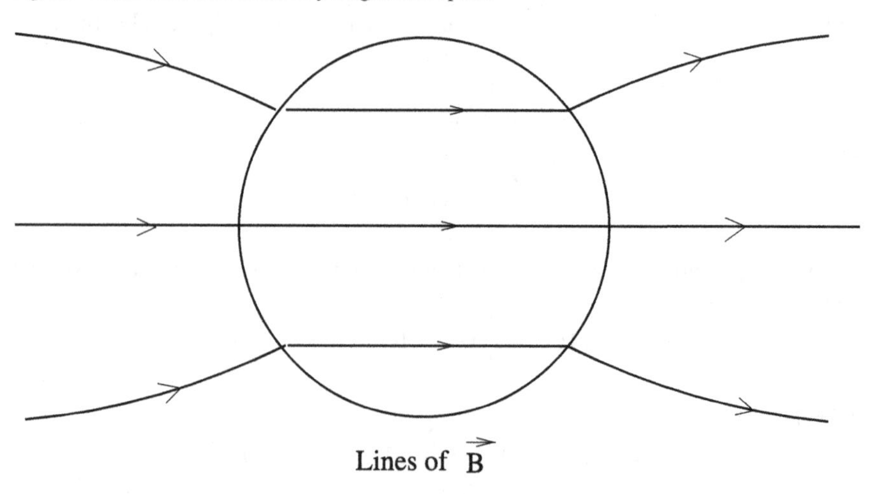

Fig. 8.6 Lines of **B** for a uniformly magnetized sphere

The calculation of **B** by $\mathbf{B} = \nabla \times \mathbf{A}$ and **H** yields results identical to those obtained by using the scalar potential.

Problems

1. Find **B** and **H** at all points for a uniformly magnetized spherical shell.
2. A cylindrical hole of radius a is bored parallel to the axis along the cylindrical conductor of radius b. If the current is uniformly distributed throughout the cross-section of the conductor and the centre of the hole is at a distance c from the axis of the cylinder, find the magnetic field at different points.
3. Find the magnetic moment of a sphere carrying a uniform surface charge density and rotating with angular velocity ω.
4. Calculate the magnetic field due to a circular coil carrying current using the vector potential. (Calculate only for points on the normal to the plane of the coil through the centre.)
5. Calculate the different multipole moments for a circular coil carrying current using (1) the scalar potential and (2) the vector potential.
6. A permanent magnet is in the form of (a) a bar of rectangular cross-section and (b) a cylinder of circular cross-section. In either case, the magnetization is constant and in the direction of the length. Determine **B** and **H** at all points on the axis.
7. Investigate the field in case of a spherical shell of matter placed in vacuum, there being an asymptotically uniform magnetic field. The shell has no permanent magnetization but has a high permeability. Show that the hollow region is practically free of any magnetic field.
8. Show that the dipole moment for a closed plane coil carrying current is of magnitude IA where I is the current and A the area of the coil (independent of the shape of the coil).
9. Calculate the scalar and vector potentials due to a rectangular coil carrying current and hence calculate the dipole and quadrupole moments.
10. Prove that if the magnetic field be solely due to permanent magnets in a bounded region, then $\int \mathbf{B} \cdot \mathbf{H}\, dv = 0$.

Chapter 9
Electromagnetic Induction and Energy of the Magnetic Field

An observable electric current may either be due to a macroscopic motion of charges, which we call a convection current and represented by $\rho\mathbf{u}$, where ρ is the charge density and $\mathbf{u}$ the velocity vector of the charge or it may be due to a drift of electrons through a conductor which on the macroscopic scale is uncharged. In the second case, we call it conduction current and it lasts so long as there is an electric field in the conductor (which of course indicates a basically non-static condition). Ohm's law states that for a given conductor, the current density vector is a linear function of the components of the electric intensity, i.e.,

$$j_i = \sum_k \sigma_{ik} E_k \tag{9.1}$$

where σ_{ik} is called the conductivity tensor and is a characteristic property of the material of the conductor at a particular temperature. (It depends also on other conditions like pressure, presence of the magnetic fields, etc.) For isotropic conductors, Eq. (9.1) reduces to the simple form

$$\mathbf{j} = \sigma \mathbf{E} \tag{9.2}$$

If I be the current flowing through a conductor of cross-sections S and length l, Eq. (9.2) yields

$$I = \frac{1}{\int dl}\sigma S \int E\, dl = \frac{\sigma S}{l} V = \frac{V}{R} \tag{9.3}$$

where R is called the resistance of the conductor and varies directly as the length and inversely as the cross-section of the conductor. For a closed loop of wire carrying current, $V = \oint \mathbf{E} \cdot d\mathbf{l} \neq 0$, i.e., the electric field is no longer irrotational. The integral $\oint \mathbf{E} \cdot dl$ is called the electromotive force in the circuit.

A. K. Raychaudhuri, *Classical Theory of Electricity and Magnetism*, Texts and Readings in Physical Sciences 21, https://doi.org/10.1007/978-981-16-8139-4_9

As a source of electromotive force, we have the simple cell for example. Faraday discovered that there is another source of electromotive force—whenever the magnetic flux through a closed coil of wire changes, a current flows through the coil. The way in which the flux change is brought about is of no importance—it may be due to the approach or recession of a permanent magnet or another coil carrying current or due to the increase or decrease of current in a neighbouring coil or by a motion of the closed coil itself. The important thing was that the current in the closed coil lasted only as long as the flux was changing. More quantitatively, the electromotive force in the coil was found to be given by the rate of change of flux, i.e.,

$$\oint \mathbf{E} \cdot d\mathbf{l} = -\frac{d}{dt} \oint \mathbf{B} \cdot \mathbf{ds} = -\int \frac{\partial \mathbf{B}}{\partial t} \cdot \mathbf{ds} \tag{9.4}$$

where the integral on the right side is over the surface bounded by the coil and we have assumed the coil to be at rest. The occurrence of the negative sign gives the direction of the electromotive force which is such that it opposes the cause to which it is due (e.g., if it is due to the approach of a second coil carrying current, the induced current in our coil will be such that there will be a repulsion between the two coils—this is sometimes referred to as Lenz's law). Using Stokes' theorem, we convert the line integral to a surface integral and obtain

$$\nabla \times \mathbf{E} = -\dot{\mathbf{B}} \tag{9.5}$$

Notes:

1. Subject to the condition that the boundary of the surface coincides with the coil, the surface of integration in (9.4) is otherwise arbitrary. This is justified, as $\nabla \cdot \mathbf{B} = 0$, the surface integral of $\mathbf{B}$ over any closed surface vanishes.
2. Faraday observed the generation of electromotive force or electrical currents with the change of flux only with conducting coils. Maxwell advanced the idea that the coil has no essential role except that it facilitated the detection and measurement of the induced electromotive force. Thus, according to Maxwell, an electric field, which is not irrotational, originates whenever $\mathbf{B}$ changes and Eq. (9.5) is valid irrespective of the presence of the coil.

9.1 Law of Induction in a Moving Coil or Medium

Whenever the coil (in a Faraday-type experiment) or the medium is in motion, the change of flux through the coil or a specific circuital line in the medium will undergo changes which may be split up into two parts: (a) due to intrinsic time dependence of $\mathbf{B}$ and (b) due to the coil or the medium going to a region where $\mathbf{B}$ is different due to spatial non-homogeneity of $\mathbf{B}$. Such situations are common in hydrodynamics and one takes care of this by considering two differential coefficients: one giving the time rate of change at a fixed point in space and it is indicated by $\partial/\partial t$ and the other giving the change when we confine ourselves to a fixed element of the moving matter and this is indicated by d/dt. In general, for any function $F(x, y, z, t)$, we have

$$\begin{aligned} F(x+dx, y+dy, z+dz, t+dt) &- F(x, y, z, t) \\ &= \frac{\partial F}{\partial x}dx + \frac{\partial F}{\partial y}dy + \frac{\partial F}{\partial z}dz + \frac{\partial F}{\partial t}dt \\ &= \left((\mathbf{u}\cdot\nabla)F + \frac{\partial F}{\partial t}\right)dt \end{aligned}$$

$$\text{or} \quad \frac{dF}{dt} = \frac{\partial F}{\partial t} + (\mathbf{u}\cdot\nabla)F$$

Hence, if $\mathbf{E}'$ be the electric field in the moving system

$$\begin{aligned} \oint \mathbf{E}'\cdot d\mathbf{l} &= -\int \frac{d\mathbf{B}}{dt}\cdot d\mathbf{s} = -\int\left(\frac{\partial \mathbf{B}}{\partial t} + (\mathbf{u}\cdot\nabla)\mathbf{B}\right)\cdot d\mathbf{s} \\ &= -\int \frac{\partial \mathbf{B}}{\partial t}\cdot d\mathbf{s} + \int (\nabla\times(\mathbf{u}\times\mathbf{B}))\cdot d\mathbf{s} - \int \mathbf{u}(\nabla\cdot\mathbf{B})\cdot d\mathbf{s} \end{aligned}$$

where we have used the vector identity

$$\nabla\times(\mathbf{u}\times\mathbf{B}) = (\mathbf{B}\cdot\nabla)\mathbf{u} - (\mathbf{u}\cdot\nabla)\mathbf{B} + \mathbf{u}(\nabla\cdot\mathbf{B}) - \mathbf{B}(\nabla\cdot\mathbf{u})$$

and taken $\mathbf{u}$ to be a constant vector. Now using Stokes' theorem to convert the line integral on the left to a surface integral and taking account of the solenoidal character of $\mathbf{B}$, we get

$$\oint(\nabla\times\mathbf{E}')\cdot d\mathbf{s} = -\oint \dot{\mathbf{B}}\cdot d\mathbf{s} + \oint(\nabla\times(\mathbf{u}\times\mathbf{B}))\cdot d\mathbf{s}$$

$$\text{or} \quad \nabla\times\left(\mathbf{E}' - \mathbf{u}\times\mathbf{B}\right) = -\dot{\mathbf{B}} \tag{9.6}$$

However, $\mathbf{E}' - \mathbf{u}\times\mathbf{B}$ is the effective electric intensity from the point of view of a stationary observer—for a charge e moving with velocity $\mathbf{u}$ (i.e., with the medium) experiences a force $\mathbf{u}\times\mathbf{B}$ due to the magnetic field. Hence, the stationary observer will say that the electric intensity is $\mathbf{E}$ where

$$e\mathbf{E}' = e\mathbf{E} + e(\mathbf{u}\times\mathbf{B}) \tag{9.7}$$

Using (9.7) in (9.6), we get back Eq. (9.5)

$$\nabla\times\mathbf{E} = -\dot{\mathbf{B}}$$

9.2 Energy of the Magnetostatic Field

Even when a current is being driven in a stationary condition, work is done as the charge is moving under the action of a force $e\mathbf{E}$. This work appears as the joule heat at the rate of σE^2 per unit volume and is an irreversible process, associated with an increase of entropy. When, however, the current is not stationary as when a circuit with an electromotive force after being switched on, the current grows from zero to an asymptotic steady value, an additional amount of work has to be done to produce the magnetic field associated with the current. During the growth, the situation is obviously non-stationary but we may consider the process to occur infinitely slowly so that the use of the stationary state field equations remains justified. (In effect, we are neglecting radiation of electromagnetic energy in our energy balance equation.)

We start with Ohm's law

$$\mathbf{j} = \sigma\left(\mathbf{E} + \mathbf{E}_i\right) \tag{9.8}$$

where $\mathbf{E}$ is the impressed electric field (e.g., due to a battery of cells) and $\mathbf{E}_i$ is the induction electric field as given by Eq. (9.5). The work done by the impressed electric field is $(\mathbf{E}\cdot d\mathbf{l})\,(\mathbf{j}\cdot d\mathbf{s}\,dt)$ where $\mathbf{j}\cdot d\mathbf{s}\,dt$ is the amount of charge flowing through $d\mathbf{s}$ in time dt, the multiplication by $\mathbf{E}$ gives the force on this charge and further multiplication by the displacement gives the work done. Hence, from Eq. (9.8), the work done is

$$\int_v \mathbf{E}\cdot\mathbf{j}\,dvdt = \int \frac{j^2}{\sigma}dvdt - \int \mathbf{E}_i\cdot\mathbf{j}\,dvdt$$

The first term on the right is simply the joule heat, and the second term exists only so long as the magnetic field is changing with time. Replacing $\mathbf{j}$ by $\nabla\times\mathbf{H}$ and using the vector identity

$$\nabla\cdot(\mathbf{A}\times\mathbf{B}) = -\mathbf{A}\cdot(\nabla\times\mathbf{B}) + \mathbf{B}\cdot(\nabla\times\mathbf{A})$$

we get

$$-\int \mathbf{E}_i\cdot\mathbf{j}\,dv\,dt = \int \nabla\cdot(\mathbf{H}\times\mathbf{E}_i)dv\,dt - \int (\nabla\times\mathbf{E}_i)\cdot\mathbf{H}\,dv\,dt$$

The first integral on the right, being a divergence integral, may be converted into a surface integral over the sphere at infinity and vanishes; in the absence of the radiation field, $\mathbf{E}_i$ and $\mathbf{H}$ both vanish in this case faster than $1/r^2$. (The absence of radiation is a consequence of our assumption of quasi-stationary conditions.) The second term on the right, thus, represents the energy of the magnetic field, and using Eq. (9.5), we have

$$U_{\text{mag}} = \int_v \int_B \mathbf{H}\cdot d\mathbf{B}\,dv \tag{9.9}$$

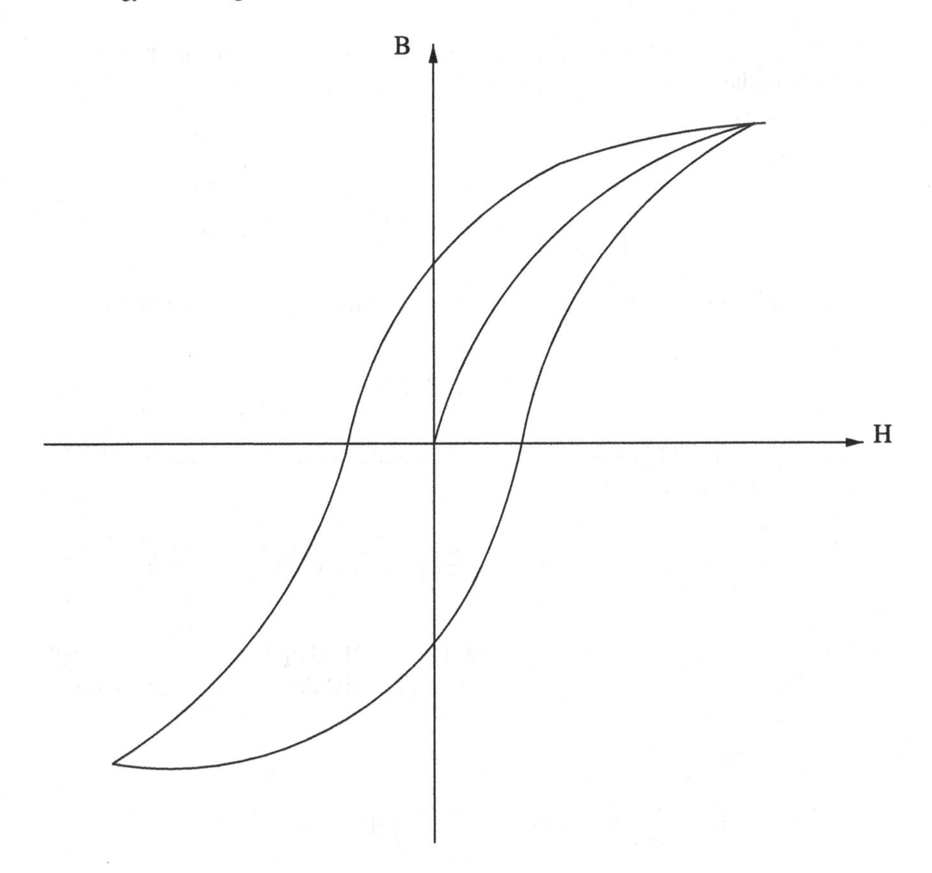

Fig. 9.1 Hysteresis curve for a ferromagnetic material

This expression for the magnetic field energy is integrable only if there is a functional relation between **H** and **B**. For para- and dia-magnetics, one may assume the simple relation $\mathbf{B} = \mu\mathbf{H}$ with μ a constant depending on the nature of the material but not on **H** (or **B**). The energy density is then simply

$$\frac{1}{2}\mathbf{B}\cdot\mathbf{H} = \frac{1}{2}\mu H^2$$

which is of the same form as the electrostatic field energy density. In the case of ferromagnetics, however, there is no linear relation—indeed **B** is not even uniquely determined by **H** and a typical form of the variation of **B** with **H** for a ferromagnet is of the form shown. The curve is called the hysteresis curve and is shown in Fig. 9.1. Thus, as **H** goes through a cycle of changes, we get a loop of non-vanishing area in the hysteresis curve and it represents the energy dissipated as heat when the ferromagnetic specimen goes through a cycle of magnetization. This is also, of course, an irreversible effect and is distinct from the joule heat.

We can express the energy as associated with current circuits. Equation (9.9) may be written (in the absence of ferromagnets)

$$\begin{aligned}U_{\text{mag}} &= \frac{1}{2}\int \mathbf{B}\cdot\mathbf{H}\,dv = \frac{1}{2}\int \mathbf{H}\cdot(\nabla\times\mathbf{A})\,dv \\ &= \frac{1}{2}\int \nabla\cdot(\mathbf{A}\times\mathbf{H})\,dv + \frac{1}{2}\int \mathbf{A}\cdot(\nabla\times\mathbf{H})\,dv\end{aligned}$$

Again the first integral vanishes because of the conditions at infinity and we have

$$U_{\text{mag}} = \frac{1}{2}\int \mathbf{A}\cdot\mathbf{j}\,dv$$

Considering the current $\mathbf{j}$ to be distributed in a number of distinct coils, such that for any coil $\mathbf{j}\,dv = I d\mathbf{l}$, we have

$$U_{\text{mag}} = \frac{1}{2}\sum_k \oint \mathbf{A}\cdot(I_k d\mathbf{l}_k) = \sum_k \frac{I_k}{2}\int(\nabla\times\mathbf{A})\cdot\mathbf{ds}_k = \frac{1}{2}\sum_k I_k\Phi_k$$

where we have used Stokes' theorem and $\Phi_k = (\int \mathbf{B}\cdot\mathbf{ds})_k$ is the flux linked with the k-th circuit. As the induction or the vector potential due to a number of current carrying coils is additive, if Φ_{ki} be the flux through the k-th coil due to the current in the i-th coil, the flux Φ_k through the k-th coil is

$$\begin{aligned}\Phi_k &= \sum_{i(\neq k)} \Phi_{ki} + L_k I_k = \sum_{i(\neq k)} \int \mathbf{B}_{ki}\cdot\mathbf{ds}_k + L_k I_k \\ &= \sum_{i(\neq k)} \oint A_{ki}\cdot d\mathbf{l}_k + L_k I_k \\ &= \frac{\mu_0}{4\pi}\sum_{i(\neq k)} \oint I_i \frac{d\mathbf{l}_i\cdot d\mathbf{l}_k}{\mid \mathbf{r}_i - \mathbf{r}_k \mid} + L_k I_k\end{aligned}$$

where the integration is over the entire lengths of the circuits i and k and $|\mathbf{r}_i - \mathbf{r}_k|$ is the distance between the elements $d\mathbf{l}_i$ and $d\mathbf{l}_k$. The prime over the summation indicates that the term $i = k$ is to be omitted. The flux linked with any coil due to the current in itself is taken care of separately by the term $L_k I_k$, where L_k is called the self inductance of the coil. Substituting this expression for Φ_k, the energy expression becomes

$$\begin{aligned}U_{\text{mag}} &= \frac{\mu_0}{8\pi}\sum_i \sum_{k(k\neq i)} I_i I_k \int\int \frac{d\mathbf{l}_i\cdot d\mathbf{l}_k}{\mid \mathbf{r}_i - \mathbf{r}_k \mid} + \frac{1}{2}\sum_k L_k I_k^2 \\ &= \frac{1}{2}\sum_k\sum_i L_{ik} I_i I_k \qquad (9.10)\end{aligned}$$

where the functions L_{ik}s are defined as

$$L_{ik} = \frac{\mu_0}{4\pi}\oint\oint \frac{d\mathbf{l}_i \cdot d\mathbf{l}_k}{\mid \mathbf{r}_i - \mathbf{r}_k \mid} \qquad \text{for } i \neq k$$
$$L_{ik} = L_k = L_i \qquad \text{for } i = k \qquad (9.11)$$

Obviously $L_{ik} = L_{ki}$. For $i \neq k$, L_{ik} is called the mutual inductance of the i-th and the k-th circuits. Note that the mutual inductances depend on the 'geometry' of the coils and their position and orientation but do not depend on the current strengths. They may be defined either in terms of the flux linked for unit current or, because of Faraday's law, as the induced electromotive force per unit rate of change of current.

The expression (9.10) for the magnetic energy is of the same form as that for kinetic energy in generalized velocities q_i

$$T = \frac{1}{2}\sum a_{ik}\dot{q}_i\dot{q}_k$$

As in that case, the condition of positive definiteness of the energy puts some restriction on the possible value of L_{ik}s. Thus, the self inductances must all be positive, and for any two coils, one must have the inequality

$$L_i L_k > L_{ik}^2$$

The force or torque which can be calculated directly from the force expression

$$\mathbf{F} = \int \mathbf{j} \times \mathbf{B}\, dv$$

may be obtained also from the energy expression. Thus, if q_i be a generalized coordinate connected with the position or orientation of a coil, the corresponding 'force' is

$$Q_i = \left(\frac{\partial U}{\partial q_i}\right)_{I=\text{constant}} \qquad (9.12)$$

where, of course, if the generalized coordinate be an angl coordinate, the corresponding generalized force is a torque. The unusual sign in the force formula (9.12) deserves some comment. As any coordinate q_i changes, an induced electromotive force is brought into play in the coils opposing this change. To maintain the currents I_k constant, one has to change the impressed electromotive forces, and thus, an additional term in the energy change is introduced. Thus, taking into account the induced electromotive force $-\sum \frac{\partial \Phi_k}{\partial q_i}\dot{q}_i$ in the k-th circuit, the energy conservation equation is

$$\sum_i Q_i\dot{q}_i + \sum\left(\frac{\partial U_{\text{mag}}}{\partial q_i}\right)_I \dot{q}_i - \sum_i\sum_k\left(\frac{\partial \Phi_k}{\partial q_i}\right)_I \dot{q}_i I_k = 0$$

But

$$\sum_k \left(\frac{\partial \Phi_k}{\partial q_i}\right)_I I_k = 2\left(\frac{\partial U_{\text{mag}}}{\partial q_i}\right)_I$$

Hence

$$\sum_i Q_i q_i - \sum \left(\frac{\partial U_{\text{mag}}}{\partial q_i}\right)_I \dot{q}_i = 0$$

Thus, we get Eq. (9.12) with the sign indicated. We may check the correctness of our calculations as also the consistency of our theory by deriving the force between two current circuits by putting the energy expression as given in the first line of Eq. (9.10) in Eq. (9.12)

$$F_{12} = \frac{\mu_0}{4\pi}\oint\oint I_1 I_2 (d\mathbf{l}_1 \cdot d\mathbf{l}_2)\,\nabla_1 \left(\frac{1}{\mid \mathbf{r}_1 - \mathbf{r}_2 \mid}\right)$$

where $\mathbf{F}_{12}$ is the force on the coil 1 due to coil 2 and ∇_1 represents the gradients with respect to the coordinates of coil 1. This expression is the same as that obtained from Eqs. (8.3) and (8.4) of Chap. 8.

9.3 Calculation of Self and Mutual Inductances

We may either use the flux formula $\oint \mathbf{B}\cdot\mathbf{ds}$ in case of a closed coil or calculate the energy of the magnetic field or use Eq. (9.11) to obtain the values of the inductances. However, not only the convenience of calculation but also other considerations may decide in favour of some particular choice. As we have seen in electrostatics, when the sources are spread out to infinity, sometimes the integral $\int \rho dv'/|\mathbf{r}-\mathbf{r}'|$ diverges but the electric intensity integral

$$\mathbf{E}(\mathbf{r}) = \frac{1}{4\pi\epsilon_0}\int \rho(\mathbf{r}')\frac{\mathbf{r}-\mathbf{r}'}{|\mathbf{r}-\mathbf{r}'|^3}\,dv'$$

converges and we can find a function whose gradient is the intensity. Here also, similar cases arise when the sources extend to infinity—the integral in (9.11) may diverge but the flux integral may converge to a finite value. This is the case in our Example 3.

Example 1. Self inductance of a long straight wire.

Here, we shall suppose that the field is cut off at a finite distance from the wire to obtain a finite value of the inductance. If the current be uniformly distributed over the cross-section of the wire, a circle of radius r will enclose a current Ir^2/a^2 where I is the total current carried by the wire and a the radius of the wire. Therefore, for the field $\mathbf{H}$ at a distance r $Ir(2\pi a^2)$, the energy per unit length for the region within the material of the wire is

$$\frac{1}{2}\int_0^a (\mathbf{H}\cdot\mathbf{B})\, 2\pi r\, dr = \frac{\mu_0 I^2}{4\pi a^4}\int_0^a r^3 dr = \frac{\mu_0 I^2}{16\pi}$$

assuming that the permeability is independent of $\mathbf{H}$. The field energy per unit length of the outside space is

$$\frac{\mu_0}{2}\int_a^r \left(\frac{I}{2\pi r}\right)^2 2\pi r\, dr = \frac{\mu_0 I^2}{4\pi}\log\left(\frac{r}{a}\right)$$

where the field has been artificially cut off at r. Using the expression $U_{\text{mag}} = \frac{1}{2}LI^2$, we have the self inductance per unit length

$$L = \frac{\mu_0}{8\pi} + \frac{\mu_0}{2\pi}\log\left(\frac{r}{a}\right)$$

At high frequencies of alternating current, the current becomes concentrated on the surface of the wire and L falls to $\frac{\mu_0}{2\pi}\log(r/a)$.

Example 2. Mutual inductance of two parallel, circular coils, the line joining their centres being perpendicular to the plane of the coils.
The distance between two elements r_{12} of coordinates $(r = a_1, z = 0, \phi = \theta_0)$ and $(r = a_2, z = d, \phi = \theta_0 + \theta)$ is

$$r_{12}^2 = a_1^2 + a_2^2 - 2a_1 a_2 \cos\theta + d^2$$

(The rectangular cartesian coordinates of the two elements are $(x_1 = a_1\cos\theta_0,$ $y_1 = a_1\sin\theta_0, z_1 = 0)$ and $(x_2 = a_2\cos(\theta_0+\theta), y_2 = a_2\sin(\theta_0+\theta), z_2 = d)$. Hence, $r_{12}^2 = (x_2 - x_1)^2 + (y_2 - y_1)^2 + d^2$. Substituting the values of x_1, x_2, y_1, y_2, we get the above expression.)

The mutual inductance, according to (9.11), is thus

$$L_{12} = \frac{\mu_0}{4\pi}\oint_\theta\oint_{\theta_0}\frac{a_1 a_2\cos\theta}{\sqrt{a_1^2 + a_2^2 - 2a_1 a_2\cos\theta + d^2}}\, d\theta_0\, d\theta$$

The integral cannot be evaluated in closed form. However, in case the two coils are of nearly equal radii and the separation $d \ll a$, the integral has been approximately evaluated to give

Fig. 9.2 A long wire and a coplanar circular coil

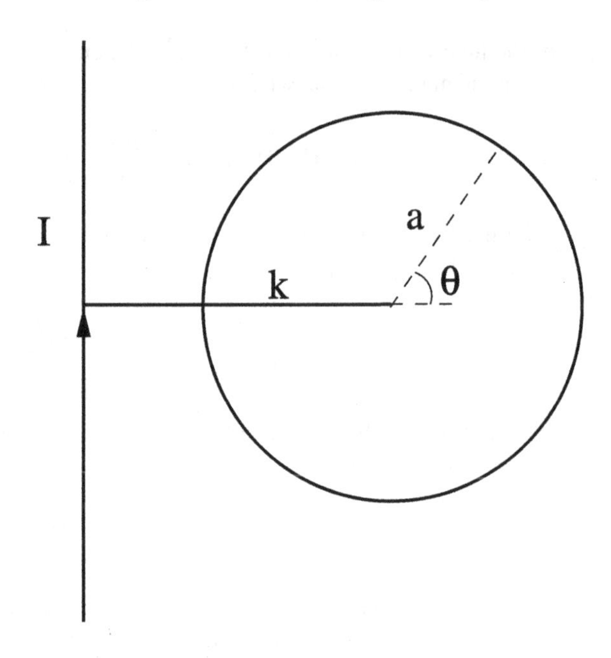

$$L_{12} = \mu_0 a \left(\log \left(\frac{8a}{b} \right) - 2 \right)$$

$$\text{where } a = \frac{a_1 + a_2}{2}, \qquad b^2 = 4a^2 \frac{(a_1 - a_2)^2 + d^2}{4a^2 + d^2}$$

Example 3. A long wire and a coplanar circular coil of radius a, and the centre of the circular coil at a distance k from the long wire.
In this case (Fig. 9.2), as we have already mentioned, the integral in (9.11) diverges. However, we may calculate the flux through the circular coil. The field due to the long wire is everywhere normal to the plane of the circle and considering an element lying between r and $r + dr$, θ and $\theta + d\theta$, the flux through the element is $\frac{\mu_0 I}{2\pi(k + r\cos\theta)} r dr d\theta$. Hence, the total flux is

$$\frac{\mu_0 I}{2\pi} \int_0^a \int_0^{2\pi} \frac{r dr\, d\theta}{k + r\cos\theta} = \mu_0 I \left(k - \sqrt{k^2 - a^2} \right)$$

and consequently, the mutual inductance is

$$\mu_0 \left(k - \sqrt{k^2 - a^2} \right)$$

9.4 Current in Circuits with Inductance, Resistance and Capacitance

If I be the current at any instant in a circuit with an inductor L, resistor R and capacitor C, and Q be the charge on the capacitor at the instant, the potential difference between the capacitor plates is Q/C, and from Faraday's law, the inductor gives rise to an electromotive force $-LdI/dt$, so that if the impressed electromotive force be E, Ohm's law gives

$$E - L\frac{dI}{dt} - \frac{q}{C} = RI \tag{9.13}$$

If E be independent of time, the equation is readily integrated (note that $I = dQ/dt$) to give

$$Q = EC + Ae^{\alpha_+ t} + Be^{\alpha_- t}$$

where A and B are arbitrary constants of integration to be fixed by initial conditions and

$$\alpha_\pm = -\frac{R}{2L} \pm \sqrt{\frac{R^2}{4L^2} - \frac{1}{LC}} = -\frac{R}{2L} \pm \beta \text{ (say)}$$

Thus, the change of Q will be oscillatory or aperiodic accordingly as R is less or greater than $2\sqrt{L/C}$. In case of charging of the capacitor, at $t = 0$, $Q = Q_0 = 0$ and we can write the solution formally as

$$Q = \begin{cases} CE - Ae^{-\frac{Rt}{2L}} \sin(\beta t + \theta) & \text{where } A\sin\theta = CE \text{ (oscillatory case)} \\ CE - Ae^{-\frac{Rt}{2L}} \sinh(\beta t + \theta) & \text{where } A\sinh\theta = CE \text{ (aperiodic case)} \end{cases}$$

Thus, even if the charging be oscillatory, the oscillations are damped out because of the irreversible joule heat production. In reality, there is an additional source of damping due to the radiation of electromagnetic waves which is, however, being neglected in the quasi-stationary approximation. In either case, the charge of the capacitor asymptotically assumes the equilibrium value CE. In the oscillatory case, the charge attains values above the equilibrium value occasionally. The case of discharge of a capacitor may be discussed similarly taking the initial value of Q as CE and putting the impressed electromotive force as zero. Again the charge asymptotically vanishes but this may be either by damped oscillations or in an aperiodic manner. In all these cases, the current is obtained by differentiating the expression for Q with respect to time.

The case of a circuit containing only an inductor and a resistor is of some interest—the differential equation is now simply

$$L\frac{dI}{dt} + RI = E \tag{9.14}$$

and the solution in the case of growth and decay of current is, respectively,

$$I = \begin{cases} \frac{E}{R}\left(1 - r^{-Rt/L}\right) & \text{(growth)} \\ I_0 r^{-Rt/L} & \text{(decay)} \end{cases}$$

9.4.1 The Case of an Alternating Impressed Electromotive Force

We shall assume the impressed electromotive force to be sinusoidally varying with time—in a sense this does not involve any loss of generality, for any periodic function may be Fourier decomposed into sinusoidal components, and thus, the solution may be obtained as a superposition of those due to sinusoidal electromotive forces. For a sinusoidal electromotive force, the differential equation reads

$$L\frac{d^2Q}{dt^2} + R\frac{dQ}{dt} + \frac{Q}{C} = E_0 \sin \omega t \tag{9.15}$$

The general solution of this equation is a sum of the general solutions of the equation obtained by putting the right-hand side to zero and a particular integral of the complete equation. The former is just the discharge equation we have already studied and as it is damped out usually in a fairly short time. It is called a transient, and the effective solution is, thus, the particular integral. In the present case, the latter and the corresponding current through the circuit are

$$Q = -\frac{E_0}{\sqrt{R^2\omega^2 + \left(\frac{1}{C} - L\omega^2\right)^2}} \cos\left[\omega t + \tan^{-1}\frac{1}{C}\left(L\omega - \frac{1}{C\omega}\right)\right] \tag{9.16}$$

$$I = \frac{E_0}{\sqrt{R^2 + \left(\frac{1}{C\omega} - L\omega\right)^2}} \sin\left[\omega t + \tan^{-1}\frac{1}{C}\left(L\omega - \frac{1}{C\omega}\right)\right] \tag{9.17}$$

One can describe the situation by saying that the role of the resistor is taken over by the expression

$$Z \equiv \sqrt{R^2 + \left(\frac{1}{C\omega} - L\omega\right)^2} \tag{9.18}$$

which is called the impedance of the circuit and the current, which is of the same period as the electromotive force, lags in phase by $\tan^{-1}\frac{1}{C}\left(L\omega - \frac{1}{C\omega}\right)$.

The simple case where there is no capacitor may be considered a little further. The work done by the applied electromotive force in time dt is

$$\frac{E_0^2}{\sqrt{R^2 + L^2\omega^2}} \sin \omega t \, \sin\left(\omega t + \tan^{-1}\frac{L\omega}{R}\right) dt$$

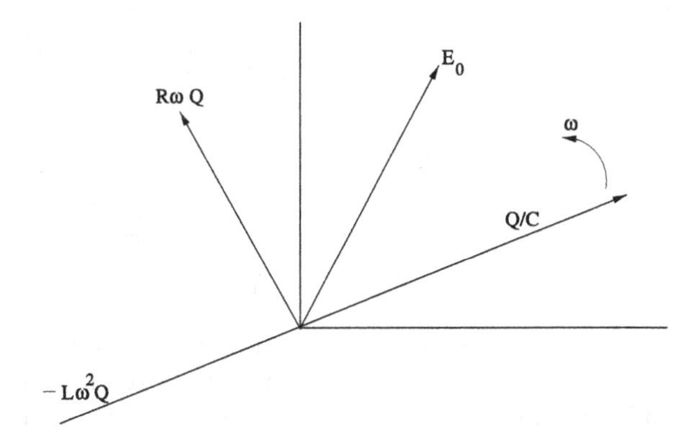

Fig. 9.3 Vector diagram for LCR circuit with alternating voltage

so that the average rate of work done during a complete period is

$$\frac{1}{2}E_0 I_0 \cos\left(\tan^{-1}\left(\frac{L\omega}{R}\right)\right)$$

which can be written as

$$E_m I_m \cos\theta$$

where E_m and I_m are the root mean square values of the electromotive force and current, respectively, and θ is the phase difference between the current and electromotive force. The potential drop across the resistors RI is in phase with the current I and hence work is done in driving the current through it: this appears as joule heat. The electromotive force across the inductor $L(dI/dt)$ has a phase difference of $\pi/2$ with the current, and hence, the time integral of the work done over a complete period vanishes. This accounts for the decrease of the rate of work done below the value $E_m I_m$.

The study of alternating currents is sometimes facilitated by what is called a vector diagram (Fig. 9.3). As is well known a complex number can be represented as a vector in a plane (Argand diagram), thus, the processes of addition and subtraction of complex numbers reduce to the rules of vector addition and subtraction. In the case of sinusoidal electromotive forces, the typical differential equation gives rise to $\sin\omega t$ and $\cos\omega t$ terms which must separately cancel out. Representing the electromotive force by the real (or imaginary) part of $\exp(i\omega t)$ and using the vector representation, the above corresponds to the separate cancellation of real and imaginary parts, i.e., of two orthogonal components of the vector equation. Again a differentiation of $\cos(\omega t)$ or $\sin(\omega t)$ with respect to t multiplies the constant factor by ω and changes the phase by $\pi/2$. In the vector diagram, this means a change of length by the factor ω and an counterclockwise turning of the vector through $\pi/2$. Thus, considering Eq. (9.15), $L\frac{d^2Q}{dt^2}$ and Q/C give two vectors in the same direction with lengths $L\omega^2 Q$ and Q/C, respectively, while the term RdQ/dt gives a vector of length $R\omega Q$ and at right angles to the previous vector. Thus, the resultant will be of length

$$\omega Q_0 \sqrt{R^2 + \left(L\omega - \frac{1}{C\omega}\right)^2}$$

This must be equal to the length of the vector on the right-hand side of Eq. (9.15); hence,

$$Q = \frac{E_0}{\omega\sqrt{R^2 + \left(L\omega - \frac{1}{C\omega}\right)^2}}$$

Again the phase of the resultant of the left-hand side terms is ahead of that of Q by $\cot^{-1}\frac{1}{R}\left(\frac{1}{C\omega} - L\omega\right)$. These are the results given in (9.16). As time progresses, the entire system of vectors turns in the anticlockwise direction with angular velocity ω.

Problems

1. An electromotive force is maintaining a steady current in a circuit, so that $\oint \mathbf{E} \cdot d\mathbf{l} \neq 0$. Hence $\nabla \times \mathbf{E} \neq 0$. Therefore $\dot{\mathbf{B}} \neq 0$. As conditions are stationary, $\dot{\mathbf{B}}$ will be constant in time and so $\mathbf{B}$ will monotonically increase with time whenever there is an electromotive force driving a steady current. Point out the fallacy in the above argument.
2. Calculate the self inductance per unit length of a long solenoid. Calculate also the mutual inductance for a secondary solenoid of N total turns wound over the central region of the long solenoid.
3. The parallel railway lines have a connection through a millivoltmeter. What will be the reading of the millivoltmeter when a train moves on the rails with a velocity of 60 km per hour? The separation between the rails is 2 m and the vertical component of the earth's magnetic field at the place is 0.1 Oersted.
4. A circular coil of radius r and resistance R is spinning about one of its diameters with angular velocity ω. If the magnetic field B is perpendicular to the axis of rotation, show that the average rate of dissipation of energy due to induced currents is $\frac{1}{2R}\pi^2 r^4 \omega^2 B^2$.
5. Two circuits having inductors L_1 and L_2 and resistance R_1 and R_2, respectively, are so placed that their mutual inductance is M. A sinusoidal electromotive force is impressed on the first circuit. Set up the differential equations and derive the integrals. Discuss the limiting case when the resistances vanish and $M^2 \approx L_1 L_2$.
6. A circular coil of radius R carries a current $I_0 \sin \omega t$. At a distance $\sqrt{3}R$, a coil of radius $r \ll R$ is placed, initially at rest. Calculate the motion of the smaller coil if the bigger coil be fixed. Take the resistance of the smaller coil to be Ω.
7. Fermi considered the electron as a distribution of charge and magnetic dipoles in a finite spherical region of radius r. Show that if the mass of the electron is to be interpreted as electromagnetic field energy, then the contribution of the magnetic energy dominates and the radius of the electron is $\sim 10^{-12}$ cm (magnetic dipole moment is $\frac{he}{4\pi m}$). If, however, the electric field energy alone is considered, the radius is $\sim 10^{-13}$ cm.

Chapter 10
Maxwell's Equations, Electromagnetic Energy and Momentum

So far we have come across the following equations for the electromagnetic field variables:

$$\nabla \cdot \mathbf{D} = \rho \tag{10.1}$$

$$\nabla \times \mathbf{E} = -\dot{\mathbf{B}} \tag{10.2}$$

$$\nabla \cdot \mathbf{B} = 0 \tag{10.3}$$

$$\nabla \times \mathbf{H} = \mathbf{j} \tag{10.4}$$

However, it is easy to see that this system of equations is notconsistent. Suppose for example that the current density $\mathbf{j}$ is solely due to a mass motion of the charges with velocity $\mathbf{u}$ (i.e., we consider a convection current). Then $\mathbf{j} = \rho\mathbf{u}$ and taking the divergence of Eq. (10.4), we get

$$\nabla \cdot (\rho\mathbf{u}) = 0 \tag{10.5}$$

However, the conservation of charges requires

$$\frac{\partial \rho}{\partial t} + \nabla \cdot (\rho\mathbf{u}) = 0 \tag{10.6}$$

Hence, Eqs. (10.5) and (10.6) are consistent only in case the charge density does not change with time—a condition we have associated with the stationary situation. To get over this difficulty, Maxwell modified Eq. (10.4) by the addition of a new term so that it became

$$\nabla \times \mathbf{H} = \dot{\mathbf{D}} + \mathbf{j} \qquad \boxed{\nabla \times \mathbf{H} = \frac{1}{c}\dot{\mathbf{D}} + \frac{4\pi}{c}\mathbf{j} \text{ (Gaussian units)}} \tag{10.7}$$

A. K. Raychaudhuri, *Classical Theory of Electricity and Magnetism*, Texts and Readings in Physical Sciences 21, https://doi.org/10.1007/978-981-16-8139-4_10

Taking the divergence of (10.7) and using (10.1), we get Eq. (10.6). The question of consistency may be put into a slightly different form. Let us count the number of Eqs. (10.2) and (10.4) are vector equations and, therefore, give effectively six equations. Together with (10.1) and (10.3), we have, therefore, eight equations. If we pose the problem as one in which ρ and $\mathbf{j}$ (the sources of the field) are specified, we have six unknowns (the electric and magnetic field components $\mathbf{E}$ and $\mathbf{B}$ for the moment we assume that phenomenological relations between $\mathbf{D}$ and $\mathbf{E}$ as well as between $\mathbf{H}$ and $\mathbf{B}$ are known, so that only one pair of vectors is to be regarded as unknown). Thus, we have an over-determined system, and consistency demands that there must be $(8-6)=2$ identities between the eight equations. On taking the divergence, Eq. (10.2) gives

$$\frac{\partial}{\partial t}(\boldsymbol{\nabla}\cdot\mathbf{B})=0 \tag{10.8}$$

Equation (10.8) shows that if at a particular instant of time (10.3) is satisfied everywhere in space, it will be automatically satisfied at all subsequent instants everywhere. Thus, (10.3) becomes some sort of an initial or boundary condition. A similar situation holds for (10.7) and (10.1). Taking the divergence of (10.7) and using the conservation of charge equation

$$\frac{\partial\rho}{\partial t}+(\boldsymbol{\nabla}\cdot\mathbf{j})=0 \tag{10.9}$$

we get

$$\frac{\partial}{\partial t}(\boldsymbol{\nabla}\cdot\mathbf{D}-\rho)=0 \tag{10.10}$$

Thus, (10.1) can also be regarded as an initial condition. The system, thus, reduces to six equations for six unknowns with two restrictions on initial conditions. (A problem in which the condition at say time $t=0$ is specified—called the Cauchy data—and one is to calculate the time evolution of the variables is called the Cauchy problem. Thus, if the solution of the electromagnetic field problems from the Maxwell equations is posed as a Cauchy problem, Eqs. (10.1) and (10.3) are merely constraints which the Cauchy data must satisfy.)

Of course, matters are sometimes differently stated—one uses (10.7) and (10.1) to obtain (10.9) and says that conservation of charge follows from Maxwell's equations. Returning to Eq. (10.7), the new term introduced by Maxwell is called the displacement current and can be split up into two parts

$$\dot{\mathbf{D}}=\epsilon_0\dot{\mathbf{E}}+\dot{\mathbf{P}}$$

It is obvious that a change of polarization is associated with a movement of bound charges and so the introduction of $\dot{\mathbf{P}}$ as a current does not cause great surprise, but the term $\epsilon_0\dot{\mathbf{E}}$ would be present even in a vacuum and any so-called physical interpretation of this term seems out of question. However, it brings in a similarity in the reciprocal relation of electric and magnetic fields—a change of **B** brings in an electric field (Eq. (10.2)) and a change of **E** causes a magnetic field (Eq. (10.7)).

We have already mentioned about phenomenological relations to supplement Maxwell's equations. In the simple case of a linear homogeneous dielectric and absence of ferromagnetic materials and permanent magnets, the phenomenological relations are

$$\mathbf{D} = \epsilon \mathbf{E} \tag{10.11}$$

$$\mathbf{B} = \mu \mathbf{H} \tag{10.12}$$

where the electric permittivity ϵ and the magnetic permeability μ are related to the corresponding quantities in the free space by $\epsilon = \epsilon_0(1 + \chi_e)$ and $\mu = \mu_0(1 + \chi_e)$, with χ_e and χ_m being the electric and the magnetic susceptibilities of the medium, respectively. Considering the current to be a superposition of conduction and convection effects, we introduce Ohm's law in the form

$$\mathbf{j} = \sigma \mathbf{E} + \rho \mathbf{u} \tag{10.13}$$

The equations become simpler in the case of vacuum and no sources. We need not then distinguish between **D** and **E** or **B** and **H**. Thus,

$$\nabla \cdot \mathbf{E} = 0 \tag{10.14}$$

$$\nabla \times \mathbf{E} = -\dot{\mathbf{B}} \tag{10.15}$$

$$\nabla \cdot \mathbf{B} = 0 \tag{10.16}$$

$$\nabla \times \mathbf{B} = \epsilon_0 \mu_0 \dot{\mathbf{E}} \tag{10.17}$$

We have a remarkably simple system of equations which exhibit a symmetry between electric and magnetic fields.

Maxwell's equations are linear differential equations—this introduces considerable simplicity in mathematical discussions. Further, because of this, we may superpose two distinct solutions to obtain a solution, e.g., if $\mathbf{E}_1$, $\mathbf{D}_1$, $\mathbf{B}_1$, $\mathbf{H}_1$, ρ_1 and $\mathbf{j}_1$ constitute a solution of Maxwell's equations and the corresponding variables with subscript 2 constitute another solution, the linear combination

$$c_1\mathbf{E}_1 + c_2\mathbf{E}_2,\ c_1\mathbf{D}_1 + c_2\mathbf{D}_2,\ \ldots,\ c_1\mathbf{j}_1 + c_2\mathbf{j}_2$$

will also satisfy Maxwell's equations. This apparently introduces a non-uniqueness in the solution, since to any solution we may add any solution of the source-free equations to obtain a different solution for the same source distribution. In the static case, non-uniqueness is gotten rid of by means of boundary conditions—nontrivial solutions of source-free equations are eliminated. In the non-static situation, however, things are not that simple and as we shall see there are intriguing questions.

10.1 The Electromagnetic Waves in an Isotropic Homogeneous Dielectric

The medium is considered uncharged but to begin with we do retain a finite conductivity and consequently a conduction current

$$\nabla \cdot \mathbf{E} = \frac{1}{\epsilon}\nabla \cdot \mathbf{D} = 0 \tag{10.18}$$

$$\nabla \cdot \mathbf{B} = \mu\nabla \cdot \mathbf{H} = 0 \tag{10.19}$$

$$\nabla \times \mathbf{E} = -\dot{\mathbf{B}} = -\mu\dot{\mathbf{H}} \tag{10.20}$$

$$\nabla \times \mathbf{H} = \dot{\mathbf{D}} + \mathbf{j} = \epsilon\dot{\mathbf{E}} + \sigma\mathbf{E} \tag{10.21}$$

From Eqs. (10.20) and (10.21), we have

$$\nabla \times (\nabla \times \mathbf{E}) = -\mu(\nabla \times \dot{\mathbf{H}}) = -\mu\epsilon\ddot{\mathbf{E}} - \mu\sigma\dot{\mathbf{E}}$$

Now using (10.18) and the vector identity $\nabla \times (\nabla \times \mathbf{E}) = \nabla(\nabla \cdot \mathbf{E}) - \nabla^2\mathbf{E}$, we get

$$\nabla^2\mathbf{E} = \mu\epsilon\ddot{\mathbf{E}} + \mu\sigma\dot{\mathbf{E}} \tag{10.22}$$

Let us suppose that the second term on the right-hand side of (10.22) is negligible: this may be either due to low conductivity or a time dependence of $\mathbf{E}$ such as to make the second derivative with respect to t much greater than the first derivative. Whatever be the case, we have then the simple wave equation

$$\nabla^2\mathbf{E} = \mu\epsilon\ddot{\mathbf{E}} \tag{10.23}$$

The velocity of propagation is, thus, $v = 1/\sqrt{\mu\epsilon}$ reducing to the value $c = 1/\sqrt{\epsilon_0\mu_0}$ for vacuum. This linking up of the velocity of light with electromagnetic quantities was one of the first impressive results of Maxwell's theory.

To have an idea of the nature of the waves to which our equations have led, we consider a plane monochromatic solution of the form

$$\begin{aligned}\mathbf{E} &= \mathbf{E}_0 \exp\left(\frac{2\pi i}{\lambda}(lx + my + mz \pm vt)\right) \\ &= \mathbf{E}_0 \exp(i\mathbf{k}\cdot\mathbf{r} \pm i\omega t) \qquad \text{with } k^2 = \frac{\omega^2}{v^2}\end{aligned} \tag{10.24}$$

The equi-phase surfaces (the wavefronts) are the planes $\mathbf{k}\cdot\mathbf{r} =$ constant and the frequency is $\omega/(2\pi)$ and $|k| = 2\pi/\lambda$, where λ is the wavelength. The waves are apparently non-dispersive but that is only because we have not so far considered the possible dependence of the dielectric constant on frequency of the incident electric field. We may obtain an exactly similar wave equation for **H**. Taking the curl of Eq. (10.21) and then eliminating **E** with the help of Eq. (10.20), we obtain

$$\nabla^2 \mathbf{H} = \epsilon \ddot{\mathbf{H}} + \sigma \mu \dot{\mathbf{H}} \tag{10.25}$$

Equation (10.25) is of the same form as (10.22) and assuming similar circumstances as in going from (10.22) to (10.23), we get

$$\nabla^2 \mathbf{H} = \epsilon \mu \ddot{\mathbf{H}} = \frac{1}{v^2} \ddot{\mathbf{H}} \tag{10.26}$$

and we again take a plane wave solution

$$\mathbf{H} = \mathbf{H}_0 e^{i(\mathbf{k}' \cdot \mathbf{r} \pm \omega' t)} \tag{10.27}$$

The velocities of **E** and **H** waves are obviously identical. In writing out (10.27), we are allowing for the possibility that the wavefronts and frequencies may be different. However, the coupling between the two fields that exists because of Maxwell's equations constrains these to be identical. By substituting from (10.27) and (10.24) in (10.20) and (10.21), we get (neglecting the conduction term)

$$i\mathbf{k} \times \mathbf{E} = \pm i \mu \omega' \mathbf{H} \tag{10.28}$$

$$i\mathbf{k}' \times \mathbf{H} = \pm i \epsilon \omega \mathbf{E} \tag{10.29}$$

Note that with the plane waveform, the operator $\nabla = i\mathbf{k}$ and $\frac{\partial}{\partial t} = \pm i\omega$. As (10.27) and (10.28) are to hold at all times and all positions of space, we get $\mathbf{k} = \mathbf{k}'$ and $\omega = \omega'$. The equations show further that the three vectors **E**, **H** and **k** are mutually orthogonal. Thus, the electromagnetic field vectors lie in the wavefront and the propagation vector is normal to the front. With this, Eqs. (10.18) and (10.19) which read

$$\mathbf{k} \cdot \mathbf{E} = 0, \qquad \mathbf{k} \cdot \mathbf{H} = 0$$

are trivially satisfied. Also, from (10.28) or (10.29), we have (as $k^2 = \omega^2/v^2$)

$$\sqrt{\epsilon}\, |\mathbf{E}| = \sqrt{\mu}\, |\mathbf{H}|$$

so that if we take over the expressions of magnetostatic and electrostatic field energy densities, that we have obtained previously to be valid in this non-static case as well, we may say that the field energy is partitioned equally between the magnetic and electric fields. (This result, obtained by neglecting the conduction term, will not hold for good conducting materials, as we shall see later.)

10.2 Energy Flux and the Poynting Vector

We go back to Eqs. (10.2) and (10.7). Taking the scalar product of (10.7) with **E** and that of (10.2) with **H**, we get on subtracting

$$\mathbf{E}\cdot\nabla\times\mathbf{H}-\mathbf{H}\cdot\nabla\times\mathbf{E}=\left(\mathbf{E}\cdot\dot{\mathbf{D}}+\dot{\mathbf{B}}\cdot\mathbf{H}\right)+\mathbf{E}\cdot\mathbf{j}$$
$$\text{or}\quad \nabla\cdot(\mathbf{H}\times\mathbf{E})=\frac{1}{2}\frac{\partial}{\partial t}(\mathbf{E}\cdot\mathbf{D}+\mathbf{B}\cdot\mathbf{H})+\mathbf{E}\cdot\mathbf{j}$$

since the left-hand side in the first line above reduces to $\nabla\cdot(\mathbf{H}\times\mathbf{E})$. From our knowledge of stationary electromagnetic fields, the right-hand side is apparently the time rate of change of electromagnetic field energy density plus the rate of generation of joule heat per unit volume. Integrating the above equation over a certain region of space and converting the volume integral on the left-hand side into a surface integral, we get the even more significant equation

$$-\oint(\mathbf{E}\times\mathbf{H})\cdot\mathbf{ds}=\frac{1}{2}\frac{\partial}{\partial t}\int(\mathbf{E}\cdot\mathbf{D}+\mathbf{B}\cdot\mathbf{H})\,dv+\int\mathbf{E}\cdot\mathbf{j}\,dv \tag{10.30}$$

This equation is interpreted as the principle of conservation of energy—the loss of energy in the region of integration is just the flux of energy through the boundary of the region represented by the surface integral on the left. Thus, the vector $\mathbf{E}\times\mathbf{H}$ apparently gives the amount of energy flowing through a unit area per unit time, the element of area being held normal to the direction of $\mathbf{E}\times\mathbf{H}$. This vector is known as the Poynting vector. There are, however, a few intriguing points:

1. We have not proved that the expression $\frac{1}{2}(\mathbf{E}\cdot\mathbf{D}+\mathbf{B}\cdot\mathbf{H})$ gives the energy density in case of non-static fields. In the case of static fields, this expression for the energy density was deduced on the assumption that the field intensities vanish at infinity at least as fast as $1/r^2$, but in the case of radiation fields, the field intensities vanish as $1/r$ and consequently $\int_\infty \mathbf{E}\cdot\mathbf{D}\,dv$ would blow up.
2. There exist realizable static fields where $\mathbf{E}\times\mathbf{H}$ does not vanish but because the field is static, there is no question of an energy flux. As an example, consider a long solenoid carrying current and a capacitor oriented so that they give electric and magnetic fields orthogonal to one another. The usual answer to this difficulty is that one must consider the surface integral over an entire closed surface as the energy flux through that surface and must not consider the value of $\mathbf{E}\times\mathbf{H}$ as giving the flux at each point separately. Such an argument has an apparent justification in the form of Eq. (10.30). However, this does not solve the following puzzle.
3. Consider a long straight wire carrying current. Because of the continuity condition, there must be an electric field parallel to the axis of the wire (say the z-axis). The magnetic field lines are circles in the perpendicular plane, so that $\mathbf{E}\times\mathbf{H}$ turns out to be in the radial direction and towards the axis of the wire. Thus, our formula would give a continuous flux of energy into the wire through its entire surface. Of course, inside the wire too, there would be such a flux towards the axis. Indeed the

argument is independent of the form of the wire—in its immediate neighbourhood, there is always a non-vanishing $\mathbf{E} \times \mathbf{H}$ directed inwards.

Note that in these cases, Eq. (10.30) shows that the flux of energy exactly accounts for the Joule heat production.

10.3 The General Stress Tensor and Momentum of Radiation

We have seen in electrostatics that consistent with the requirement of field theory, force in electrostatics may be represented as derived from surface forces—the so-called Maxwell stress tensor. In a similar way, in magnetic fields, we may construct a tensor whose divergence gives the force density. We now add up these two stress tensors to write (using the convention that summation over a repeated index is to be understood)

$$T_{ik} = -\frac{1}{2}\left(E_l D_l + H_l B_l\right)\delta_{ik} + \left(E_i D_k + H_i B_k\right) \tag{10.31}$$

and assert that this is the general stress tensor for any electromagnetic field—static or non-static. This means that we hope to obtain the force density by calculating its divergence. We have

$$\begin{aligned}\frac{\partial T_{ik}}{\partial x_k} = -\frac{1}{2}\big(&E_{l,i}D_l + E_l D_{l,i} + H_{l,i}B_l + H_l B_{l,i}\\ &-2E_{i,k}D_k - 2E_i D_{k,k} - 2H_{i,k}B_k - 2H_i B_{k,k}\big)\end{aligned} \tag{10.32}$$

where a comma followed by an index indicates differentiation with respect to the corresponding coordinate.[1] Thus, $\nabla \cdot \mathbf{A} = A_{i,i}$. Using now

$$\begin{aligned}[\mathbf{D} \times (\nabla \times \mathbf{E})]_x &= D_y(\nabla \times \mathbf{E})_z - D_z(\nabla \times \mathbf{E})_y\\ &= D_y\left(\frac{\partial E_y}{\partial x} - \frac{\partial E_x}{\partial y}\right) - D_z\left(\frac{\partial E_x}{\partial z} - \frac{\partial E_z}{\partial x}\right)\\ &= D_k\left(\frac{\partial E_k}{\partial x} - \frac{\partial E_x}{\partial x_k}\right)\end{aligned}$$

where the term $D_x\left(\frac{\partial E_x}{\partial x} - \frac{\partial E_x}{\partial x}\right)$, which is zero, was added to get the final expression. We have

$$\begin{aligned}E_{l,i}D_l + E_l D_{l,i} - 2E_{i,k}D_k &= 2(E_{k,i} - E_{i,k})D_k + E^2\epsilon_{,i}\\ &= 2\,[\mathbf{D} \times (\nabla \times \mathbf{E})]_i + E^2\epsilon_{,i}\end{aligned}$$

[1] For example, $\frac{\partial f}{\partial x} = f_{,1}$, $\frac{\partial f}{\partial y} = f_{,2}$ and $\frac{\partial f}{\partial z} = f_{,3}$. In general, $\frac{\partial f}{\partial x_i} = f_{,i}$.

Similarly,

$$H_{l,i} B_l + H_l B_{l,i} - 2H_{i,k} B_k = 2\,[\mathbf{B} \times (\nabla \times \mathbf{H})]_i + H^2 \mu_{,i}$$

Using these expressions and also all four Maxwell's equations, we get from (10.32)

$$\frac{\partial T_{ik}}{\partial x_k} = E_i \rho + (\mathbf{j} \times \mathbf{B})_i - \frac{1}{2} E^2 \epsilon_{,i} - \frac{1}{2} H^2 \mu_{,i} + \frac{\partial}{\partial t} (\mathbf{D} \times \mathbf{B})_i \tag{10.33}$$

Besides the terms which occur as force densities in the case of electrostatic and magnetic fields, we have the additional term $\frac{\partial}{\partial t}(\mathbf{D} \times \mathbf{B})_i$ which is closely related to the Poynting vector. This term is independent of the existence of charges or currents and is non-vanishing even in a vacuum provided the flux of electromagnetic energy exists. We interpret it in the following manner—the part

$$E_i \rho + (\mathbf{j} \times \mathbf{B})_i - \frac{1}{2} E^2 \epsilon_{,i} - \frac{1}{2} H^2 \mu_{,i}$$

causes a momentum change of matter carrying charges and currents. The new term must also indicate some momentum change and the only candidate that is present here is the electromagnetic field. Thus, we are led to assign a momentum to the electromagnetic field and its relation with the energy flux is consistent with the ideas of the special theory of relativity.[2]

Considering a plane wave, we have shown that **E**, **H** and **k** are mutually orthogonal. Hence, the Poynting vector as well as the momentum vector coincide in direction with the propagation vector. These results have been obtained considering a plane wave. However, a wave of arbitrary form can be considered plane over an infinitesimal region, and thus, these characteristics hold good for electromagnetic waves in general.[3]

10.4 The Pressure of Radiation

As the radiation field carries momentum, if it is absorbed or scattered (reflection is also a case of scattering) by a material obstacle, there will be a momentum transfer from the field to the obstacle which will, thus, experience a force—thus, we go to the idea of radiation pressure on any matter on which radiation is incident.

We have seen that the waves are transverse, i.e., the electric and magnetic vectors are normal to the direction of propagation. This does not completely specify their directions, e.g., if z be the direction of propagation, **E** and **H** may have any direction in the xy-plane, subject to the restriction that **E** is perpendicular to **H**.

[2] According to the special theory of relativity, energy has inertia and thus energy flux is associated with momentum.

[3] See, however, the case of wave guides, where boundary conditions lead to a different situation.

Now if the situation be such that **E** and **H** directions change in a random fashion, so that there is no distinction between different directions in the xy-plane, we say that the wave is unpolarized. If, however, **E** is always in a particular direction (say along x) and consequently **H** is always in the y-direction, we say that the wave is plane (or linearly) polarized with electric vector in the x-direction. Similarly, we have cases of circular or elliptical polarization when the electric vector sweeps out a circle or ellipse, respectively. (Recall the result that two simple harmonic vibrations in inclined directions combine to form an elliptic motion, and in special cases, the ellipse may degenerate to a straight line or circle.)

To calculate the pressure of radiation, we consider a plane polarized wave proceeding in the z-direction with its electric vector in the x-direction and magnetic vector in the y-direction. The stress tensor (10.30) is then diagonal and has the following non-vanishing components:

$$T_{xx} = T_{yy} = \frac{1}{2}\left(E_x D_x - H_y B_y\right)$$
$$T_{zz} = -\frac{1}{2}\left(E_x D_x + H_y B_y\right)$$

Thus, the stresses are purely normal but, unlike fluid pressure, are anisotropic. If the radiation falling on the xy-plane is absorbed, the amount of momentum destroyed per unit area per unit time is c time the momentum density of radiation, which is $\frac{1}{\sqrt{\epsilon\mu}}\langle \mathbf{D}\times\mathbf{B}\rangle_{\text{av}} = \sqrt{\epsilon\mu}\left\langle E_x H_y\right\rangle_{\text{av}}$. Thus, for a directed beam of radiation, the pressure is equal to the energy density.

For diffuse radiation (black body radiation), however, $\left\langle E_x^2\right\rangle_{\text{av}} = \left\langle E_y^2\right\rangle_{\text{av}} = \left\langle E_z^2\right\rangle_{\text{av}} = \frac{1}{3}\left\langle E^2\right\rangle_{\text{av}}$, etc. The Maxwell stress tensor reduces to the form of that of a perfect fluid

$$T_{ik} = -\frac{1}{3}\left\langle\frac{1}{2}\left(\epsilon E^2 + \mu H^2\right)\right\rangle_{\text{av}} \delta_{ik} = -p\,\delta_{ik}$$

Thus, the pressure is isotropic and is of magnitude one-third of the energy density.

Problems

1. Show that black body radiation undergoing reversible adiabatic expansion satisfies the equation $\rho V^{4/3} =$ constant where V is the volume and ρ the density of radiation.
2. Investigate the condition that

$$\mathbf{E} = \mathbf{a}\cos\omega t + \mathbf{b}\sin\omega t, \quad \mathbf{j} = \mathbf{p}\cos\omega t + \mathbf{q}\sin\omega t, \quad \mathbf{H} = 0$$

   may be a solution of the Maxwell equations.
3. Obtain a complete solution of Maxwell's equations for an infinite straight wire carrying a current, taking account of the phenomenological relation $\mathbf{j} = \sigma\mathbf{E}$. Calculate $(\mathbf{E}\times\mathbf{H})$ and interpret the situation physically.

4. Verify that the field

$$\mathbf{E}(\mathbf{r}, t) = \frac{e}{4\pi\epsilon_0} \frac{\mathbf{R} - \mathbf{u}R/c}{\left(\mathbf{R} - \frac{1}{c}\mathbf{u} \cdot \mathbf{R}\right)^3} \left(1 - \frac{u^2}{c^2}\right), \qquad \mathbf{B} = \frac{1}{c}\mathbf{u} \times \mathbf{E}$$

where $\mathbf{R} = \mathbf{r} - \mathbf{r}'$ and $\mathbf{u}$ is a constant vector whose magnitude is small compared to the velocity of light c and satisfies the Maxwell equations with $\rho(\mathbf{r}', t) = e\,\delta\left(\mathbf{r}' - \mathbf{u}t\right)$ and $\mathbf{j} = \rho\mathbf{u}$.

5. Show that for an infinite solenoid carrying current, if one takes into account the phenomenological relation $\mathbf{j} = \sigma\mathbf{E}$, there is no stationary solution.
6. If the non-vanishing components of the fields in polar coordinates be given by

$$H_\varphi = \frac{\sin\theta}{r}\left(\frac{f}{c^2} + \frac{f}{cr}\right)$$

$$E_\varphi = 2\cos\theta\left(\frac{f}{cr^2} + \frac{f}{r^3}\right) \quad E_\theta \quad = \frac{f}{c^2 r}\sin\theta + \left(\frac{f}{cr^2} + \frac{f}{r^3}\right)\sin\theta$$

find the energy crossing a sphere of large radius per unit time.

7. Study a spherical wave solution of Maxwell's equations with particular stress on the possible orientations of the electric and magnetic fields.

Chapter 11
Reflection and Refraction of Electromagnetic Waves

When an electromagnetic wave falls on the interface between two media having different values of ϵ and μ, a part of the energy is sent back into the first medium but in an altered direction of propagation (the reflected wave) and a part penetrates into the second medium (refracted wave). In reality, matters are more complicated—there will be absorption, i.e., derangement of the energy to heat and irregular scattering. However, right now we shall not take these into consideration. We shall take for simplicity the case of a plane wave (a parallel beam in the language of geometrical optics) and the interface also will be taken to be a plane surface labelled $z = 0$, so that the region $z > 0$ constitutes medium 1 and $z < 0$ constitutes medium 2 (Fig. 11.1).

The unit vector normal to the interface at the point of incidence will be indicated by $\mathbf{n}$; thus, its only non-vanishing component is in the z-direction. In general, unprimed objects will refer to the incident wave, singly primed to the reflected wave and doubly primed to the refracted wave. As the boundary conditions (right now we are not considering their exact form) are to be satisfied all over the interface and at all times, the three wave disturbances must depend on x, y and t in the same manner at $z = 0$, i.e.,

$$e^{i(\mathbf{k}\cdot\mathbf{r}-\omega t)} = e^{i(\mathbf{k}'\cdot\mathbf{r}-\omega' t)} = e^{i(\mathbf{k}''\cdot\mathbf{r}-\omega'' t)}$$

are identities at $z = 0$. Hence,

$$\omega = \omega' = \omega'' \tag{11.1}$$

$$\mathbf{k}\cdot\mathbf{r} = \mathbf{k}'\cdot\mathbf{r} = \mathbf{k}''\cdot\mathbf{r} \tag{11.2}$$

for any vector $\mathbf{r}$ normal to $\mathbf{n}$. From (11.1), we get that the frequencies of the three waves are identical. (This is not a trivial result, since the velocities of the waves being different in the two media, either the frequency or the wavelength or both must differ in the incident and refracted waves—the identity of frequencies settles this question.) From (11.2), we get first that $\mathbf{k}$, $\mathbf{k}'$, $\mathbf{k}''$ and $\mathbf{n}$ lie in the same plane, and secondly

A. K. Raychaudhuri, *Classical Theory of Electricity and Magnetism*, Texts and Readings in Physical Sciences 21, https://doi.org/10.1007/978-981-16-8139-4_11

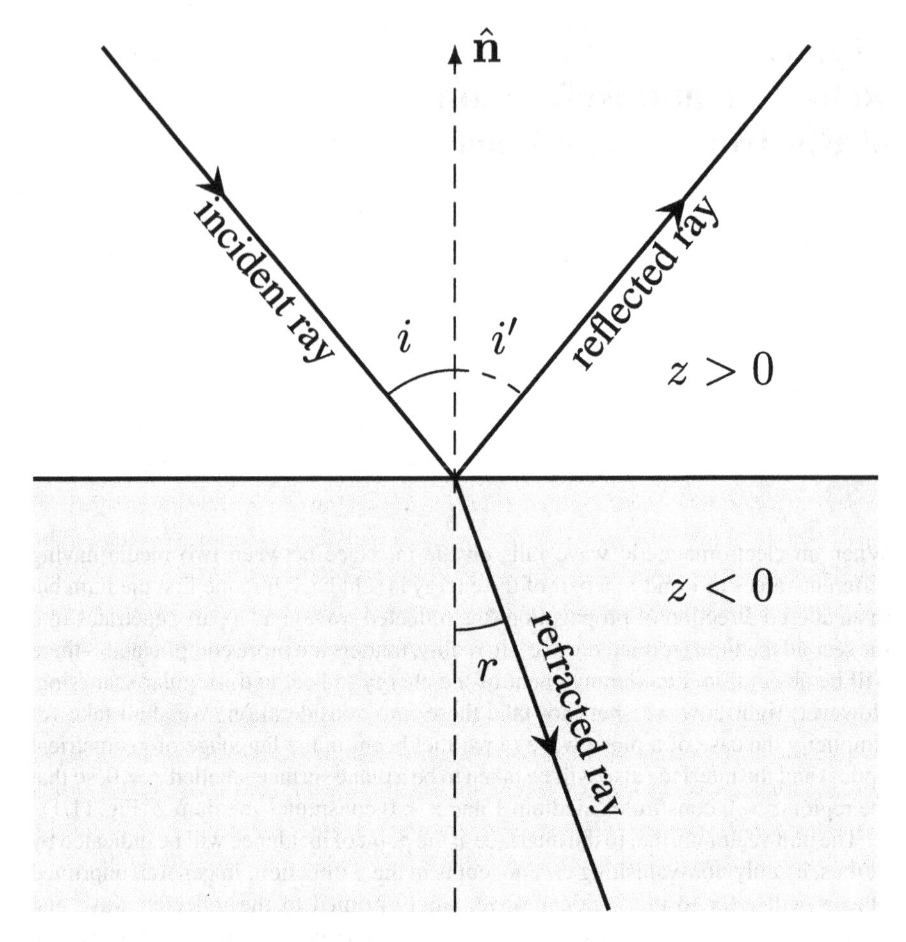

Fig. 11.1 Geometry of reflection and refraction

$$|\mathbf{k}| \sin i = |\mathbf{k}'| \sin i' = |\mathbf{k}''| \sin r \tag{11.3}$$

where i, i' and r are the angles of incidence, reflection and refraction, as indicated in the figure. As $|\mathbf{k}| = \omega/v$,

$$|\mathbf{k}| = |\mathbf{k}'| \quad \text{and} \quad \frac{|\mathbf{k}''|}{|\mathbf{k}|} = \frac{v'}{v''} = \sqrt{\frac{\epsilon''\mu''}{\epsilon'\mu'}}$$

Thus, $i = i'$ and

$$\frac{\sin i}{\sin r} = \sqrt{\frac{\epsilon''\mu''}{\epsilon'\mu'}} = \frac{n''}{n'} \tag{11.4}$$

where n, the ratio of the velocity of electromagnetic waves in vacuum to the velocity in a medium, is called the refractive index of the medium. Combining all this, we have the laws of reflection and refraction.

1. The incident ray, the reflected ray and the refracted ray as well as the normal to the interface lie in one and the same plane. The frequency of the three waves remains the same (and so the wavelengths of the incident and refracted waves are different).
2. The angles of incidence and reflection are equal and the sine of the angle of incidence bears a constant ratio with the sine of the angle of refraction—the value of the ratio depends on the nature of the two media and in general on the frequency of the incident wave (Snell's law), as we shall see afterwards.

So far we have not used the particular nature of the waves except in expressing the velocity of the waves in terms of ϵ and μ. Hence excepting that, all other conclusions would hold quite generally irrespective of the nature of the wave. We now use the specific boundary conditions of electromagnetic fields. We recall the following relations:

$$\mathbf{E} = \mathbf{E}_0 e^{i(\mathbf{k}\cdot\mathbf{r}-\omega t)}, \qquad \mathbf{B} = \mathbf{B}_0 e^{i(\mathbf{k}\cdot\mathbf{r}-\omega t)} \tag{11.5}$$

$$\mathbf{B} = \sqrt{\mu\epsilon}\,\frac{\mathbf{k}\times\mathbf{E}}{|\mathbf{k}|}, \qquad \frac{|k''|}{\sqrt{\mu''\epsilon''}} = \frac{|k|}{\sqrt{\mu'\epsilon'}} \tag{11.6}$$

Thus, the boundary conditions are

$$\epsilon'(\mathbf{E}_0 + \mathbf{E}_0') \cdot \mathbf{n} = \epsilon'' \mathbf{E}_0'' \cdot \mathbf{n} \qquad \text{(continuity of } \mathbf{D}\cdot\mathbf{n}) \tag{11.7}$$

$$(\mathbf{k} \times \mathbf{E}_0 + \mathbf{k}' \times \mathbf{E}_0') \cdot \mathbf{n} = (\mathbf{k}'' \times \mathbf{E}_0'') \cdot \mathbf{n} \qquad \text{(continuity of } \mathbf{B}\cdot\mathbf{n}) \tag{11.8}$$

$$(\mathbf{E}_0 + \mathbf{E}_0') \times \mathbf{n} = \mathbf{E}_0'' \times \mathbf{n} \qquad \text{(continuity of } \mathbf{E}\times\mathbf{n}) \tag{11.9}$$

$$\frac{1}{\mu'}(\mathbf{k} \times \mathbf{E}_0 + \mathbf{k}' \times \mathbf{E}_0') \times \mathbf{n} = \frac{1}{\mu''}(\mathbf{k}'' \times \mathbf{E}_0'') \times \mathbf{n} \qquad \text{(continuity of } \mathbf{H}\times\mathbf{n}) \tag{11.10}$$

We shall break up our calculations into two parts—first, we shall consider that the electric vector **E** is in the plane containing **k** and **n** (this plane is called the plane of incidence). Correspondingly, **B** in this case will be normal to the plane of incidence. In the second case, the situation will be reversed—**E** will be normal to the plane of incidence and **B** in this plane.

Case I: In this case, **E** is in the plane of incidence (Fig. 11.2).
Since **E** is in the plane of **k** and **n**, so **E** is of the form $a\mathbf{k} + b\mathbf{n}$ and **B** is in the perpendicular plane, so $\mathbf{B}\cdot\mathbf{n} = 0$. It is easy to check that these conditions imposed on the incident wave automatically leads to the same conditions for reflected and refracted waves.
Of the boundary conditions, (11.8) is now trivially satisfied (both sides zero), while (11.7) and (11.10) are identical in view of Snell's law, Eq. (11.4). We, thus, get only two independent equations from the four boundary conditions. From (11.7) and (11.9), we get

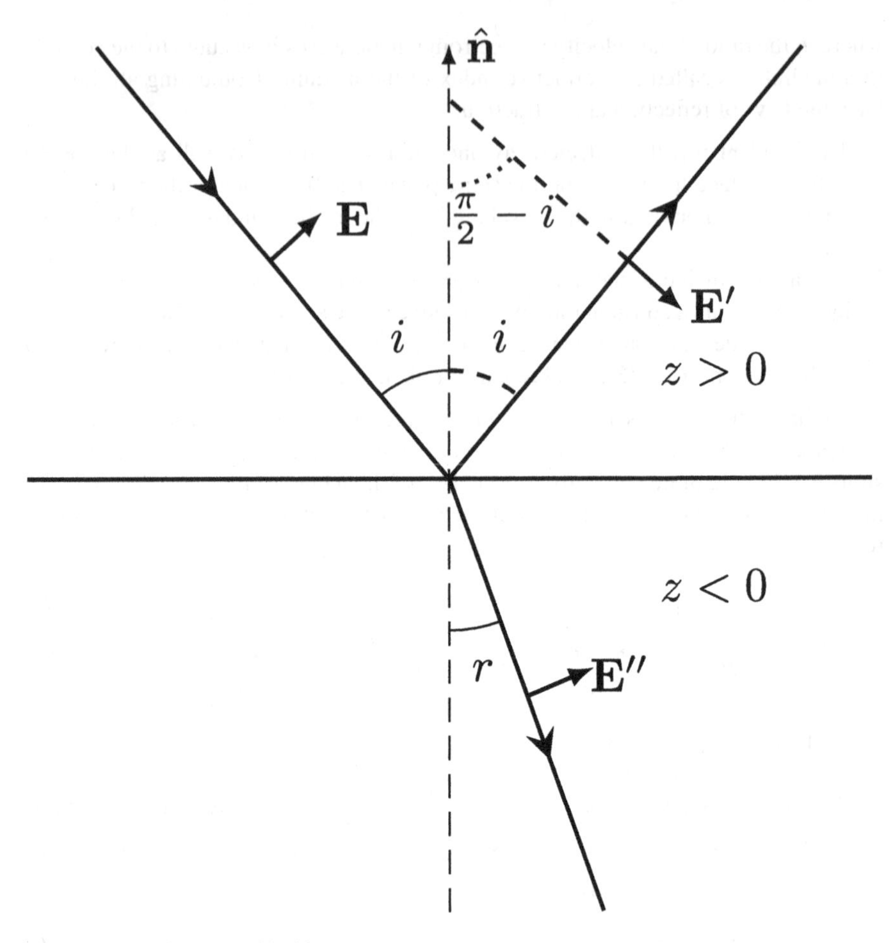

Fig. 11.2 **E** in the plane of incidence

$$\epsilon'(E_0 + E_0') \sin i = \epsilon'' E_0'' \sin r \tag{11.11}$$
$$(E_0 - E_0') \cos i = E_0'' \cos r \tag{11.12}$$

Using (11.4), Eqs. (11.11) and (11.12) become

$$\sqrt{\frac{\epsilon'}{\mu'}}\,(E_0 + E_0') = \sqrt{\frac{\epsilon''}{\mu''}}\,E_0'' \tag{11.13}$$

$$\left(E_0 - E_0'\right) \cos i = E_0''\sqrt{1 - \frac{n'^2}{n''^2}\sin^2 i} \tag{11.14}$$

One can solve (11.13) and (11.14) to determine E_0' and E_0 in terms of E_0'' and the angle of incidence i.

$$\frac{E_0''}{E_0} = \frac{2n'n''\cos i}{\frac{\mu'}{\mu''}n''^2\cos i + n'\sqrt{n''^2 - n'^2\sin^2 i}} \tag{11.15}$$

$$\frac{E_0'}{E_0} = \frac{\frac{\mu'}{\mu''}n''^2\cos i - n'\sqrt{n''^2 - n'^2\sin^2 i}}{\frac{\mu'}{\mu''}n''^2\cos i + n'\sqrt{n''^2 - n'^2\sin^2 i}} \tag{11.16}$$

These formulae are known after Fresnel who deduced them on the basis of the ether theory. Equation (11.16) shows that with μ' and μ'' nearly equal, the reflected amplitude will vanish if $\tan i = \frac{n''}{n'}$. This angle is called Brewster's angle, and for a wave incident at this particular angle, the reflected wave will have no electric vector component in the plane of incidence. Thus, if the original radiation be unpolarized, the reflected radiation in this case will be completely polarized with the electric vector perpendicular to the plane of incidence.

The formulae (11.15) and (11.16) are complicated; however, it is somewhat interesting to consider the case of normal incidence when the formulae are considerably simpler. Taking $\mu' = \mu'' = \mu_0$, we get in this case

$$\frac{E_0''}{E_0} = \frac{2n'}{n' + n''}, \qquad \frac{E_0'}{E_0} = \frac{n'' - n'}{n' + n''} \tag{11.17}$$

Case II: In this case, $\mathbf{E}$ is perpendicular to the plane of incidence (Fig. 11.3). Since $\mathbf{E}$ is perpendicular to the plane of incidence, i.e., $\mathbf{E} \cdot \mathbf{n} = 0$, $\mathbf{B}$ is in the plane of incidence.

Equation (11.7) is now trivially satisfied, while (11.8) and (11.9) are identical. From the remaining two boundary equations, utilizing Eq. (11.4), we obtain after some algebraic calculation (taking $\mu' = \mu'' = \mu_0$)

$$\frac{E_0''}{E_0} = \frac{2n'\cos i}{n'\cos i + \sqrt{n''^2 - n'^2\sin^2 i}} \tag{11.18}$$

$$\frac{E_0'}{E_0} = \frac{n'\cos i - \sqrt{n''^2 - n'^2\sin^2 i}}{n'\cos i + \sqrt{n''^2 - n'^2\sin^2 i}} \tag{11.19}$$

Again for normal incidence

$$\frac{E_0''}{E_0} = \frac{2n'}{n' + n''}, \qquad \frac{E_0'}{E_0} = \frac{n' - n''}{n' + n''} \tag{11.20}$$

Thus, at normal incidence, reflection and refraction do not discriminate between the two cases, i.e., the state of polarization.

It is of interest to compare the intensities of different waves by which we mean the flux of energy through unit area per unit time, the element of area being normal to the propagation vector. Thus, it is given by the product of energy density and the velocity of propagation. Let R be the ratio of the amount of energy reflected from unit area of the boundary to that incident on the same area. It is called the reflection coefficient. Again let T be the ratio of the amount of energy refracted to the amount incident—T is called the transmission coefficient. Then, in terms of i and r, taking $\mu' = \mu'' = \mu_0$, we get for Case I (see (11.15) and (11.16))

$$R_1 = \frac{E_0'^2 \cos i'}{E_0^2 \cos i} = \left(\frac{n'' \cos i - n' \cos r}{n'' \cos i + n' \cos r}\right)^2 = \frac{\tan^2(i-r)}{\tan^2(i+r)} \tag{11.21}$$

$$T_1 = \frac{\epsilon'' E_0''^2 v'' \cos r}{\epsilon' E_0^2 v' \cos i} = \frac{n'' E_0''^2 \cos r}{n' E_0^2 \cos i} = \frac{\sin 2i \sin 2r}{\sin^2(i+r)\cos^2(i-r)} \tag{11.22}$$

The cosines appear in R and T for the inclinations of the directions of propagation to the normal. From Eqs. (11.21) and (11.22), we have $R_1 + T_1 = 1$, which simply shows that the energy is being conserved and absorption, etc., are being neglected. For Case II, we have from (11.18) and (11.19) again

$$R_2 = \frac{E_0'^2 \cos i'}{E_0^2 \cos i} = \frac{\sin^2(i-r)}{\sin^2(i+r)} \tag{11.23}$$

$$T_2 = \frac{n'' E_0''^2}{n' E_0^2 \cos i} = \frac{\sin 2i \sin 2r}{\sin^2(i+r)} \tag{11.24}$$

For normal incidence, formulae (11.21), (11.22), (11.23) and (11.24) become

$$R_n = \left(\frac{n' - n''}{n' + n''}\right)^2, \quad T_n = \frac{4n'^2}{(n' + n'')} \tag{11.25}$$

irrespective of the state of polarization. As both the numerator and the denominator vanish in (11.21), we have to adopt a limiting procedure or use (11.20). In general, as the two components of **E** (parallel and perpendicular to the plane of incidence) are reflected and refracted in different proportions, any unpolarized wave will give rise to partial polarization for both the reflected and refracted waves. The only exceptions occur in the case of normal incidence when there is no polarization and incidence at Brewster's angle and when the reflected wave is completely polarized. The refracted wave is, however, never completely polarized.

Formula (11.25) provides a method for the determination of the refractive index of opaque materials. Alternatively, formula (11.25) may be subject to observational verification if the reflection coefficient be measured and the refractive index is obtained from dispersion theory. We give below (Table 11.1) the relevant results for rock salt obtained from Czerny's experiments (Zeits. für Physik **65**, 600 (1930), quoted in R.W. Wood's *Physical Optics*, Third edition, McMillan, New York, p. 532 (1936)). The agreement between the observed and calculated values of the reflection coefficient is quite impressive, especially so when we consider that we have neglected all absorption effects.

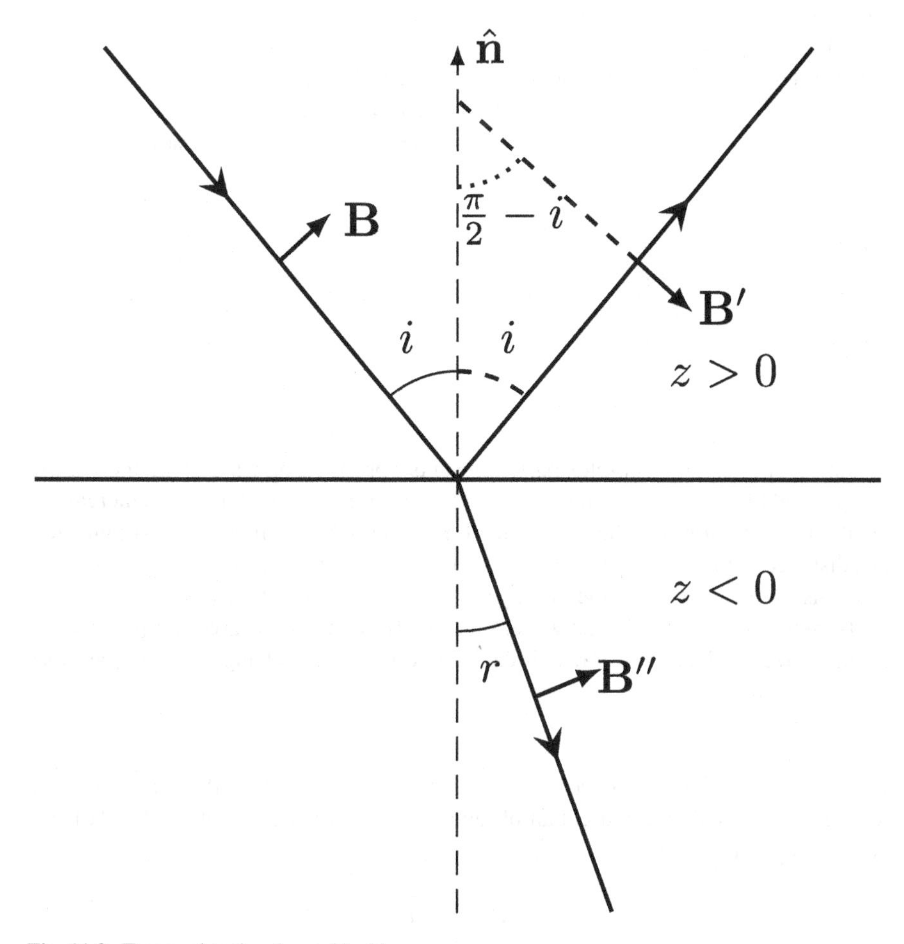

Fig. 11.3 **E** normal to the plane of incidence

11.1 Total Internal Reflection

If $n' > n''$, then the angle r is no longer real for angles of incidence i exceeding $\sin^{-1}(n''/n')$, and for $\sin r$, it then exceeds unity. One says that there is in those cases no refracted wave in the second medium and the incident energy is completely reflected back into the first medium. We shall investigate this phenomenon a little more. The equation for the refracted disturbance is of the form (if xz be the plane of incidence, z being normal to the interface)

$$\begin{aligned}\mathbf{E}'' = \mathbf{E}_0 e^{i(\mathbf{k}''\cdot r - \omega t)} &= \mathbf{E}_0 e^{ik''(x \sin r + z \cos r)} e^{-i\omega t} \\ &= \mathbf{E}_0 e^{ik''(x \sin r - \omega t)} e^{-k'' z \sqrt{(n'/n'')^2 \sin^2 i - 1}}\end{aligned}$$

Table 11.1 Refractive index of rock salt

Wavelength (in cm $\times 10^4$)	Refractive index n	Reflection coefficient at normal incidence R	
		Observed	Calculated
300	2.454	17.7%	17.9%
117	2.688	20.9%	19.9%
94	2.922	24.0%	24.3%
83	3.198	27.4%	27.5%
70	4.259	38.4%	39.0%
63	6.149	51.9%	56.5%

As $\sin r$ is still real, the complex part represents a propagating wave along the x-axis, i.e., parallel to the interface but its amplitude decreases exponentially with increasing depth of penetration into the second medium. The wave is effectively extinguished in a distance of the order $1/k''$, i.e., of the order of wavelength. Note that this wave is not transverse for although the propagation is along the x-axis, E_x does not vanish.

Turning now to the reflected wave, Eqs. (11.16) and (11.19) give complex values for the reflected electric vector. This can be interpreted as a change of phase because the wave disturbance

$$E = E_0 e^{i(kr-\omega t+\delta)} \tag{11.26}$$

with a phase factor δ may be written as $E = E_0 e^{i\delta} e^{i(kr-\omega t)}$, where the phase has been reduced to zero and the apparent amplitude $E_0 e^{i\delta}$ is complex. Equations (11.16) and (11.19) allow us to write

$$\frac{E_0'}{E_0} = \frac{a - ib}{a + ib} \tag{11.27}$$

where a in either case is $\cos i$ and $b = n^{\pm 1}\sqrt{n^2 \sin^2 i - 1}$ the positive and negative sign in the index occurring for electric vector parallel and perpendicular, respectively, to the plane of incidence.[1] Equation (11.28) may be written as

$$E_0' = E_0 e^{i\delta} \tag{11.28}$$

$$\text{where} \quad \delta_{\pm} = \tan^{-1}\left(-\frac{2n^{\pm 1}\sqrt{n^2 \sin^2 i - 1}\,\cos i}{\cos^2 i - n^{\pm 2}(n^2 \sin^2 i - 1)}\right) \tag{11.29}$$

Thus, for both the cases, the reflected intensity will be the same as the incident intensity consistently with the idea of total reflection but the reflection brings in,

[1] We have introduced a new notation: $n \equiv n'/n''$ stands for the refractive index of the first medium *relative* to the second. This is appropriate, for one studies total reflection with the first medium, say, glass and the second medium is air. Thus, n is the refractive index of glass with the second medium understood.

in general, a phase difference between the two components. This does not happen in the case of ordinary reflection. The phase difference vanishes for $i = \pi/2$ and $i = \sin^{-1}(1/n)$. To understand the effect of this phase difference, suppose that the incident wave is plane polarized so that the parallel and perpendicular components or the electric vector are of the form $a\cos(\omega t)$ and $b\cos(\omega t)$. In the reflected wave, they will be of the form $a\cos(\omega t)$ and $b\cos(\omega t + \delta)$. They will now combine to form an elliptic vibration in general. Thus, according to theory, while in ordinary reflection, plane polarized wave, after reflection, remains plane polarized, in total reflection, we shall have, in general, elliptically polarized wave after reflection of a plane polarized wave.

11.2 Reflection at Conducting Surfaces (Metallic Reflection)

We recall Eqs. (10.21) and (10.25) from the last chapter

$$\nabla^2\mathbf{E} = \mu\epsilon\ddot{\mathbf{E}} + \mu\sigma\dot{\mathbf{E}} \tag{11.30}$$

$$\nabla^2\mathbf{H} = \epsilon\ddot{\mathbf{H}} + \mu\sigma\dot{\mathbf{H}} \tag{11.31}$$

Before going to the discussion of wave phenomena, let us consider the approach of a good conductor to the equilibrium state in which there is no electric field in the body of the conductor (electrostatic condition). Consider for a moment the field $\mathbf{E}$ to be constant in space—Eq. (11.30) then gives

$$\epsilon\ddot{\mathbf{E}} + \sigma\dot{\mathbf{E}} = 0$$

$$\text{or} \qquad \mathbf{E} = \mathbf{E}_0 e^{-(\sigma/\epsilon)t}$$

The field, thus, decreases exponentially with time—the e-folding time being ϵ/σ. This time, called the relaxation time, has a small value for good conductors. If we take copper, for example, $\sigma = 5 \times 10^7\ (\Omega\text{m})^{-1}$ and for the moment taking the dielectric constant to be of the order of unity so that $\epsilon \simeq \epsilon_0 = 8.85 \times 10^{-12}\,\text{C}^2/\text{N} - \text{m}^2$, the relaxation time comes out as $\sim 10^{-19}$ s which is smaller than the time period corresponding to even optical frequencies ($\sim 10^{-15}$ s). We may also study the decay of charge density within the body of the conductor. The charge conservation equation $\dot{\rho} + \nabla \cdot \mathbf{j} = 0$, on substituting $\mathbf{j} = \sigma\mathbf{E}$, becomes $\dot{\rho} + \sigma\nabla \cdot \mathbf{E} = 0$, where we have taken the conductivity to be constant. Now using $\nabla \cdot \mathbf{E} = \nabla \cdot \mathbf{D}/\epsilon = \rho/\epsilon$, we have finally

$$\dot{\rho} + \frac{\sigma}{\epsilon}\rho = 0$$

which integrates to give

$$\rho = \rho_0 e^{-\sigma t/\epsilon}$$

Thus, the relaxation time for the charge density is also quite small in good conductors. (What happens to the charge? Unlike the field, it cannot just disappear. The answer is that in the body of the conductor, the charge disappears only to be concentrated at the surface of the conductor where it results in a discontinuity of the normal component of the displacement.)

We return to the consideration of waves. For a monochromatic wave, $\mathbf{E} = \mathbf{E}_0(x, y, z)e^{i\omega t}$ where $\mathbf{E}_0$ is a function of the space coordinates only. Then Eq. (11.30) reduces to

$$\nabla^2 \mathbf{E}_0 = -\omega^2 p^2 \mathbf{E}_0 \tag{11.32}$$

$$\text{where} \quad p^2 = \mu\epsilon - i\frac{\sigma\mu}{\omega} \tag{11.33}$$

Formally p^2 in Eq. (11.32) plays the role of a complex dielectric constant. To simplify the discussion, let the wave be plane polarized with the vector $\mathbf{E}$ in the y-direction and $\mathbf{H}$ in the z-direction, x being the direction of propagation. Thus, from (11.32),

$$E_y = E_0 e^{i\omega t - ip\omega x} \tag{11.34}$$

$$H_z = H_0 e^{i\omega t - ip\omega x} \tag{11.35}$$

Substituting in the equation $\nabla \times \mathbf{E} = -\dot{\mathbf{B}}/c$, we have

$$\mu H_0 = pE_0 \tag{11.36}$$

As p is complex, the electric and magnetic vectors are not in phase. Writing

$$p = n - i\alpha \tag{11.37}$$

We get from (11.33)

$$n^2 - \alpha^2 = \mu\epsilon, \quad \text{and} \quad n\alpha = \frac{\sigma\mu}{2\omega} \tag{11.38}$$

so that

$$n^4 - \mu\epsilon n^2 - \frac{\sigma^2\mu^2}{4\omega^2} = 0$$

$$\text{or} \quad n^2 = \frac{\mu}{2}\left(\epsilon + \sqrt{\epsilon^2 + \frac{\sigma^2}{\omega^2}}\right) \tag{11.39}$$

$$\text{and} \quad \alpha^2 = \frac{\mu}{2}\left(-\epsilon + \sqrt{\epsilon^2 + \frac{\sigma^2}{\omega^2}}\right) \tag{11.40}$$

The field vectors, thus, have the form

$$E_y = E_0 e^{-\omega\alpha x} e^{i\omega(t-nx)} \tag{11.41}$$

$$\begin{aligned} B_z &= p E_0 e^{-\omega\alpha x} e^{i\omega(t-nx)} \\ &= \sqrt{n^2+\alpha^2}\, E_0 e^{-\omega\alpha x} e^{i\omega(t-nx-\theta)} \end{aligned} \tag{11.42}$$

where we have written

$$\begin{aligned} p &= \sqrt{n^2+\alpha^2}\, e^{-i\theta} \\ \theta &= \tan^{-1}\left(\frac{\alpha}{n}\right) \end{aligned} \tag{11.43}$$

Thus, both the electric and magnetic field vectors are damped out—the depth in which they are reduced by a factor $1/e$ is $(1/\alpha\omega)$. For metals, the conductivity term is usually much greater than the dielectric constant. Hence, we have to a fairly high degree of approximation

$$\alpha\omega = \sqrt{\frac{\sigma\mu\omega}{2}} = \sqrt{\frac{\pi c\sigma\mu}{\lambda_0}} \tag{11.44}$$

or the depth of penetration is $d \sim \sqrt{\frac{2}{\sigma\mu\omega}}$. Again taking the case of copper: for a wavelength of 1 cm, the skin depth is about 40 μm while for $\lambda_0 = 600$ nm, the depth of penetration comes out as ~0.02 μm. Although both B and E are damped out, the energy may be shown to be associated primarily with the magnetic field. We have from Eqs. (11.41) and (11.42), for the ratio between the electric and magnetic energy densities,

$$\begin{aligned} \frac{U_{\text{mag}}}{U_{\text{el}}} &= \frac{B^2/\mu}{E^2/\epsilon} = \frac{n^2+\alpha^2}{\mu\epsilon} \\ &= \sqrt{1+\frac{\sigma^2}{\omega^2\epsilon^2}} = \sqrt{1+\left(\frac{\sigma\lambda_0}{2\pi c\epsilon}\right)^2} \end{aligned} \tag{11.45}$$

where we have used (11.39) and (11.40). Equation (11.45) is exact, considering the case of metals like copper (11.45) gives to a good approximation

$$\frac{U_{\text{mag}}}{U_{\text{el}}} = \frac{\lambda_0\sigma}{2\pi c\epsilon}$$

Substituting values for copper we find that in the optical region, i.e., for wavelengths $\sim 10^3$ nm, electrical energy constitutes only about 0.3% of the total energy.

In the region where the conductivity term dominates over the dielectric constant, (11.39) and (11.40) may be approximated by

$$n = \alpha = \sqrt{\frac{\mu\sigma}{2\omega}} \tag{11.46}$$

Also, the phase difference between the electric and magnetic fields is $\theta = \pi/4$. The effect of high conductivity is, thus, two-fold—an extinction of the field within a short depth which makes the metal opaque, this effect increases with frequency and explains the concentration of high frequency currents in the surface region of a conducting wire (the skin effect). Secondly, there is an effective increase in the dielectric constant and consequently the refractive index n. Formally one can forget the conductivity and introduce instead a complex dielectric constant p^2/μ (see Eq. (11.33)). To find the reflection coefficient, we must use the boundary conditions as before. For simplicity, we consider the incidence to be normal—so that consistent with (11.34) and (11.35), $z = 0$ defines the boundary plane. For a plane polarized wave of the types (11.34) and (11.35), the incident, the reflected and the refracted waves are given by

$$E_y = \eta H_z = E_0 e^{i\omega(t-x/c)} \qquad \text{(incident wave in air)}$$
$$-E'_y = \eta H'_z = E'_0 e^{i\omega(t+x/c)} \qquad \text{(reflected wave in air)}$$
$$E''_y = E''_0 e^{i\omega(t-px/c)}$$
$$H''_z = \frac{p}{\mu} E''_0 e^{i\omega(t-px/c)} \qquad \text{(refracted wave in metal)}$$

(cf. (11.41) and (11.41)) and where $\eta = \sqrt{\mu/\epsilon}$. Note that we have appended a negative sign to E'_y to take care of the fact that the reflected wave is proceeding in a direction opposite to the incident wave. With our choice of polarization, the normal components of **D** and **B** vanish and so one pair of boundary conditions is satisfied trivially; the other pair gives

$$E_0 - E'_0 = E''_0, \qquad E_0 + E'_0 = \frac{p\eta}{\mu} E''_0$$

so that

$$E'_0 = \frac{p\eta - \mu}{p\eta + \mu} E_0, \qquad E''_0 = \frac{2\mu}{p\eta + \mu} E_0$$

Therefore, using (11.37), the reflection coefficient is

$$R = \frac{|E'_0|^2}{E_0^2} = \frac{(p\eta - \mu)(p^*\eta - \mu)}{(p\eta + \mu)(p^*\eta + \mu)} = \frac{(n\eta - \mu)^2 + \alpha^2\eta^2}{(n\eta + \mu)^2 + \alpha^2\eta^2} \tag{11.47}$$

Using (11.46), we get for $n\eta \gg \mu$

$$R = \frac{n\eta - \mu}{n\eta + \mu} = 1 - \frac{2\mu}{n} = 1 - 2\sqrt{\frac{2\omega\epsilon}{\sigma}} \tag{11.48}$$

Thus, if the frequency be not very large ($2\omega\epsilon \ll \sigma$), R approaches unity—the incident energy is almost completely reflected back into the air and there is little energy that penetrates into the metal. This conclusion has been confirmed experimentally, but, for shorter wavelengths as in the optical region, the observed value of R decreases and the metal begins to be transparent especially in the case of alkali metals.

This can be understood on the basis of a classical free electron theory, originally advanced by Drude. According to this theory, the electrons in the metal execute a forced oscillatory motion under the influence of the electric vector in the incident wave. As the electrons are free, there is no restoring force but the motion is impeded due to collision with atoms. This brings in a damping term proportional to the velocity, and thus, the equation of motion is of the form

$$\ddot{x} + b\dot{x} = \frac{q}{m} E_0 e^{i\omega t} \tag{11.49}$$

The radiation damping is proportional to the third-order derivative of the displacement—in (11.49), the damping is being taken proportional to the velocity as in the cases under consideration, collision damping dominates over radiation damping. The steady-state integral of (11.49) is

$$x = Ae^{-i\omega t}, \qquad A = -\frac{qE_0/m}{\omega^2 + ib\omega}$$

The complex nature of A indicates a phase difference between the driving force and the oscillating electrons. This displacement of the electrons gives an induced polarization and one obtains for the dielectric constant (which also is complex now)

$$\epsilon = 1 - Nq^2 m\epsilon_0(\omega^2 + ib\omega)$$

where N is the number density of electrons, each carrying a charge q. The complex dielectric constant will play the role of p^2/μ (cf. 11.33) and, as in (11.37), we write

$$\sqrt{\epsilon\mu} = n - i\alpha$$

so that

$$n^2 - \alpha^2 = \mu\left(1 - \frac{Nq^2}{m\epsilon_0}\frac{1}{\omega^2 + b^2}\right)$$
$$n\alpha = -\frac{Nq^2}{m\epsilon_0}\frac{b}{\omega^3 + b^2\omega}$$

The reflection coefficient is, thus, still given by (11.47)

$$R = \frac{(n\eta - \mu)^2 + \alpha^2}{(n\eta + \mu)^2 + \alpha^2} \tag{11.50}$$

We may consider two limiting cases—in the first case of fairly long wavelengths, the polarization term is large and α is large—so that according to (11.47) the reflection coefficient tends towards unity. This is the situation already considered. In the second case of short wavelength (high frequency), the polarization term is small, α is thus small, and thus, the reflection coefficient tends to the form

$$R = \left(\frac{n\eta - \mu}{n\eta + \mu} \right)^2$$

As n depends on ω, ϵ and σ, this gives a low value of the reflection coefficient for high frequencies. Correspondingly, the transmission assumes an appreciable value. This is consistent with small values of α and the metal then loses its strong reflecting power and becomes transparent.

Lastly, we note that in case the reflecting power is unity and the incidence is normal, the incident and reflected waves together form a system of stationary waves. These stationary waves have been observed experimentally.

Problems

1. The incident wave is elliptically polarized. What will be the changes in the state of polarization brought about by reflection and refraction?
2. Find the condition that an incident plane polarized wave after reflection at a glass-air interface may become circularly polarized. The refractive index of glass is 1.5.
3. How would you physically explain the presence of electromagnetic wave in the second medium in case of total reflection? Show that these disturbances do not transport any energy.

Chapter 12
Wave Guides and Cavity Resonators

We now study the case of electromagnetic waves in metallic pipes whose ends may be open (wave guides) or closed (cavity resonators). The basis is still Maxwell's equations but the boundary conditions at the metal-dielectric interface introduce some constraints. We have seen in the last chapter that the electromagnetic fields can penetrate inside the metals only up to a small depth $\sim \sqrt{2/(\sigma\mu\omega)}$. Ideally, if the conductivity σ tends to infinity, there would be no electric field inside the conductor and we go over to the electrostatic condition that the electric field tangential to the metal-insulator surface vanishes. We shall in the major part of our discussion confine ourselves to this idealized situation but this leaves out the conduction currents that actually flow near the surface, and the consequent Joule heat generated means an attenuation of the electromagnetic waves (wave guides) or a diminution of the stored energy (cavity resonators). Hence, to evaluate these effects, we shall have to consider the case of finite conductivity and penetration of fields into the body of the conductor.

We write out Maxwell's equations—the material in the pipe being air, we take $\epsilon = \epsilon_0$ and $\mu = \mu_0$

$$\nabla \cdot \mathbf{E} = 0 \tag{12.1}$$

$$\nabla \times \mathbf{E} = -\dot{\mathbf{B}} \tag{12.2}$$

$$\nabla \cdot \mathbf{B} = 0 \tag{12.3}$$

$$\nabla \times \mathbf{B} = \mu_0\epsilon_0\dot{\mathbf{E}} \tag{12.4}$$

The equations lead, as before, to the wave equations

$$\Box\, \mathbf{E} \equiv \nabla^2\mathbf{E} - \frac{1}{c^2}\frac{\partial^2\mathbf{E}}{\partial t^2} = 0 \tag{12.5}$$

$$\Box\, \mathbf{B} \equiv \nabla^2\mathbf{B} - \frac{1}{c^2}\frac{\partial\partial^2\mathbf{B}}{\partial\partial t^2} = 0 \tag{12.6}$$

A. K. Raychaudhuri, *Classical Theory of Electricity and Magnetism*, Texts and Readings in Physical Sciences 21, https://doi.org/10.1007/978-981-16-8139-4_12

We consider a cylindrical pipe with its axis along the z-axis which is also the direction of propagation of the waves. As we are talking of propagation, the case is of wave guides rather than of cavities. Thus, the fields that we are considering will be of the form

$$\mathbf{E} = \mathbf{E}^0(x, y)\, e^{i(kz-\omega t)} \tag{12.7}$$

$$\mathbf{B} = \mathbf{B}^0(x, y)\, e^{i(kz-\omega t)} \tag{12.8}$$

the superscript 0 indicating the amplitude. Here, the fields are not constants over the constant z-planes which are the wavefronts. This is a necessary consequence of the boundary conditions. We can call such waves plane fronted, as distinct from plane waves—by the latter term, we shall mean those waves in which the field vectors are the same everywhere on the plane wavefront.

As the boundary conditions and consequently the forms (12.7, 12.8) introduce a distinction between the transverse and longitudinal directions, it will be convenient to write out Maxwell's equations explicitly, taking into account (12.7, 12.8) and the prescription $\frac{\partial}{\partial t} \to -i\omega$ and $\frac{\partial}{\partial z} \to ik$

$$\frac{\partial E_x^0}{\partial x} + \frac{\partial E_y^0}{\partial y} + ikE_z^0 = 0 \tag{12.9}$$

$$\frac{\partial E_z^0}{\partial y} - ikE_y^0 = i\omega B_x^0 \tag{12.10}$$

$$ikE_x^0 - \frac{\partial E_z^0}{\partial x} = i\omega B_y^0 \tag{12.11}$$

$$\frac{\partial E_y^0}{\partial x} - \frac{\partial E_x^0}{\partial y} = i\omega B_z^0 \tag{12.12}$$

$$\frac{\partial B_x^0}{\partial x} + \frac{\partial B_y^0}{\partial y} + ikB_z^0 = 0 \tag{12.13}$$

$$\frac{\partial B_z^0}{\partial y} - ikB_y^0 = -i\omega\mu_0\epsilon_0 E_x^0 \tag{12.14}$$

$$ikB_x^0 - \frac{\partial B_z^0}{\partial x} = -i\omega\mu_0\epsilon_0 E_y^0 \tag{12.15}$$

$$\frac{\partial B_y^0}{\partial x} - \frac{\partial B_x^0}{\partial y} = -i\omega\mu_0\epsilon_0 E_z^0 \tag{12.16}$$

The four curl equations (12.10), (12.11), (12.14) and (12.15) involve the transverse components E_x^0, E_y^0, B_x^0 and B_y^0 linearly, and thus, we may solve for them in terms of the transverse derivative of the longitudinal field components E_z^0 and B_z^0 as

$$E_x^0 = \frac{i}{\frac{\omega^2}{c^2} - k^2}\left(k\frac{\partial E_z^0}{\partial x} + \omega\frac{\partial B_z^0}{\partial y}\right) \tag{12.17}$$

$$E_y^0 = -\frac{i}{\frac{\omega^2}{c^2} - k^2}\left(\omega\frac{\partial B_z^0}{\partial x} - k\frac{\partial E_z^0}{\partial y}\right) \tag{12.18}$$

$$B_x^0 = -\frac{i}{\frac{\omega^2}{c^2} - k^2}\left(\frac{\omega}{c^2}\frac{\partial E_z^0}{\partial y} - k\frac{\partial B_z^0}{\partial x}\right) \tag{12.19}$$

$$B_y^0 = \frac{i}{\frac{\omega^2}{c^2} - k^2}\left(\frac{\omega}{c^2}\frac{\partial E_z^0}{\partial x} + k\frac{\partial B_z^0}{\partial y}\right) \tag{12.20}$$

where we have used $c^2 = 1/(\mu_0\epsilon_0)$. We next spell out the specific boundary conditions. Firstly, the components of **E** tangential to the surface vanishes, i.e.,

$$\mathbf{n} \times \mathbf{E} = 0 \tag{12.21}$$

One tangential direction everywhere is obviously parallel to the z-axis, so that at the surface

$$E_z = 0 \tag{12.22}$$

Suppose that at one point the orthogonal tangential direction is parallel to the x-axis, then at that point, E_x also vanishes and hence $B_y^0 = 0$ from (12.11). As the y-direction is normal to the surface at this point,

$$\mathbf{n} \cdot \mathbf{B} = 0 \tag{12.23}$$

Using these in Eq. (12.14),

$$\frac{\partial B_z^0}{\partial y} = 0 \rightarrow (\mathbf{n} \cdot \nabla)B_z^0 = 0 \tag{12.24}$$

It is clear that Eqs. (12.23) and (12.24) are merely consequences of Eqs. (12.17)–(12.20). In practice, we shall take Eqs. (12.22) and (12.24) as our boundary conditions. This will have the advantage that only the longitudinal components are involved. The field equations ensure that Eq. (12.21) will then be satisfied.

With Eqs. (12.7) and (12.8), Eqs. (12.5) and (12.6) reduce to

$$\left(\frac{\partial^2}{\partial x^2} + \frac{\partial^2}{\partial y^2}\right)\mathbf{E} + \left(\frac{\omega^2}{c^2} - k^2\right)\mathbf{E} = 0$$
$$\left(\frac{\partial^2}{\partial x^2} + \frac{\partial^2}{\partial y^2}\right)\mathbf{B} + \left(\frac{\omega^2}{c^2} - k^2\right)\mathbf{B} = 0 \tag{12.25}$$

We may confine our attention to the equations for E_z^0 and B_z^0, for once these are determined, Eqs. (12.17–12.20) would determine the transverse field components.

Equations (12.25) are Schrödinger- like equations in two dimensions for E_z^0 and B_z^0. In general, acceptable solutions will exist only for discrete values of $(\frac{\omega^2}{c^2} - k^2)$. However, as the boundary conditions for E_z^0 and B_z^0 are different (namely, Eqs. (12.22) and (12.24), respectively), the eigenvalues will be different for the two equations. But before going into that study, let us see whether we can have completely transverse waves in pipes similar to those in free unbounded space.

12.1 Transverse Electromagnetic Waves (TEM Waves)

By TEM waves, we mean $E_z^0 = B_z^0 = 0$. We have then, from Eqs. (12.12) and (12.9), $\mathbf{E}_0$ has zero divergence and curl, so that

$$\mathbf{E}^0 = -\nabla\Psi(x, y) \tag{12.26}$$

$$\nabla^2\Psi(x, y) = 0 \tag{12.27}$$

With the boundary condition that the potential function Ψ is a constant over the surfaces of the cylinder, this leads to the conclusion that Ψ is constant everywhere, i.e., the electric field vanishes everywhere. (This is just the familiar electrostatic theorem that the space inside a hollow conductor is field free.) However, if inside the hollow there are other conductors at a different potential from the outer one as in the spherical or cylindrical capacitor, the field need not vanish. Similarly, transverse waves can exist in the annular gap between two coaxial pipes. Assuming for the moment that we have such TEM waves, we have from Eqs. (12.26), (12.27) and (12.25)

$$k^2 = \frac{\omega^2}{c^2} \tag{12.28}$$

showing that the waves have the velocity c and are non-dispersive. We have now from Eqs. (12.11), (12.15) and (12.28)

$$E_x^0 = \pm cB_y^0, \qquad E_y^0 = \mp cB_x^0$$

which indicates that here also, as in the case of free space waves

$$E^2 - c^2B^2 = \mathbf{E}\cdot\mathbf{B} = 0$$

To sum up, TEM waves exist only in annular regions and in case they exist their characteristics are the same as that of waves in free space.

In view of our remarks following Eq. (12.25), the modes divide naturally into two types—one in which the magnetic field alone is transverse (called TM modes) and the other in which the electric field alone is transverse (called TE modes).

12.2 TM and TE Modes

The transverse electric (TE) and transverse magnetic (TM) modes are as follows:

(a) In TM modes, $B_z = 0$ and $E_z \neq 0$. We have from Eqs. (12.14), (12.15), (12.17) and (12.18)

$$[\text{TM}] \quad \mathbf{B}^0 = \frac{\omega}{c^2 k} \hat{\mathbf{z}} \times \mathbf{E}^0) \tag{12.29}$$

$$E_\alpha^0 = \frac{ik}{(\omega^2/c^2) - k^2} \nabla_\alpha E_z^0 \tag{12.30}$$

where $\hat{\mathbf{z}}$ indicates a unit vector in the direction of the z-axis and the index α above runs over the transverse coordinates x and y.

(b) In TE modes, $E_z = 0$ and $B_z \neq 0$. Therefore,

$$[\text{TE}] \quad \mathbf{E}^0 = -\frac{\omega}{k} \hat{\mathbf{z}} \times \mathbf{B}^0 \tag{12.31}$$

$$B_\alpha^0 = \frac{ik}{(\omega^2/c^2) - k^2} \nabla_\alpha B_z^0 \tag{12.32}$$

which follow from Eqs. (12.10), (12.11), (12.19) and (12.20).

Thus, in the TM (respectively TE) case, the transverse components of $\mathbf{E}^0$ (respectively $\mathbf{B}^0$) are determined in terms of E_z^0 (respectively B_z^0) from Eqs. (12.30) and (12.32). Using these next, the transverse components of $\mathbf{B}^0$ (respectively $\mathbf{E}^0$) are determined from Eqs. (12.29) (or (12.31)). One of the equations in (12.29) or (12.31) simply indicates the TM (respectively TE) nature of the mode. Basically, the problem, thus, reduces to the integration of Eqs. (12.25) for E_z^0 (respectively B_z^0) in the TM (respectively TE) case with the appropriate boundary conditions (12.22) and (12.24). One of these (12.24) (or (12.22)) is trivially satisfied in the TM (respectively TE) case. The other boundary condition leads to the requirement that $(\omega^2/c^2) - k^2$ must be positive. (We omit a formal proof. The reader may find a consideration of the following type helpful—the boundary condition of the vanishing of a function over a closed contour in the xy-plane requires that if the function is not to vanish identically, its modulus must have at least one maximum somewhere in the region within the contour, and this in view of Eqs. (12.25) would require $(\omega^2/c^2) - k^2$ to be positive. Thus,

$$\frac{\omega^2}{c^2} - k^2 = \lambda_\alpha^2 \tag{12.33}$$

where λ_α^2 is the α-th eigenvalue (say). Thus, there is a cut-off frequency ω_0 given by $\omega_0 = \lambda_\alpha c$. As for frequencies below this value, k would be imaginary and then there can be no propagation (see Eqs. (12.7) and (12.8)). Equation (12.33) further shows that the phase velocity $v_p = \omega/k$ exceeds the velocity of light and tends to infinity at the cut-off frequency ω_0.

12.3 Energy Flux, Attenuation in Wave Guides and Q of Cavities

Equations (12.29) and (12.30) show that **E** and **B** are orthogonal to each other in both TM and TE modes. As one of the two vectors is not transverse, the Poynting vector will have a transverse component. However, from Eqs. (12.17–12.20), we find that the transverse components have a phase difference of $\pi/2$ with the longitudinal components; Hence, the time average of the energy flux in the transverse directions will vanish.

From Eq. (12.29), we have for the TM mode

$$\mathbf{E}^0 \times \mathbf{B}^0 = \frac{\omega}{c^2 k}\mathbf{E}^0 \times (\hat{\mathbf{z}} \times \mathbf{E}^0) = \frac{\omega}{c^2 k}(E^0)^2\hat{\mathbf{z}} - E_z^0\mathbf{E}^0]$$

As already noted, the time average of the transverse component vanishes, while that of the longitudinal component of the Poynting vector is

$$\begin{aligned} \langle(\mathbf{E}\times\mathbf{H})_z\rangle &= \frac{1}{2}\frac{\omega}{c^2 k}\frac{1}{\mu_0}\left((E^0)^2 - (E_z^0)^2\right) \\ &= \frac{\omega}{c^2 k \mu_0}\left((E_x^0)^2 + (E_y^0)^2\right) = \frac{\omega\epsilon_0}{k}(E_t^0)^2 \end{aligned} \tag{12.34}$$

so that the flux through any section of the cylinder is

$$P = \frac{\omega\epsilon_0}{k}\int (E_t^0)^2\,dxdy \tag{12.35}$$

Using Eq. (12.29), the integral of E_t^0 over a section of the cylinder is

$$\int (E_t^0)^2 dxdy = \frac{k^2}{\left(\frac{\omega^2}{c^2} - k^2\right)^2}\int \left|\nabla_t E_z^0\right|^2\,dxdy$$

Using Green's theorem in two dimensions, this is

$$\frac{k^2}{\left(\frac{\omega^2}{c^2} - k^2\right)^2}\left(\oint E_z^0\frac{\partial E_z^0}{\partial n}\,dl - \int E_z^0\nabla_t^2 E_z^0\,dxdy\right)$$

where $\partial/\partial n$ indicates a derivative normal to the element dl of the boundary of the cylinder in some direction on the xy-plane. The first integral vanishes because of Eq. (12.22) while in the second integral, we get, using Eq. (12.25),

$$P = \frac{\omega k\epsilon_0}{\frac{\omega^2}{c^2} - k^2}\int (E_z^0)^2\,dxdy \tag{12.36}$$

Again using Eqs. (12.29), (12.35) and (12.36), the energy per unit length of the cylinder is

$$
\begin{aligned}
U &= \frac{1}{2}\int \left(\epsilon_0\left((E_t^0)^2 + (E_z^0)^2\right) + \frac{1}{\mu_0}(B_t^0)^2\right) dxdy \\
&= \frac{1}{2}\int \left(\frac{\epsilon_0\omega^2}{c^2k^2}(E_t^0)^2 + \frac{\omega^2}{\mu_0 c^4 k^2}(E_t^0)^2\right) dxdy \\
&= \frac{\epsilon_0\omega^2}{c^2k^2}\int (E_t^0)^2\, dxdy
\end{aligned}
\tag{12.37}
$$

Now, comparing Eqs. (12.35), (12.36) and (12.37), we can introduce a concept of a flow velocity of energy

$$
v = \frac{P}{U} = \frac{c^2k}{\omega} = \frac{c^2}{v_p} \tag{12.38}
$$

It turns out that v, thus, introduced is the same as v_g, the group velocity $v_g = \partial\omega/\partial k$. Note that $v_g v_p = c^2$, as in the case of the de Broglie waves. All the above calculations of energy flux and energy per unit length can be gone through in an exactly similar manner for the TE modes leading to Eqs. (12.35), (12.36) and (12.37), where only **E** is replaced by **B**. Therefore, Eq. (12.37) holds in general.

It is of some interest to note from the second step in Eq. (12.37) that considering the entire transverse section, the energy is equally partitioned between the magnetic and electric fields, but unlike in free space (or isotropic dielectric), the equality does not hold good at each point separately.

For the infinite conductivity case considered so far, there is no dissipation of energy, and the wave proceeds without attenuation in a wave guide , and the electromagnetic oscillations in a cavity suffer no decay. If the conductivity is finite, the electric field penetrates to some extent (called the skin depth) inside the conductor and the Joule heat generated means a dissipation of electromagnetic energy into heat. This dissipation may be evaluated by calculating the flux of electromagnetic energy into the body of the conductor. In the case of infinite conductivity, this flux vanishes as **E** has no component tangential to the surface of the conductor. In the case of finite conductivity, we have seen in Chap. 11 (vide Eqs. (11.37), (11.44) and (11.47)) that the tangential component of **E** (we write E_t) is orthogonal to the tangential component of **H** (written H_t). Moreover, for high enough conductivity, their magnitudes are related as

$$
E_t = \frac{1+i}{2}\sqrt{\frac{\mu\omega}{\sigma}}\,H_t \tag{12.39}
$$

Thus, the average of the time rate of loss of energy per unit area of the conductor is given by

$$
\frac{1}{4}\sqrt{\frac{\mu\omega}{\sigma}}\,|H_t|^2 \tag{12.40}
$$

We can evaluate the above expression by taking H_t^0 to have the value corresponding to infinite conductivity. This, of course, will introduce some error but as H_t^0 is being multiplied by E_t^0 which is itself a small quantity, the error introduced is of a higher order of smallness.

Expression (12.40) may be used both in the cases of wave guides and cavities. In the case of wave guides, one introduces an attenuation constant K as follows:

$$K = \frac{\text{power loss per unit length of the guide}}{\text{power transmitted through the guide}}$$

Hence, the flux of energy is multiplied by the factor e^{-Kz}. In the case of cavities, one introduces a quantity Q which, for any particular mode, is defined as follows:

$$\begin{aligned} Q &= \frac{2\pi \times (\text{energy in the cavity pertaining to the mode})}{\text{energy dissipated as heat during one cycle of the mode}} \\ &= \frac{\text{energy of the mode}}{\text{energy dissipated in time } 1/\omega} \end{aligned} \tag{12.41}$$

where ω is the angular frequency of the mode. There is an intimate connection between Q and the sharpness of response of the cavity.[1] If U represents the total energy in the cavity for a mode of angular frequency ω, then

$$\begin{aligned} Q &= \frac{\omega U}{\langle dU/dt \rangle} \\ \text{or} \quad U &= U_0 e^{-\omega t/Q} \end{aligned} \tag{12.42}$$

Hence, the time dependence of the field intensities will be of the form

$$E = E_0\, e^{-\frac{\omega_0 t}{2Q}}\, e^{-i\omega_0 t}$$

The above form is valid for $0 \leq t \leq \alpha$ in order to have the field intensities and energy finite. Thus, the variation with time is no longer sinusoidal and we may use the Fourier analysis to get

[1] The reader, who is conversant with the problem of sharpness of resonance in case of mechanical vibrations, may recall the relation between the sharpness of resonance and damping in those cases.

$$E(t) = \int_{-\infty}^{\infty} E_\omega e^{i\omega t} d\omega$$

$$\text{or} \quad E_\omega = \frac{1}{2\pi} \int_{-\infty}^{\infty} E(t) e^{i\omega t} dt$$

$$= \frac{E_0}{2\pi} \int_0^{\infty} dt \exp\left(-\frac{\omega_0 t}{2Q} + i(\omega - \omega_0)t\right)$$

$$\text{or} \quad |E_\omega|^2 = \frac{E_0^2}{4\pi^2 \left(\frac{\omega_0^2}{4Q^2} + (\omega - \omega_0)^2\right)} \tag{12.43}$$

Thus, compared to the energy at frequency ω_0, the energy will be reduced to half the value at $\omega = \omega_0 \pm \frac{\omega_0}{2Q}$. Hence, the half width of the E_ω-ω curve is ω_0/Q. We now consider the complete solutions for simple forms of transverse sections.

12.3.1 Rectangular Transverse Section

Let the sides of the cross-section of the cavity/guide be α and β. We solve Eqs. (12.25) for E_z^0 and H_z^0 subject to the boundary condition (12.22) for E_z^0 and (12.24) for H_z^0 for TM(TE) modes. Thus,

$$E_z^0 = \sum_{m,n} A_{mn} \sin\frac{m\pi x}{\alpha} \sin\frac{n\pi y}{\beta} \qquad \text{(TM)}$$

$$H_z^0 = \sum_{m,n} B_{mn} \cos\frac{m\pi x}{\alpha} \cos\frac{n\pi y}{\beta} \qquad \text{(TE)}$$

where m and n are integers and in either case

$$\frac{\omega_{mn}^2}{c^2} = k^2 + \pi^2\left(\frac{m^2}{\alpha^2} + \frac{n^2}{\beta^2}\right)$$

The cut-off frequency is given by

$$\omega_0 = \begin{cases} \pi c/\alpha & \text{if } \alpha > \beta \\ \pi c/\beta & \text{if } \alpha < \beta \end{cases}$$

and occurs for the pair of values (0, 1) for (m, n). The above cut-off frequencies are for TE mode, as with either m or n vanishing, the TM mode disappears. As there is no restriction on k, the permitted frequencies form a continuous spectrum. The transverse components of the fields are readily calculated from Eqs. (12.29) to (12.32).

So much for wave guides. If the ends are also closed by conductors, we have a cavity with the additional constraint that the boundary conditions must be satisfied over these ends as well, i.e., for $z = 0, \gamma$. Indeed the resonator is now simply a rectangular parallelepiped and there is no meaningful difference between the three directions x, y and z. It does not make much sense to speak of transverse and longitudinal directions and TM and TE modes. We may just write down the form of the electric and magnetic fields for the normal modes

$$\begin{aligned} E_x(l, m, n) &= A \cos\frac{l\pi x}{\alpha} \sin\frac{m\pi y}{\beta} \sin\frac{n\pi z}{\gamma} e^{-i\omega t} \\ E_y(l, m, n) &= B \sin\frac{l\pi x}{\alpha} \cos\frac{m\pi y}{\beta} \sin\frac{n\pi z}{\gamma} e^{-i\omega t} \\ E_z(l, m, n) &= C \sin\frac{l\pi x}{\alpha} \sin\frac{m\pi y}{\beta} \cos\frac{n\pi z}{\gamma} e^{-i\omega t} \end{aligned} \tag{12.44}$$

where l, m and n must all be integers and any particular set determine a mode with frequency

$$\frac{\omega^2}{c^2} = \pi^2\left(\frac{l^2}{\alpha^2} + \frac{m^2}{\beta^2} + \frac{n^2}{\gamma^2}\right) \tag{12.45}$$

Further because of $\nabla \cdot \mathbf{E} = 0$

$$A\frac{l}{\alpha} + B\frac{m}{\beta} + C\frac{n}{\gamma} = 0 \tag{12.46}$$

Also,

$$\begin{aligned} B_x &= D \sin\frac{l\pi x}{\alpha} \cos\frac{m\pi y}{\beta} \cos\frac{n\pi z}{\gamma} e^{-i\omega t} \\ B_y &= E \cos\frac{l\pi x}{\alpha} \sin\frac{m\pi y}{\beta} \cos\frac{n\pi z}{\gamma} e^{-i\omega t} \\ B_z &= F \cos\frac{l\pi x}{\alpha} \cos\frac{m\pi y}{\beta} \sin\frac{n\pi z}{\gamma} e^{-i\omega t} \end{aligned} \tag{12.47}$$

Finally, in view of $\nabla \times \mathbf{E} = -\dot{\mathbf{B}}$, we have

$$\begin{aligned} \omega D &= \pi\left(\frac{Cm}{\beta} - \frac{Bn}{\gamma}\right) \\ \omega E &= \pi\left(\frac{An}{\gamma} - \frac{Cl}{\alpha}\right) \\ \omega F &= \pi\left(\frac{Bl}{\alpha} - \frac{Am}{\beta}\right) \end{aligned} \tag{12.48}$$

The allowed frequencies now form a discrete spectrum—in fact, the reader will recognize a three-dimensional form of what are called stationary waves with nodes or antinodes at the boundary faces. The similarity of the field expressions with those of the displacement in the case of a vibrating string may also be noted.

In view of (12.46), only two of the three unknowns A, B and C are independent for any particular mode, i.e., set of l, m and n—thus, there are two linearly independent directions of the electric vector in any particular mode. Again for any mode, $\mathbf{E} \cdot \mathbf{B} = 0$ as may be verified directly from Eq. (12.48). A general solution may be built up by an algebraic combination of the different modes.

12.3.2 Circular Cylindrical Cavity

If the radius of the cylinder be R, the boundary conditions appear as restrictions on the field or its derivative at (x, y) where $x^2 + y^2 = R^2$. In view of this, it is easiest to solve Eq. (12.25) by writing it in cylindrical polar coordinate (r, ϕ). For TM waves, we get

$$\frac{1}{r}\frac{\partial}{\partial r}\left(\frac{r\partial E_z^0}{\partial r}\right) + \frac{1}{r^2}\frac{\partial^2 E_z^0}{\partial \varphi^2} + \left(\frac{\omega^2}{c^2} - k^2\right) E_z^0 = 0 \tag{12.49}$$

with the boundary condition $E_z^0 = 0$ at $r = R$. Exactly this equation with identical boundary condition appears in the theory of vibrations of circular membranes fixed at the boundary, and a very exhaustive discussion is given in Rayleigh's theory of sound.

We note that as φ is an angular coordinate, E_z^0 will be periodic in φ, with period 2π. Hence, it can be expanded as a sum of the Fourier terms

$$E_z^0 = \sum E_m e^{im\varphi} \tag{12.50}$$

where m is any integer and E_m is a function of r alone that satisfies the differential equation

$$\frac{1}{r}\frac{d}{dr}\left(r\frac{dE_m}{dr}\right) + \left(\frac{\omega^2}{c^2} - k^2 - \frac{m^2}{r^2}\right) E_m = 0 \tag{12.51}$$

This is Bessel's equation which we have already met with in Chap. 6. Therefore, the solution is

$$E_m = J_m\left(r\sqrt{\frac{\omega^2}{c^2} - k^2}\right) \tag{12.52}$$

where J_m is called the Bessel function of order m. The possible frequencies are determined by the boundary condition $E_z^0 = 0$ at $r = R$ or

$$J_m\left(R\sqrt{\frac{\omega^2}{c^2}-k^2}\right)=0 \tag{12.53}$$

This gives an infinite number of possible frequencies, so that a typical normal mode can be expressed as

$$E_z^0 = A_{mn}J_m\left(r\sqrt{\frac{\omega_n^2}{c^2}-k^2}\right) \tag{12.54}$$

where $R\sqrt{\frac{\omega_n^2}{c^2}-k^2}$ is the n-th root of Eq. (12.53), i.e., the n-th zero of the Bessel function J_m.

For the TE modes, the differential equation remains the same as (12.49) in form, with E_z^0 replaced by H_z^0. Thus, we have

$$H_z^0 = B_{mn}J_m\left(r\sqrt{\frac{\omega_n^2}{c^2}-k^2}\right)$$

however, the boundary condition (12.24) now gives $dH_z^0/dr = 0$, so that the allowed frequency spectrum is different.

If the ends of the cylinder are closed so as to form a cavity instead of a guide, we have the additional boundary condition for the TM modes that the electric vector must vanish at the ends, say at $z = 0$ and L, where L is the length of the cylinder. This leads to $k = p\pi/L$; consequently, the resonating frequencies are

$$\frac{\omega_n^2}{c^2}=\frac{\chi_{mn}^2}{R^2}+\frac{p^2\pi^2}{L^2}$$

where χ_{mn} is the n-th root of (12.52) and p is an integer. Thus, by altering the length L, the cavity may be tuned to different frequencies. The discussion is similar for the TE modes, except that the boundary condition is now $\partial H_z^0/\partial z = 0$ at $z = 0$ and $z = L$. That again leads to the same relation for k as in TM modes.

Problems

1. Find the scalar and vector potentials for the field given by Eqs. (12.47) and (12.48).
2. Complete the calculations of the field components for a circular cylindrical cavity and show that the time averaged energy for the electric and magnetic fields are equal.
3. Calculate Q for a rectangular cavity for the TM and TE modes of lowest frequencies.
4. Write down the Maxwell stress components for the lowest TM mode of a rectangular cavity and calculate the force on the different walls.

5. Show that the total number of modes having frequencies in the interval ν and $\nu + d\nu$ in a rectangular cavity of sides α, β and γ is $8\pi\alpha\beta\gamma\nu^2 d\nu/c^3$. (Assume that the wavelength λ is small compared to the linear dimensions of the cavity.)

Chapter 13
Electromagnetic Waves in Anisotropic Media

We shall be interested in crystalline media, other than those belonging to the cubic type, in which the dielectric constant may not be considered as a scalar, but must be replaced by a tensor, i.e.,

$$D_i = \epsilon_{ik} E_k \tag{13.1}$$

In the above equation and in all that follows, we shall assume the Einstein convention that a summation is to be understood over any repeated tensor index. We have seen previously that if the energy of the electromagnetic field is to be a state function, i.e., if it is to depend only on the field strengths and not in any way on the history as to the path by which the fields have come to their final value, then the dielectric tensor must be symmetric, i.e., $\epsilon_{ik} = \epsilon_{ki}$. We shall assume this condition to hold good. In principle, the permeability would also be a tensor but in the absence of ferromagnetic materials, the permeability does not differ appreciably from its value in vacuum, so we shall take $\mathbf{B} = \mu_0 \mathbf{H}$. The Maxwell equations still hold good

$$\nabla \cdot \mathbf{D} = 0 \tag{13.2}$$

$$\nabla \times \mathbf{E} = -\dot{\mathbf{B}} \tag{13.3}$$

$$\nabla \cdot \mathbf{B} = 0 \tag{13.4}$$

$$\nabla \times \mathbf{H} = \dot{\mathbf{D}} \tag{13.5}$$

Proceeding as in the case of isotropic dielectrics, we obtain

$$-\nabla^2 \mathbf{H} = \nabla \times \frac{\partial \mathbf{D}}{\partial t} \tag{13.6}$$

$$\nabla \times (\nabla \times \mathbf{E}) = -\mu_0 \frac{\partial^2 \mathbf{D}}{\partial t^2} \tag{13.7}$$

Consider a plane wave

A. K. Raychaudhuri, *Classical Theory of Electricity and Magnetism*, Texts and Readings in Physical Sciences 21, https://doi.org/10.1007/978-981-16-8139-4_13

$$\mathbf{E} = \mathbf{E}_0\, e^{i(\mathbf{k}\cdot\mathbf{r}-\omega t)} \tag{13.8}$$

Then $D_i = \epsilon_{il} E_l = \epsilon_{il} E_{0l} e^{i(\mathbf{k}\cdot\mathbf{r}-\omega t)}$, i.e.,

$$\mathbf{D} = \mathbf{D}_0\, e^{i(\mathbf{k}\cdot\mathbf{r}-\omega t)} \tag{13.9}$$

As in the case of isotropic media, here also the coupling between the fields requires that **B** (or **H**) should have the same form

$$\mathbf{B} = \mu_0 \mathbf{H} = \mu_0 \mathbf{H}_0\, e^{i(\mathbf{k}\cdot\mathbf{r}-\omega t)}$$

Note that for the plane waveform $\exp i(\mathbf{k}\cdot\mathbf{r}-\omega t)$, the operator $\nabla \equiv i\mathbf{k}$ and $(\partial/\partial t) \equiv -i\omega$. Using (13.8) and (13.9), we now get from Eq. (13.7)

$$\mathbf{k}\times(\mathbf{k}\times\mathbf{E}) = -\omega^2\mu_0\mathbf{D} \tag{13.10}$$

$$\text{or} \quad (\mathbf{k}\cdot\mathbf{E})\mathbf{k} - k^2\mathbf{E} = -\omega^2\mu_0\mathbf{D}$$

$$\begin{aligned} \text{i.e.,} \quad k_l E_l k_i - k^2 E_i &= -\omega^2\mu_0 D_i \\ &= -\omega^2\mu_0\epsilon_{il} E_l \end{aligned} \tag{13.11}$$

where the last expression is in components in the tensor notation.

Introducing the unit vector $\mathbf{n} = \mathbf{k}/|k|$ in the direction of $\mathbf{k}$, Eq. (13.11) reads

$$\left(n_i n_l - \delta_{il} + \frac{\omega^2\mu_0}{k^2}\epsilon_{il}\right) E_l = 0 \tag{13.12}$$

The above is a set of linear homogeneous equations in E_l. Hence, if the E_ls are not to vanish identically, the determinant of the coefficients must vanish

$$\begin{vmatrix} n_1^2 - 1 + \frac{\omega^2}{k^2}\mu_0\epsilon_{11} & n_1 n_2 + \frac{\omega^2}{k^2}\mu_0\epsilon_{12} & n_1 n_3 + \frac{\omega^2}{k^2}\mu_0\epsilon_{13} \\ n_1 n_2 + \frac{\omega^2}{k^2}\mu_0\epsilon_{12} & n_2^2 - 1 + \frac{\omega^2}{k^2}\mu_0\epsilon_{22} & n_2 n_3 + \frac{\omega^2}{k^2}\mu_0\epsilon_{23} \\ n_1 n_3 + \frac{\omega^2}{k^2}\mu_0\epsilon_{13} & n_2 n_3 + \frac{\omega^2}{k^2}\epsilon_{23} & n_1^2 - 1 + \frac{\omega^2}{k^2}\mu_0\epsilon_{33} \end{vmatrix} = 0 \tag{13.13}$$

For a given **n**, the above is an equation to determine ω^2/k^2. Since ω/k is the phase velocity of the waves and the propagation vector **k** is normal to the wavefronts, which are surfaces of equal phase, this equation gives the possible values of the phase velocity in any particular direction. As the dielectric tensor ϵ_{ik} is symmetric, it may be diagonalized by an orthogonal transformation—the axes of reference are then called principal axes of the dielectric tensor and the corresponding diagonal elements (which are the only non-vanishing elements now) are called principal dielectric constants.

Thus, with these axes $\epsilon_{jk} = 0$ if $i \neq k$. Let us write $\epsilon_{11} = \epsilon_1$, $\epsilon_{22} = \epsilon_2$, $\epsilon_{33} = \epsilon_3$. Finally, we introduce the abbreviations

$$\alpha \equiv 1 - \frac{\omega^2}{k^2}\mu_0\epsilon_1 = 1 - \frac{v^2}{c_1^2}$$
$$\beta \equiv 1 - \frac{\omega^2}{k^2}\mu_0\epsilon_2 = 1 - \frac{v^2}{c_2^2}$$
$$\gamma \equiv 1 - \frac{\omega^2}{k^2}\mu_0\epsilon_3 = 1 - \frac{v^2}{c_3^2}$$

(where $v = \omega/k$ and $c_i = 1/\sqrt{\mu_0\epsilon_i}$) to simplify Eq. (13.13) to

$$\begin{vmatrix} n_1^2 - \alpha & n_1n_2 & n_1n_3 \\ n_1n_2 & n_2^2 - \beta & n_2n_3 \\ n_1n_3 & n_2n_3 & n_3^2 - \gamma \end{vmatrix} = 0$$

Evaluating the determinant, after further simplification, we get

$$(n_1^2 - \alpha)\left(\beta\gamma - n_2^2\gamma - n_3^2\beta\right) + n_1^2n_2^2\gamma + n_1^2n_3^2\beta = 0$$
$$\text{or} \quad \alpha\beta\gamma - n_1^2\beta\gamma - n_2^2\alpha\gamma - n_3^2\alpha\beta = 0$$
$$\text{or} \quad 1 - \frac{n_1^2}{\alpha} - \frac{n_2^2}{\beta} - \frac{n_3^2}{\gamma} = 0 \qquad \text{if } \alpha\beta\gamma \neq 0$$
$$\text{or} \quad n_1^2\left(1 - \frac{1}{\alpha}\right) + n_2^2\left(1 - \frac{1}{\beta}\right) + n_3^2\left(1 - \frac{1}{\gamma}\right) = 0$$

Substituting the values of α, β and γ, we get (taking $v \neq 0$)

$$\frac{n_1^2}{v^2 - c_1^2} + \frac{n_2^2}{v^2 - c_2^2} + \frac{n_3^2}{v^2 - c_3^2} = 0 \tag{13.14}$$

Equation (13.14) is a quadratic in v^2 and shows that for any $\mathbf{n}$ (i.e., the direction of wave normal), the phase velocity has two distinct values. Therefore, one has two refracted rays in an anisotropic medium corresponding to an incident ray. As an example of two distinct velocities for any $\mathbf{n}$, consider the direction of a principal axis $(n_1, n_2, n_3) = (1, 0, 0)$, then the two possible velocities from (13.14) are c_2 and c_3.

13.1 The Relation Between the Directions of D, B, k, etc.

For an isotropic media, $\mathbf{D}$ (which is in the same direction as $\mathbf{E}$), $\mathbf{B}$ and $\mathbf{k}$ are mutually orthogonal and the direction of energy flux vector $\mathbf{E} \times \mathbf{H}$ coincides with the wave normal $\mathbf{k}$. Things are, however, much more complicated in case of anisotropic media. Equation (13.10) shows that $\mathbf{D}$ is normal to $\mathbf{k}$. Similarly, Eq. (13.4) shows that $\mathbf{B}$ is also normal to $\mathbf{k}$. Equation (13.6) gives

$$k^2\mathbf{B} = -i\mu\omega\mathbf{k}\times\mathbf{D}$$

showing that **B** is orthogonal to **D** as well. Thus, **B**, **D** and **k** continue to be mutually orthogonal. However, since **E** and **D** are, in general, not in the same direction now, **E** is no longer orthogonal to **k**. On the other hand, Eq. (13.3) shows that **E** is perpendicular to **B**. Thus, **E** lies in the plane containing **D** and **k**.

The validity of Poynting's theorem may be demonstrated for the case of anisotropic media as well. From Eqs. (13.3) and (13.5), we have

$$\begin{aligned}\nabla\cdot(\mathbf{E}\times\mathbf{H}) &= \mathbf{E}\cdot(\nabla\times\mathbf{H}) - \mathbf{H}\cdot(\nabla\times\mathbf{E})\\ &= \mathbf{E}\cdot\frac{\partial\mathbf{D}}{\partial t} + \mathbf{H}\cdot\frac{\partial\mathbf{B}}{\partial t}\end{aligned}$$

with ϵ_{ik} symmetric, $\frac{\partial}{\partial t}(E_i D_i) = \frac{\partial}{\partial t}(\mathbf{E}_i\epsilon_{ik}\mathbf{E}_k) = 2\mathbf{E}\cdot\dot{\mathbf{D}}$. Hence,

$$\nabla\cdot(\mathbf{E}\times\mathbf{H}) = \frac{\partial}{\partial t}(\mathbf{E}\cdot\mathbf{D}+\mathbf{H}\cdot\mathbf{B})$$

Thus, $\mathbf{E}\times\mathbf{H}$ continues to represent the energy flux. We shall call the direction of the energy flux the ray vector—it is, thus, normal to **E** and **H** and lies on the plane spanned by the wave normal **k** and the displacement vector **D**, but normal to neither of the two in general.

13.2 The Relation Between the Vectors in the Two Waves with the Same Wave Normal

We have seen that for any **n**, there are two values of the phase velocity v. We now go to investigate the relation between the vectors in these two waves. Calling the two velocities v_1 and v_2 (note the change in notation) and indicating the corresponding vectors by superscripts (1) and (2), respectively, Eq. (13.12) gives for the two waves

$$\left(n_i n_l - \delta_{il} + v_1^2\mu_0\epsilon_{il}\right)E_l^{(1)} = 0 \tag{13.15}$$

$$\left(n_i n_l - \delta_{il} + v_2^2\mu_0\epsilon_{il}\right)E_i^{(2)} = 0 \tag{13.16}$$

Contracting (13.15) with $E_i^{(2)}$ and (13.16) with $E_l^{(1)}$ and subtracting, we get

$$(v_1^2 - v_2^2)\mu_0\epsilon_{il}E_l^{(1)}E_i^{(2)} = 0 \tag{13.17}$$

which may be written as

$$(v_1^2 - v_2^2)\mu_0\mathbf{E}^{(1)}\cdot\mathbf{D}^{(2)} = (v_1^2 - v_2^2)\mu_0\mathbf{E}^{(2)}\cdot\mathbf{D}^{(1)} = 0$$

If the two velocities be different, the vectors $\mathbf{D}^{(2)}$ and $\mathbf{E}^{(1)}$ are orthogonal (and of course so are $\mathbf{D}^{(1)}$ and $\mathbf{E}^{(2)}$). Again from (13.11), we obtain

$$-v_1^2\mu_0\mathbf{D}^{(1)}\cdot\mathbf{D}^{(2)} = (\mathbf{n}\cdot\mathbf{E}^{(1)})(\mathbf{n}\cdot\mathbf{D}^{(2)}) - \mathbf{E}^{(1)}\cdot\mathbf{D}^{(2)}$$

Both the terms on the right-hand side vanish—hence $\mathbf{D}^{(1)}$ and $\mathbf{D}^{(2)}$ are normal to one another. Thus, the pairs $\mathbf{D}^{(2)}$ and $\mathbf{B}^{(1)}$, as well as $\mathbf{D}^{(1)}$ and $\mathbf{B}^{(2)}$, are aligned. The two waves are, thus, polarized in perpendicular directions.

For any $\mathbf{n}$, and corresponding to v_1, say, Eq. (13.12) would determine the direction of $\mathbf{E}$ (and hence of $\mathbf{D}$) but not its magnitude. Thus, the relative intensities in the two waves may have arbitrary values. Therefore, the waves proceeding in any direction will have definite polarizations. As the directions of $\mathbf{E}$ and $\mathbf{H}$ are different for the two waves, the ray directions would be different. Determining the ray direction, once the direction of $\mathbf{E}$ is determined from (13.12), is of course possible but we shall not get into that exercise here.

It may be noted that the validity of Snell's laws of refraction follows from the independence of the phase velocity relative to the direction of propagation. As this is not true for anisotropic media, the laws of refraction do not hold, in general, for any of the two refracted waves.

The two schematic diagrams Fig. 13.1 show the relative orientations of $\mathbf{k}$, $\mathbf{E}$, $\mathbf{D}$, $\mathbf{B}$ as well as the Poynting vector for the two rays, for which the z-axis is taken in the direction of $\mathbf{k}$.

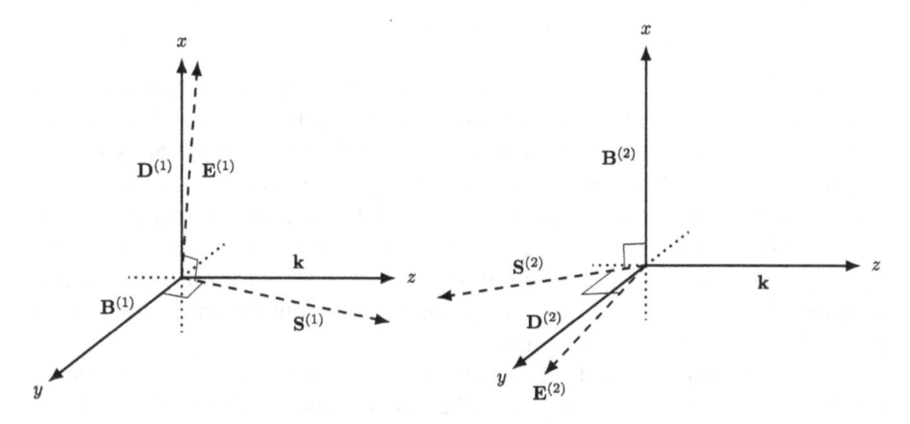

Fig. 13.1 The field directions for the two rays corresponding to a given wave normal are shown. In both cases, the corresponding **D** and **B** are orthogonal to **k** and are orthogonal to each other. The fields **E** and **B** are also orthogonal. The Poynting vector is no longer parallel to the wave vector but is perpendicular to both **E** and **B**

13.3 Crystal Classes and Optic Axes

According to the symmetry of crystalline lattices, crystals may be classified into different types showing different optical behaviours. The cubic crystals do not show any anisotropy, i.e., $\epsilon_{11} = \epsilon_{22} = \epsilon_{33}$; thus, there is no double refraction. The hexagonal and tetragonal types have an axis of symmetry such that all directions normal to it are equivalent. If the axis of symmetry be numbered 2, then $\epsilon_1 = \epsilon_3 \neq \epsilon_2$ and $c_1 = c_3 \neq c_2$. These crystals are called uniaxial crystals while the less symmetric ones are named biaxial crystals. The justification of these names will be clear presently.

In case of uniaxial crystals, the assumption $\alpha\beta\gamma \neq 0$ leading to Eq. (13.14) is not correct. Hence, we go back to the original equation $\alpha\beta\gamma - n_1^2\beta\gamma - n_2^2\alpha\gamma - n_3^2\alpha\beta = 0$. With $\alpha = \gamma$, this gives $\alpha(\alpha\beta - n_1^2\beta - n_2^2\alpha - n_3^2\beta) = 0$. Hence, either

$$\alpha = 0, \quad v = c_1 \qquad \text{or} \qquad \alpha\beta - \alpha\cos^2\theta - \beta\sin^2\theta = 0 \tag{13.18}$$

where $\cos\theta = n_2$, θ being the angle between the wave normal and the symmetry axis. Substituting the values of α and β, we get

$$\left(1 - \frac{v^2}{c_1^2}\right)\left(1 - \frac{v^2}{c_2^2}\right) - \left(1 - \frac{v^2}{c_1^2}\right)\cos^2\theta - \left(1 - \frac{v^2}{c_2^2}\right)\sin^2\theta = 0$$

which simplifies to

$$v^2 = c_1^2\cos^2\theta + c_2^2\sin^2\theta \tag{13.19}$$

Of the two velocities, $v_1 = c_1$ is independent of $\mathbf{n}$; hence, for this, Snell's law holds good. It is called the ordinary ray. The velocity of the other ray varies between the extremes c_1 and c_2. In particular, if the wave normal coincides with the symmetry axis, $\theta = 0$, both the velocities are equal. One has then only one ray—this particular direction is called the optic axis, and as there is only one such axis, crystals of this type are called uniaxial. For $\mathbf{n}$ in the direction of the optic axis, Eq. (13.12) simply gives $E_2 = 0$ leaving E_1/E_3 undetermined, which shows that the wave is transverse but unpolarized. As $\epsilon_1 = \epsilon_3$, the directions of $\mathbf{E}$ and $\mathbf{D}$ coincide and so do those of the ray direction and the wave normal.

For the more general case, $\mathbf{E}_1$, $\mathbf{E}_2$ and $\mathbf{E}_3$ are all different. To find the directions in which the velocities of the two waves will be identical, one investigates the condition for the roots of the quadratic (13.14) to be identical. This would give a relation between n_1, n_2 and n_3, of which only two are independent (because of $n_1^2 + n_2^2 + n_3^2 = 1$). One may, thus, think that there would be an infinity of directions with identical roots. As a matter of fact, the constraints of reality of (c_1, c_2, c_3) and (n_1, n_2, n_3) select out only two particular directions. Thus, there are two optic axes and hence the name biaxial crystals. To investigate mathematically, we write Eq. (13.14) in the form

$$\frac{n_1^2}{p_1}+\frac{n_2^2}{p_2}+\frac{n_3^2}{p_3}=0 \tag{13.20}$$

where $p_i = v^2 - c_i^2$. If Eq. (13.20) is to give a double root, the first differential coefficient of left-hand side with respect to v must vanish

$$\frac{n_1^2}{p_1^2}+\frac{n_2^2}{p_2^2}+\frac{n_3^2}{p_3^2}=0 \tag{13.21}$$

Equations (13.20) and (13.21) are two homogeneous equations for three quantities; hence, we have

$$\frac{n_1^2}{p_1^2(p_2-p_3)}=\frac{n_2^2}{p_2^2(p_3-p_1)}=\frac{n_3^2}{p_3^2(p_1-p_2)} \tag{13.22}$$

This last equation cannot be satisfied as the denominators cannot all be of the same sign. However, the equation has been obtained on the assumption $p_1 p_2 p_3 \neq 0$; therefore, one of them must vanish for the roots to be equal. Let $p_2 = 0$, then Eq. (13.20) will require $n_2 = 0$ and

$$n_1^2(v^2-c_3^2)+n_3^2(v^2-c_1^2)=0 \tag{13.23}$$

which will determine the other root. For the roots to be identical, v must be equal to c_2 (given by $p_2 = 0$). Hence, from (13.23) with $n_1^2 = \cos^2\theta$ and $n_3^2 = \sin^2\theta$

$$\tan^2\theta = \frac{c_2^2-c_3^2}{c_1^2-c_2^2} \tag{13.24}$$

This relation can only be satisfied if c_2 lies between c_1 and c_3. Hence, we have the following result. The two optic axes lie in the plane containing the axes of greatest and least dielectric constant and the axes bisect the angles between the optic axes. The angle made by the optic axes with the ϵ-axes being

$$\tan^{-1} \pm\sqrt{\frac{c_2^2-c_3^2}{c_1^2-c_2^2}}$$

13.4 Conical Refraction

In case the wave normal coincides with the optic axis, one of Eqs. (13.12), namely that with $i = 2$, is empty and the ratio of E_1 and E_3 may be determined from any of the other two equations. The value of E_2 remains undetermined; consequently, the direction of the Poynting vector is not unique. This gives rise to the phenomenon known as internal conical refraction, which we will now analyse.

The Poynting vector $\mathbf{S}$, giving the ray direction for wave normal along optic axis, is

$$\begin{aligned}\mathbf{S} = \mathbf{E} \times \mathbf{H} &= \frac{1}{\mu_0 \omega}\mathbf{E} \times (\mathbf{k} \times \mathbf{E}) = \frac{k}{\mu_0 \omega}\mathbf{E} \times (\mathbf{n} \times \mathbf{E}) \\ &= \frac{k}{\mu_0 \omega}\left(E^2\mathbf{n} - (\mathbf{E} \cdot \mathbf{n})\mathbf{E}\right) \end{aligned} \tag{13.25}$$

where Eq. (13.3) was used. From Eq. (13.12) using $(n_1, n_2, n_3) = (\cos\theta, 0, \sin\theta)$ and $v = c_2$, after a little algebra

$$E_1\left(\sin^2\theta - \frac{c_2^2}{c_1^2}\right) = E_3 \cos\theta \sin\theta \tag{13.26}$$

$$S^2 = \frac{k^2}{\mu_0^2 \omega^2}\left(E^4 - E^2(\mathbf{E} \cdot \mathbf{n})^2\right)$$

where the latter follows from (13.25). Hence, the unit vector $\mathbf{m} = (m_1, m_2, m_3)$ along the direction of the ray is

$$\mathbf{m} = \left(\frac{E^2\cos\theta - (\mathbf{E} \cdot \mathbf{n})E_1}{\sqrt{E^4 - E^2(\mathbf{E} \cdot \mathbf{n})^2}}, -\frac{(\mathbf{E} \cdot \mathbf{n})E_2}{\sqrt{E^4 - E^2(\mathbf{E} \cdot \mathbf{n})^2}}, \frac{E^2\sin\theta - (\mathbf{E} \cdot \mathbf{n})E_3}{\sqrt{E^4 - E^2(\mathbf{E} \cdot \mathbf{n})^2}}\right)$$

It now follows

$$\mathbf{m} \cdot \mathbf{n} = m_1 n_1 + m_3 n_3 = \frac{E^2 - (\mathbf{E} \cdot \mathbf{n})^2}{\sqrt{E^4 - E^2(\mathbf{E} \cdot \mathbf{n})^2}} \tag{13.27}$$

$$\begin{aligned} m_1 n_1 c_3^2 + m_3 n_3 c_1^2 &= \frac{E^2(c_3^2\cos^2\theta + c_1^2\sin^2\theta)}{\sqrt{E^4 - E^2(\mathbf{E} \cdot \mathbf{n})^2}} \\ &= \frac{E^2 c_2^2}{\sqrt{E^4 - E^2(\mathbf{E} \cdot \mathbf{n})^2}} \end{aligned} \tag{13.28}$$

where, in obtaining the last relation, we have used Eq. (13.24). Now, from (13.27) and (13.28), we get

$$(m_1 n_1 + m_3 n_3)(m_1 n_1 c_3^2 + m_3 n_3 c_1^2) = c_2^2 \tag{13.29}$$

Hence, if the coordinates of the end point of a ray are (x, y, z) and $m_1 = x/\sqrt{x^2 + y^2 + z^2}$, etc., from Eq. (13.29),

$$(x n_1 c_3^2 + z n_3 c_1^2)(x n_1 + z n_3) = c_2^2(x^2 + y^2 + z^2)$$

This equation represents a cone. Therefore, when the wave normal is along the optic axis, there is an infinite number of rays (i.e., directions of flux of energy) which lie on the surface of a cone. Hence, an unpolarized wave, incident from outside at such an angle that the refracted wave normal in the crystal is along the optic axis, gives rise to a cone of rays, each constituent ray of the cone being plane polarized.

13.5 External Conical Refraction

To understand the phenomenon of external conical refraction, we shall have to find the relation between the ray direction, i.e., the direction of the Poynting vector, and the wave normal. The Poynting vector is given by

$$\mathbf{S} = \frac{E^2}{\mu_0 v}(\mathbf{n} - (\mathbf{n}\cdot\mathbf{e})\mathbf{e}) \tag{13.30}$$

where $\mathbf{e}$ is the unit vector in the direction of $\mathbf{E}$ and $v = \omega/k$. Now Eq. (13.12) can be written as

$$n_i(\mathbf{n}\cdot\mathbf{E}) - E_i + \frac{v^2}{c_i^2}E_i = 0$$

$$\text{or} \qquad e_i = \frac{n_i(\mathbf{n}\cdot\mathbf{e})}{1 - \frac{v^2}{c_i^2}} \tag{13.31}$$

From the equations above, we get

$$S_i = \frac{E^2}{\mu_0 v}\left(1 - \frac{(\mathbf{n}\cdot\mathbf{e})^2 i}{1 - \frac{v^2}{c_i^2}}\right) n_i \tag{13.32}$$

If ψ be the angle between the ray direction $\mathbf{S}$ and the wave normal $\mathbf{n}$, then from (13.30),

$$\cos\psi = \frac{S_i n_i}{\sqrt{S_i S_i}} = \sqrt{1 - (\mathbf{e}\cdot\mathbf{n})^2}, \qquad \mathbf{e}\cdot\mathbf{n} = \sin\psi$$

a result might have been obtained directly, since $\mathbf{S}$, $\mathbf{E}$ and $\mathbf{n}$ all lie in the same plane, all three being orthogonal to $\mathbf{B}$, and $\mathbf{S}$ is normal to $\mathbf{E}$. Substituting now in Eq. (13.32)

$$\frac{S_i}{|S|} = n_i\left(1 - \frac{\sin^2\psi}{1 - \frac{v^2}{c_i^2}}\right)\sqrt{1 - (\mathbf{n}\cdot\mathbf{e})^2} = n_i\,\frac{c_i^2\cos^2\psi - v^2}{(c_i^2 - v^2)\cos\psi}$$

Introducing the ray velocity $v_r = v\sec\psi$ (see Fig. 13.2), the above becomes

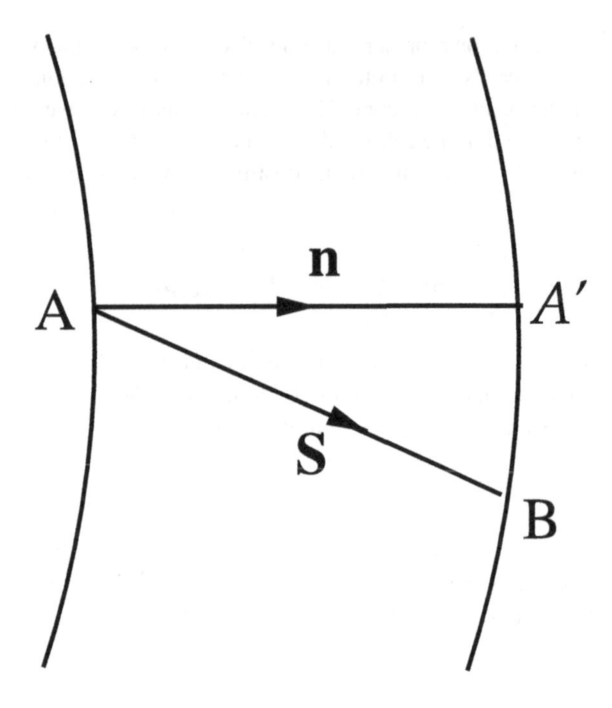

Fig. 13.2 Figure shows the relationship between the ray velocity and wave normal for external conical refraction

$$\frac{S_i}{|S|} = \frac{n_i(c_i^2 - v_r^2)}{c_i^2 - V^2}\frac{v}{v_r} \tag{13.33}$$

$$\frac{v_r}{v} = \frac{\mathrm{AB}}{\mathrm{AA'}} = \sec\psi$$

This determines, in general, two values of n_i for a given S_i. However, n_2 becomes indeterminate if $S_2 = 0$ and $V_r = c_2$. Thus, when the ray coincides with this direction (called the ray axis), then at the point of exit of the ray from the crystal, the wave normals lie on a cone. A single ray in the crystal, on emergence, produces a ring on a screen, the diameter of the ring increasing as the distance of the screen from the point of emergence increases. This phenomenon is known as external conical refraction.

Problems

1. Show that for an electrostatic field in a crystal, the electrostatic potential obeys a Poisson-type equation where the lengths along different axes are modified by suitable scale factors.
2. Show that the electrostatic field in a crystal is not a central field. Find the equations of the equipotentials for a point charge.
3. Derive the following expression for the phase velocity:

$$
\begin{aligned}
2v^2 = {} & n_1^2(c_2^2 + c_3^2) + n_2^2(c_3^2 + c_1^2) + n_3^2(c_1^2 + c_2^2) \\
& \pm \sqrt{\begin{aligned} & n_1^4(c_2^2 - c_3^2)^2 + n_2^4(c_3^2 - c_1^2)^2 + n_3^4(c_1^2 - c_2^2)^2 \\ & - 2n_1^2n_2^2(c_2^2 - c_3^2)(c_3^2 - c_1^2) - 2n_2^2n_3^2(c_3^2 - c_1^2)(c_1^2 - c_2^2) \\ & - 2n_3^2n_1^2(c_1^2 - c_2^2)(c_2^2 - c_3^2) \end{aligned}}
\end{aligned}
$$

4. If ψ_1 and ψ_2 are the angles which the wave normal makes with the optic axes, show that the phase velocities are

$$
\begin{aligned}
2v_1^2 &= c_1^2 + c_3^2 + (c_1^2 - c_3^2)\cos(\psi_1 - \psi_2) \\
2v_2^2 &= c_1^2 + c_3^2 + (c_1^2 - c_3^2)\cos(\psi_1 + \psi_2)
\end{aligned}
$$

5. If a surface is such that the normal vector to it at any point gives the phase velocity for the wave normal in that direction is called a normal surface. Show that a normal surface consists of two sheets, and study the intersections of these surfaces by the principal planes of the crystal.
6. Show that if v_r be the ray velocity for the unit vector **s** in the ray direction

$$
\frac{s_1^2 v_r^2}{v_r^2 - c_1^2} + \frac{s_2^2 v_r^2}{v_r^2 - c_2^2} + \frac{s_3^2 v_r^2}{v_r^2 - c_3^2} = 1
$$

Chapter 14
Solution of Maxwell's Equations: Retarded and Advanced Potentials

Let us start by listing the Maxwell equations again

$$\nabla \cdot \mathbf{D} = \rho \tag{14.1}$$

$$\nabla \times \mathbf{E} = -\frac{\partial \mathbf{B}}{\partial t} \tag{14.2}$$

$$\nabla \cdot \mathbf{B} = 0 \tag{14.3}$$

$$\nabla \times \mathbf{H} = \mathbf{j} + \frac{\partial \mathbf{D}}{\partial t} \tag{14.4}$$

which admit (from Eqs. (14.3) and (14.2), respectively) the introduction of a vector potential $\mathbf{A}$ and a scalar potential ϕ such that

$$\mathbf{B} = \nabla \times \mathbf{A} \tag{14.5}$$

$$\mathbf{E} = -\frac{\partial \mathbf{A}}{\partial t} - \nabla \phi \tag{14.6}$$

However, there is some freedom in the choice of $\mathbf{A}$ and ϕ. We may add a term $\nabla \Psi$ to $\mathbf{A}$ and $-\partial \psi \partial t$ to ϕ without affecting the relations (14.5) and (14.6). The scalar ψ is quite arbitrary and this transformation of the potentials is known as a gauge transformation.[1]

[1] If Ψ be any solution of the Schrödinger equation

$$\nabla^2 \Psi + \frac{8\pi^2 m}{\hbar^2}(E - V)\Psi = 0$$

then $\Psi' = \Psi e^{i\alpha}$, where α is a constant, is also a solution. For a charged particle in an electromagnetic field, the gauge change $\mathbf{A}' = \mathbf{A} + \nabla \Lambda$ and $\phi' = \phi + \partial \Lambda / \partial t$ induces a change in Ψ to $\Psi' = \Psi \exp(iq\Lambda/c^2)$ (where q is the charge carried by the particle). These two have been named by Pauli as gauge transformations of the first and second kind, respectively.

A. K. Raychaudhuri, *Classical Theory of Electricity and Magnetism*, Texts and Readings in Physical Sciences 21, https://doi.org/10.1007/978-981-16-8139-4_14

We shall use this freedom to reduce the problem to a solution of the inhomogeneous wave equations. We assume that the potentials have been chosen in such a gauge that

$$\nabla \cdot \mathbf{A} + \mu_0 \epsilon_0 \frac{\partial \phi}{\partial t} = 0 \tag{14.7}$$

Equation (14.7) is known as the Lorentz gauge condition. Let us suppose that some charge and current distribution are specified, in vacuum. This will allow us to substitute $\mathbf{D} = \epsilon_0 \mathbf{E}$ and $\mathbf{B} = \mu_0 \mathbf{H}$. Eliminating the variables $\mathbf{H}$ and $\mathbf{D}$ from (14.4), by (14.5) and (14.6), we get

$$\nabla(\nabla \cdot \mathbf{A}) - \nabla^2 \mathbf{A} = \mu_0 \mathbf{j} - \frac{1}{c^2} \frac{\partial^2 \mathbf{A}}{\partial t^2} - \mu_0 \epsilon_0 \nabla \frac{\partial \phi}{\partial t}$$

Utilizing the Lorentz condition (14.7) and writing $\Box$ for the operator $\nabla^2 - \frac{1}{c^2}\frac{\partial^2}{\partial t^2}$ (called the d'Alembertian), the above becomes

$$\Box \mathbf{A} = -\mu_0 \mathbf{j} \tag{14.8}$$

$$\Box \phi = -\frac{\rho}{\epsilon_0} \tag{14.9}$$

where the latter comes from a similar elimination of $\mathbf{D}$ from Eq. (14.1) using (14.6) and (14.7). Equations (14.7–14.9) together ensure the conservation of charge

$$\frac{\partial \rho}{\partial t} + \nabla \cdot \mathbf{j} = 0 \tag{14.10}$$

Mathematically, the vector Eq. (14.8) and the scalar Eq. (14.9) are of identical form (just split up the vector equation into the three component equations to see it). We shall find the solution to this type of equation in three steps

1. Eliminate the time dependence from the equations. This will be done by using the method of the Fourier transform.
2. Solve the resulting equation by finding Green's function. In reality, the so-called solution is merely in the form of an integral.
3. Restore the time dependence and obtain the solution in the form of an integral.

We recapitulate a few properties of the Fourier transforms for ready reference. Under very general conditions (which we do not specify here), a function $F(t)$ can be represented as an infinite integral

$$F(t) = \int_{-\infty}^{+\infty} g(\omega)\, e^{i\omega t}\, d\omega \tag{14.11}$$

which may be inverted as

$$g(\omega) = \frac{1}{2\pi} \int_{-\infty}^{+\infty} F(t)\, e^{-i\omega t}\, dt \tag{14.12}$$

If $F(t)$ be real, $g^*(\omega) = g(-\omega)$, where the asterisk indicates the complex conjugate. The Dirac delta function can be represented as a Fourier integral: this as well as its reciprocal relation are as follows:

$$\delta(x) = \frac{1}{2\pi} \int_{-\infty}^{+\infty} e^{-ikx}\, dk \tag{14.13}$$

$$1 = \int_{-\infty}^{+\infty} e^{ikx}\, \delta(x)\, dx \tag{14.14}$$

In view of Eqs. (14.11) and (14.12) if

$$\int_{-\infty}^{+\infty} g(x)\, e^{-i\omega x}\, dx = 0 \tag{14.15}$$

then $g(x) = 0$ for all x. Now consider the Fourier transforms

$$\rho(\mathbf{r}, t) = \int_{-\infty}^{+\infty} \rho_\omega e^{i\omega t} d\omega \tag{14.16}$$

$$\rho_\omega(\mathbf{r}) = \frac{1}{2\pi} \int_{-\infty}^{+\infty} \rho(\mathbf{r}, t) e^{-i\omega t} dt \tag{14.17}$$

$$\phi(\mathbf{r}, t) = \int_{-\infty}^{+\infty} \phi_\omega e^{i\omega t} d\omega \tag{14.18}$$

$$\phi_\omega(\mathbf{r}) = \frac{1}{2\pi} \int_{-\infty}^{+\infty} \phi(\mathbf{r}, t) e^{-i\omega t} dt \tag{14.19}$$

Substituting these in (14.9), we get

$$\int_{-\infty}^{+\infty} \left(\nabla^2 \phi_\omega + \frac{\omega^2}{c^2} \phi_\omega + \frac{\rho_\omega}{\epsilon_0} \right) e^{i\omega t}\, d\omega = 0$$

Hence, from (14.15),

$$\nabla^2 \phi_\omega + \frac{\omega^2}{c^2} \phi_\omega + \frac{\rho_\omega}{\epsilon_0} = 0 \tag{14.20}$$

Thus, the first step of eliminating the variable t has been achieved, and we have now to find Green's function which is a solution of the equation

$$\nabla^2 G(\mathbf{r}, \mathbf{r}') + \frac{\omega^2}{c^2} G(\mathbf{r}, \mathbf{r}') = \delta(\mathbf{r} - \mathbf{r}') \tag{14.21}$$

The solution is

$$G(\mathbf{r}, \mathbf{r}') = -\frac{\exp\left(\pm i\frac{\omega}{c}|\mathbf{r} - \mathbf{r}'|\right)}{4\pi|\mathbf{r} - \mathbf{r}'|}$$

Whether this satisfies Eq. (14.21) may be verified by direct differentiation

$$\frac{\partial}{\partial x}\left(\frac{e^{ikr}}{r}\right) = \frac{ikx}{r^2}e^{ikr} + e^{ikr}\frac{\partial}{\partial x}\left(\frac{1}{r}\right)$$
$$\frac{\partial^2}{\partial x^2}\left(\frac{e^{ikr}}{r}\right) = -\frac{k^2x^2}{r^2}\frac{e^{ikr}}{r} + \frac{ik}{r^2}e^{ikr} - \frac{3ik}{r^2}\left(\frac{x^2}{r^2}\right)e^{ikr} + e^{ikr}\frac{\partial^2}{\partial x^2}\left(\frac{1}{r}\right)$$

Since the boundary condition $\phi \to 0$ as $|\mathbf{r} - \mathbf{r}'| \to \infty$ is homogeneous, we have the solution of (14.20)

$$\phi_\omega(\mathbf{r}) = \frac{1}{4\pi\epsilon_0}\int \frac{\exp\left(\pm i\frac{\omega}{c}|\mathbf{r} - \mathbf{r}'|\right)}{4\pi|\mathbf{r} - \mathbf{r}'|}\rho_\omega(\mathbf{r}')\,dv'$$

Now restoring the time dependence with the help of Eq. (14.18) and using Eq. (14.17)

$$\phi(\mathbf{r}, t) = \frac{1}{4\pi\epsilon_0}\int_{-\infty}^{\infty}\frac{d\omega}{2\pi}e^{i\omega t}\int \rho(\mathbf{r}', t')e^{-i\omega t'}\,dt'\int \frac{\exp\left(\pm i\frac{\omega}{c}|\mathbf{r} - \mathbf{r}'|\right)}{4\pi|\mathbf{r} - \mathbf{r}'|}\,dv'$$

In view of (14.13), the ω integration gives

$$\int_{-\infty}^{\infty} d\omega\, e^{i\omega(t - t' \pm |\mathbf{r} - \mathbf{r}'|/c)} = 2\pi\delta\left(t - t' \pm \frac{1}{c}|\mathbf{r} - \mathbf{r}'|\right)$$

Thus,

$$\begin{aligned}\phi(\mathbf{r}, t) &= \frac{1}{4\pi\epsilon_0}\int_{-\infty}^{\infty}dt'\int dv'\,\frac{\rho(\mathbf{r}', t')}{|\mathbf{r} - \mathbf{r}'|}\,\delta\left(t - t' \pm \frac{1}{c}|\mathbf{r} - \mathbf{r}'|\right)\\ &= \frac{1}{4\pi\epsilon_0}\int dv'\,\frac{1}{|\mathbf{r} - \mathbf{r}'|}\,\rho\left(\mathbf{r}', t \pm \frac{1}{c}|\mathbf{r} - \mathbf{r}'|\right) \qquad (14.22)\\ \mathbf{A}(\mathbf{r}, t) &= \frac{\mu_0}{4\pi}\int dv'\,\frac{1}{|\mathbf{r} - \mathbf{r}'|}\,\mathbf{j}\left(\mathbf{r}', t \pm \frac{1}{c}|\mathbf{r} - \mathbf{r}'|\right) \qquad (14.23)\end{aligned}$$

where the latter expression was obtained in an exactly similar manner from (14.8). These equations above are the desired solutions in integral form. The solutions involve an alternative choice of sign before $|\mathbf{r} - \mathbf{r}'|/c$ and hence there are two independent solutions. The general solution will be a linear combination of these two with constant coefficients adding up to unity.

More generally, we can append to (14.22) or (14.23) any solution of the homogeneous equation

$$\Box\phi = \Box\mathbf{A} = 0$$

Thus, in the case of central symmetry, the solution of the homogeneous equation is any function of the argument $t - \mathbf{r}/c$ or $t + \mathbf{r}/c$. They are known, respectively, as outgoing and incoming waves and these could be added to the solution (14.22) or (14.23). Thus, we do not have the uniqueness of the solutions in the form we had in the static case. Let us consider the two signs separately

$$\begin{aligned}\phi_{\text{ret}}(\mathbf{r}, t) &= \frac{1}{4\pi\epsilon_0}\int dv' \frac{1}{|\mathbf{r}-\mathbf{r}'|}\rho\left(\mathbf{r}', t - \frac{1}{c}|\mathbf{r}-\mathbf{r}'|\right) \\ &= \frac{1}{4\pi\epsilon_0}\int dv' \frac{\rho(\mathbf{r}', t_{\text{ret}})}{|\mathbf{r}-\mathbf{r}'|}\end{aligned} \tag{14.24}$$

It is tempting to give an interpretation of this equation on the basis of the idea that the influence from the 'source' to the field point travels through intervening space and takes a finite time in the passage. This is of course a necessary requirement of field-theoretic amplitude. Thus, the field experienced at $(\mathbf{r}, t)$ is a superposition (indeed a linear superposition because of the linearity of the Maxwell equations) of the effect of sources distributed at different points. But according to the field idea, these effects have travelled all the way from the source point to the field point and hence must have started at a time $|\mathbf{r}-\mathbf{r}'|/c$ earlier than the instant of observation at the field point. Thus, the field at the instant t will bear the imprint of the condition of the sources at $t - |\mathbf{r}-\mathbf{r}'|/c$ and that is exactly what Eq. (14.24) shows. This solution is called the retarded potential and $t - |\mathbf{r}-\mathbf{r}'|/c$ is referred to as the retarded time.

All this sounds quite sensible, but the trouble comes when we consider the other solution, which is

$$\phi_{\text{adv}}(\mathbf{r}, t) = \frac{1}{4\pi\epsilon_0}\int dv' \frac{1}{|\mathbf{r}-\mathbf{r}'|}\rho\left(\mathbf{r}', t + \frac{|\mathbf{r}-\mathbf{r}'|}{c}\right) \tag{14.25}$$

This is called the advanced potential and if we try to rephrase the language that we used in interpreting Eq. (14.24), we should say that the influence starting from the source point at a later instant $t + |\mathbf{r}-\mathbf{r}'|/c$ apparently gains the time $|\mathbf{r}-\mathbf{r}'|/c$ in travelling the distance $|\mathbf{r}-\mathbf{r}'|$ and arrives somewhat earlier than it started. For this solution, therefore, a signal can apparently travel from the present to the past and the field at $(\mathbf{r}, t)$ bears the imprint of the condition that the sources 'will' acquire at later instants. All this sounds crazy and indeed one gets a correct picture of the observed phenomena by simply taking the retarded potential solution.

However, the matter is not that simple and very fundamental questions are involved. The root of the matter lies in a time reversal symmetry. Our Eqs. (14.1–14.4) remain unchanged under the following transformations:

$$\begin{aligned} t &\to -t & \mathbf{E} &\to \mathbf{E} \\ \mathbf{r} &\to \mathbf{r} & \mathbf{D} &\to \mathbf{D} \\ \rho &\to \rho & \mathbf{B} &\to -\mathbf{B} \\ \mathbf{j} &\to -\mathbf{j} & \mathbf{H} &\to -\mathbf{H} \end{aligned}$$

Hence, corresponding to any solution progressing forward in time, we have a solution proceeding backward in time. Thus, while an accelerated charge losing energy by radiation is described by the retarded potential, we can use the advanced potential to have a solution where a charge takes up energy from infinity and is, thus, spontaneously accelerated. However, queer such things may seem, the advanced potentials have been used to deduce the radiation reaction formula in a manner that is more satisfactory than other procedures. Tied up with this problem is also the statistical mechanical problem that while mechanics admits a time reversal symmetry, the entropy principle singles out a particular direction of time flux.

We must satisfy ourselves that our solution is self consistent—we have used a gauge condition (14.7) to obtain the solution; hence, the solution itself must satisfy the gauge condition. We note that in the integrands of Eqs. (14.22) and (14.23), ρ and $\mathbf{j}$ are functions of $\mathbf{r}'$ and $t' = t - R/c$ where $R = |\mathbf{r} - \mathbf{r}'|$. Hence,

$$\left(\frac{\partial \rho}{\partial t}\right)_{\mathbf{r},\mathbf{r}'} = \frac{\partial \rho}{\partial t'}$$

$$\left(\frac{\partial \mathbf{j}}{\partial x_k'}\right)_{t,x_k} = \left(\frac{\partial \mathbf{j}}{\partial x_k'}\right)_{t'} + \left(\frac{\mathbf{j}}{\partial R}\right)_{t,\mathbf{r}'} \left(\frac{\partial R}{\partial x_k'}\right)_{t,\mathbf{r}}$$

Therefore,

$$\begin{aligned}
\frac{4\pi}{\mu_0}\nabla\cdot\mathbf{A} = \frac{4\pi}{\mu_0}\left(\frac{\partial A_k}{\partial x_k}\right)_t &= \int\left\{j_k\left(\frac{\partial(1/R)}{\partial x_k}\right)_{t,\mathbf{r}'} + \frac{1}{R}\left(\frac{\partial j_k}{\partial x_k}\right)_{t,\mathbf{r}'}\right\}dv' \\
&= -\int\left\{j_k\left(\frac{\partial(1/R)}{\partial x_k'}\right)_{t,\mathbf{r}} + \frac{1}{R}\left(\frac{\partial j_k}{\partial R}\right)_{t,\mathbf{r}'}\left(\frac{\partial R}{\partial x_k}\right)_{t,\mathbf{r}'}\right\}dv' \\
&= -\int\left\{\frac{\partial}{\partial x_k'}\left(\frac{\partial j_k}{\partial R}\right) + \frac{1}{R}\left(\frac{\partial j_k}{\partial x_k'}\right)_{t,\mathbf{r}} - \frac{1}{R}\left(\frac{\partial j_k}{\partial R}\right)\left(\frac{\partial R}{\partial x_k'}\right)_{t,\mathbf{r}}\right\}dv' \\
&= \int\frac{1}{R}\left(\frac{\partial j_k}{\partial x_k'}\right)_t dv'
\end{aligned}$$

so finally

$$\frac{\mu_0}{4\pi}\nabla\cdot\mathbf{A} + \mu_0\epsilon_0\frac{\partial\phi}{\partial t} = \frac{\mu_0}{4\pi}\int\frac{1}{R}\left(\nabla'\cdot\mathbf{j} + \frac{\partial\rho}{\partial t'}\right)dv' = 0$$

which is the same as Eq. (14.7).

14.1 The Near Field and the Far Field

We shall suppose that the sources are situated within a bounded region and the field point is outside this region. Now if the distance of the field point from the source region be small enough such that in the time taken by the electromagnetic influence to travel the distance, i.e., R/c, the source characteristics ρ and $\mathbf{j}$ do not appreciably change, and we can replace t' by t in the integrals of Eqs. (14.22) and (14.23) and obtain

$$\phi(\mathbf{r},t) = \frac{1}{4\pi\epsilon_0}\int \frac{\rho\left(\mathbf{r}',t\right)}{|\mathbf{r}-\mathbf{r}'|}\,dv'$$

$$\mathbf{A}(\mathbf{r},t) = \frac{\mu_0}{4\pi}\int \frac{\mathbf{j}\left(\mathbf{r}',t\right)}{|\mathbf{r}-\mathbf{r}'|}\,dv' \tag{14.26}$$

These formulae are rigorously true only in the stationary case and are very nearly true if the distance $R = |\mathbf{r}-\mathbf{r}'|$ be sufficiently small. These give the already studied Coulomb and Biot-Savart fields, which fall off as the inverse square of the distance. We shall refer to these as the near fields.

In the other extreme, suppose that $|\mathbf{r}-\mathbf{r}'|$ is so large that considering a particular component in the Fourier integral, the corresponding periodic time T is much smaller than the time of travel $|\mathbf{r}-\mathbf{r}'|/c$, i.e., if anticipating our results, we introduce a wavelength corresponding to the periodic time T, i.e., $\lambda = cT$, then $\lambda \ll |\mathbf{r}-\mathbf{r}'|$. We shall also suppose that the dimensions of the source region are small compared to the distance of the field point so that $|\mathbf{r}'| < |\mathbf{r}|$. Under these conditions, we shall perform an expansion in the powers of r and retain only the lowest power of $1/r$. Thus,

$$|\mathbf{r}-\mathbf{r}'| = r - \frac{\mathbf{r}\cdot\mathbf{r}'}{r} \tag{14.27a}$$

$$\frac{1}{|\mathbf{r}-\mathbf{r}'|} = \frac{1}{r} \tag{14.27b}$$

$$\frac{\partial}{\partial x_i}e^{ik|\mathbf{r}-\mathbf{r}'|} = \frac{x_i}{r}\,e^{ik|\mathbf{r}-\mathbf{r}'|} \tag{14.27c}$$

From Eq. (14.23), taking the retarded potential,

$$\mathbf{A}(\mathbf{r},t) = \int_{-\infty}^{\infty} \mathbf{A}_\omega(\mathbf{r})\,e^{i\omega t}\,d\omega \tag{14.28}$$

$$= \frac{\mu_0}{4\pi}\int \mathbf{j}_\omega(\mathbf{r}')\,\frac{e^{i\omega(t-|\mathbf{r}-\mathbf{r}'|/c)}}{|\mathbf{r}-\mathbf{r}'|}\,dv'\,d\omega$$

$$\mathbf{A}_\omega(\mathbf{r}) = \frac{\mu_0}{4\pi}\int \frac{\mathbf{j}_\omega(\mathbf{r}')}{|\mathbf{r}-\mathbf{r}'|}e^{-i\omega|\mathbf{r}-\mathbf{r}'|/c}\,dv'$$

$$= \frac{\mu_0}{4\pi r}\int \mathbf{j}_\omega(\mathbf{r}')\,e^{-i\omega|\mathbf{r}-\mathbf{r}'|/c}\,dv' \tag{14.29}$$

$$\nabla\cdot\mathbf{A}_\omega(\mathbf{r}) = \frac{\mu_0}{4\pi r}\int\left(-\frac{i\omega}{c}\right)\frac{\mathbf{r}}{r}\cdot\mathbf{j}_\omega(\mathbf{r}')\,e^{-i\omega|\mathbf{r}-\mathbf{r}'|/c}\,dv'$$

$$= -\frac{i\omega}{cr}\,\mathbf{r}\cdot\mathbf{A}_\omega(\mathbf{r}) \tag{14.30}$$

The transition from Eqs. (14.29) to (14.30) is typical as showing that for large distances, the operator ∇ applied to a Fourier transform is equivalent to $-\frac{i\omega\mathbf{r}}{cr}$. Also, since $f(\mathbf{r}, t) = \int f_\omega(\mathbf{r})e^{i\omega t}$, the time derivative $\frac{\partial f}{\partial t} = \int i\omega f_\omega(\mathbf{r})e^{i\omega t}$, which implies that $\partial f \partial t = i\omega f_\omega$, i.e., the Fourier transform of the derivative of a function is simply the product of the Fourier transform of the function with $i\omega$. Using these results in the Lorentz gauge condition $\nabla \cdot \mathbf{A} + \frac{1}{c^2}\frac{\partial \phi}{\partial t} = 0$, we get

$$\phi_\omega = \frac{c}{r}\mathbf{r} \cdot \mathbf{A}_\omega \tag{14.31}$$

We next consider the relation $\mathbf{E} = -\nabla\phi - \dot{\mathbf{A}}$ and get

$$\mathbf{E}_\omega = i\omega\left(\frac{\mathbf{r}}{r^2}(\mathbf{r} \cdot \mathbf{A}_\omega) - \mathbf{A}_\omega\right) \tag{14.32}$$

Obviously,

$$\mathbf{E}_\omega \cdot \mathbf{r} = 0 \tag{14.33}$$

Also, from $\mathbf{B} = \nabla \times \mathbf{A}$,

$$\mathbf{B}_\omega = -\frac{i\omega}{cr}\mathbf{r} \times \mathbf{A}_\omega \tag{14.34}$$

Hence, from (14.32) and (14.34),

$$\frac{1}{r}\mathbf{r} \times \mathbf{E}_\omega = -\frac{i\omega}{r}\mathbf{r} \times \mathbf{A}_\omega = c\mathbf{B}_\omega \tag{14.35}$$

showing that $\mathbf{H}$, $\mathbf{E}$ and $\mathbf{r}$ are mutually perpendicular and $|\mathbf{E}| = c|\mathbf{B}|$. Thus, the distant field has the same characteristic as those of the wave field we have already studied.

The wave or radiation field is characterized by the vanishing of $\mathbf{E} \cdot \mathbf{B}$ and $E^2 - c^2B^2$. They are scalars not only for purely spatial coordinate transformations but also are also Lorentz invariants. By this, we mean that if these two vanish in any frame of reference, they will vanish in all frames of reference in uniform relative motion. However, this holds for only pure radiation fields. In other cases, it may happen that in one frame there is a non-vanishing Poynting vector. One is, thus, led to conclude the presence of a flux of radiation, but going over to another frame in relative motion, the Poynting vector may be made to vanish, and thus, the radiation flux proves illusory.

The energy flux corresponding to a particular frequency may be easily calculated. The Poynting vector is

$$\frac{\mathbf{r}}{c\mu_0 r}|\mathbf{E}|^2$$

Now both the $\mathbf{E}_\omega$ and $\mathbf{E}_{-\omega}$ terms contribute to the same frequency ν. Hence,

$$
\begin{aligned}
|\mathbf{E}_\omega|^2 &= \left(E_\omega e^{-i\omega t} + E_\omega e^{i\omega t}\right)^2 \\
&= E_\omega^2 e^{-2i\omega t} + E_\omega^2 e^{2i\omega t} + 2|E_\omega|^2 \\
&= 2c^2\mu_0^2|H_\omega|^2
\end{aligned}
$$

The first two terms (in the last but one line) vanish on taking the time average. Using the expression (14.34), the Poynting vector is

$$
\begin{aligned}
\frac{2\omega^2\mathbf{r}}{c\mu_0 r^3}|\mathbf{r}\times\mathbf{A}_\omega|^2 &= \frac{\omega^2\mu_0\mathbf{r}}{8\pi^2 c r^3}\left(\int \frac{\mathbf{r}\times\mathbf{j}_\omega}{|\mathbf{r}-\mathbf{r}'|} e^{-i\omega|\mathbf{r}-\mathbf{r}'|/c} dv'\right)^2 \\
&= \frac{\omega^2\mu_0\mathbf{r}}{8\pi^2 c r^3}\left(\int j_{\omega\perp}\, e^{-i\omega|\mathbf{r}-\mathbf{r}'|/c} dv'\right)^2
\end{aligned}
\tag{14.36}
$$

where the symbol $\perp$ indicates the component orthogonal to $\mathbf{r}$. Note that the Poynting vector falls off as $1/r^2$ and hence its integral over the sphere at infinity does not vanish. Note also that the Poynting vector at a point is non-vanishing only if the source current has a component orthogonal to $\mathbf{r}$.

Problem

1. Show that the power radiated in the solid angle $d\Omega$ due to a current $I \sin\frac{\pi z}{l}\cos\omega t$ in a thin wire conductor extending from $z = 0$ to l is $\frac{\omega^4 l^2 I^2}{32\pi^2 c^2}\sqrt{\frac{\mu_0}{\epsilon_0}}\sin^2\theta\, d\Omega$, where θ being the angle between the z-axis and $\mathbf{r}$.

Chapter 15
Analysis of the Radiation Field

In case of static fields, the electrostatic potential is given by

$$\phi(\mathbf{r}) = \frac{1}{4\pi\epsilon_0} \int \frac{\rho(\mathbf{r}')dv'}{|\mathbf{r} - \mathbf{r}'|}$$

By performing a Taylor series expansion of $1/|\mathbf{r} - \mathbf{r}'|$ about the origin ($\mathbf{r}' = 0$), we could split up the potential into terms which were called monopole, dipole, etc., fields. In a similar manner, we shall perform a series expansion of $\exp\left(i\omega|\mathbf{r} - \mathbf{r}'|/c\right)$ (in powers of r') occurring in the expression for the potential of the distant field. As we have already seen, this distant field is a radiation field and we shall, thus, obtain a splitting up of the radiation field into a series, the separate terms of which would be identified as due to temporal variations of the electric and magnetic dipole, quadrupole and higher multipole moments of the source distribution. We shall, thus, use the words dipole, quadrupole, etc., to describe the radiation where in reality, we are describing the property of the source of radiation. However, it is important to note one significant difference—in the static case, the different multipole fields fall off as different powers of $1/r$, and in the case of the radiation field, the falling off is as $1/r$ for all types. (In fact, any different type of $1/r$ dependence is not possible for radiation fields—a slower rate of fall would make the energy flux through the sphere at infinity blow up while any faster rate would make the energy flux vanish.) However, as in the case of static fields, the angular distribution of radiation fields will have typically different behaviours for different multipoles. (In case of natural sources like atoms or nuclei, we get the radiation coming from a multitude of sources, the angular dependence is smoothed out and we observe an isotropic radiation.)

The Fourier components of the radiation field potentials can be written down using Eq. (14.24) and using $|\mathbf{r} - \mathbf{r}'| \simeq r - (\mathbf{r} \cdot \mathbf{r}')/r + \cdots$ as

A. K. Raychaudhuri, *Classical Theory of Electricity and Magnetism*, Texts and Readings in Physical Sciences 21, https://doi.org/10.1007/978-981-16-8139-4_15

$$\phi_\omega(\mathbf{r}) = \frac{1}{4\pi\epsilon_0 r} e^{-i\omega r/c} \int \rho_\omega(\mathbf{r}')\, e^{i\omega \mathbf{r}\cdot\mathbf{r}'/cr}\, dv' \tag{15.1}$$

$$\mathbf{A}_\omega(\mathbf{r}) = \frac{\mu_0}{4\pi r} e^{-i\omega r/c} \int \mathbf{j}_\omega(\mathbf{r}')\, e^{i\omega \mathbf{r}\cdot\mathbf{r}'/cr}\, dv' \tag{15.2}$$

We use the Taylor expansion of the exponential

$$e^{i\omega \mathbf{r}\cdot\mathbf{r}'/cr} = \sum_s \frac{1}{s!} \left(\frac{i\omega}{c} \frac{\mathbf{r}\cdot\mathbf{r}'}{r} \right)^s$$

and analyse the effect of the first few terms in detail. To this end, let us single out

$$\frac{e^{-i\omega r/c}}{r} \equiv f(r)$$

which gives the distance dependence—the exponential part merely makes the functions (hence the physical quantities with this dependence) oscillate, while falling off as $1/r$.

I. *The term* $s = 0$*:*
We shall see that this term arises due to oscillations of the electric dipole moment. From Eqs. (14.27), (14.29) and (15.2) above,

$$\begin{aligned} \mathbf{E}_{0\omega}(\mathbf{r}) &= i\omega\ ((\mathbf{n}\cdot\mathbf{A}_\omega)\mathbf{n} - \mathbf{A}_\omega)_0 \\ &= \frac{\mu_0}{4\pi} i\omega f(r) \int \left((\mathbf{n}\cdot\mathbf{j}_\omega(\mathbf{r}'))\mathbf{n} - \mathbf{j}_\omega(\mathbf{r}') \right) dv' \end{aligned} \tag{15.3}$$

$$\begin{aligned} \mathbf{B}_{0\omega}(\mathbf{r}) &= -\frac{i\omega}{c}\ (\mathbf{n}\times\mathbf{A}_\omega)_0 \\ &= -\frac{\mu_0}{4\pi}\frac{i\omega}{c} f(r) \int \mathbf{n}\times\mathbf{j}_\omega(\mathbf{r}')\, dv' \end{aligned} \tag{15.4}$$

Note that both $\mathbf{E}$ and $\mathbf{B}$ depend only on the component of $\mathbf{j}$ normal to $\mathbf{n}$, where $\mathbf{n} = \mathbf{r}/r$ is the unit vector in the direction towards the field point and the subscript zero indicates that we are considering the $s = 0$ term in the power series expansion.

In the stationary case, we have seen that $\int \mathbf{j}\, dv = 0$; however, the proof depended on the condition $\nabla\cdot\mathbf{j} = 0$. In this non-stationary case, the charge conservation equation reads $\nabla\cdot\mathbf{j} + \dot\rho = 0$, so that

$$\int \nabla\cdot(\mathbf{j}x_i)\, dv = \int (x_i \nabla\cdot\mathbf{j} + (\mathbf{j}\cdot\nabla)\, x_i)\, dv = \int (-x_i\dot\rho + j_i)\, dv$$

The volume integral on the left may be converted into a surface integral, the surface lying outside the bounded source distribution. Thus, it vanishes and we get

$$\int j_i\, dv = \int x_i \frac{\partial \rho}{\partial t}\, dv = \frac{\partial}{\partial t} \int \rho x_i dv \tag{15.5}$$

Thus, $\int \mathbf{j}\, dv$ represents the time rate of change of dipole moment of the charge distribution.

The easiest way of finding the nature of the angular dependence of radiation flux (i.e., the Poynting vector) is to remember that in the radiation field $|\mathbf{E}| = |\mathbf{B}|$ and then to note from Eq. (15.4) that $|\mathbf{B}| \sim \int j_\omega \sin\theta\, dv$, where θ is the angle between $\mathbf{j}$ and $\mathbf{n}$. Hence, the intensity of radiation would vary as $\sin^2\theta$ and would be as shown in Fig. 15.1. The polarization of the radiation is determined by the condition that $\mathbf{B}$ is normal to both $\mathbf{j}$ and $\mathbf{n}$, while $\mathbf{E}$ is normal to $\mathbf{B}$ and $\mathbf{n}$.

II. *The term* $s = 1$*:*

The radiation fields are

$$\mathbf{E}_{1\omega}(\mathbf{r}) = \frac{\mu_0}{4\pi} \frac{(i\omega)^2}{c} f(r) \int \left((\mathbf{n} \cdot \mathbf{j}_\omega)(\mathbf{n} \cdot \mathbf{r}')\mathbf{n} - \mathbf{j}_\omega(\mathbf{r}')(\mathbf{n} \cdot \mathbf{r}') \right) dv' \tag{15.6}$$

$$\mathbf{B}_{1\omega}(\mathbf{r}) = -\frac{\mu_0}{4\pi} \left(\frac{i\omega}{c} \right)^2 f(r) \int \left(\mathbf{n} \times \mathbf{j}_\omega(\mathbf{r}') \right) (\mathbf{n} \cdot \mathbf{r}')\, dv' \tag{15.7}$$

In the stationary case, we have seen that $\int x_i j_k\, dv = -\int x_k j_i\, dv$, assuming $\nabla \cdot \mathbf{j} = 0$. But now

$$\begin{aligned} 0 &= \int \nabla \cdot (\mathbf{j}\, x_i x_k)\, dv = \int ((\mathbf{j} \cdot \nabla)\, x_i x_k + (\nabla \cdot \mathbf{j})\, x_i x_k)\, dv \\ &= \int (j_i x_k + j_k x_i)\, dv - \int \frac{\partial \rho}{\partial t} x_i x_k\, dv \end{aligned}$$

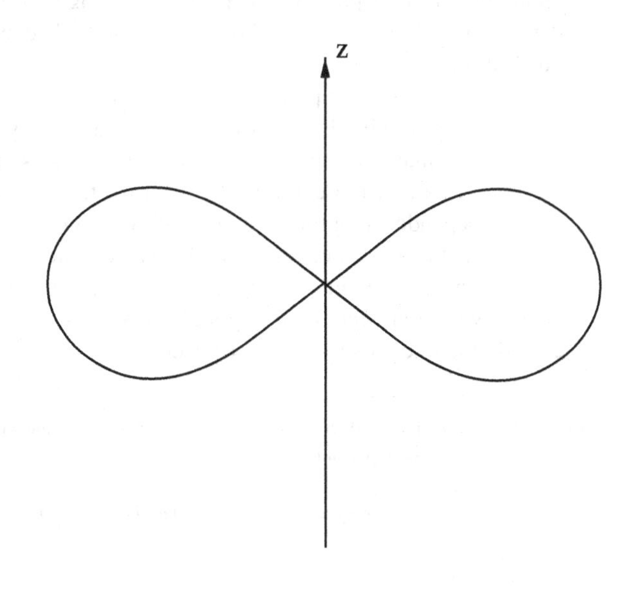

Fig. 15.1 Polar curve showing angular radiation pattern $\sim \sin^2\theta$ of an electric dipole radiation, where the current density $\mathbf{j}$ is along the z-axis

Thus, the symmetric tensor $\int (j_i x_k + j_k x_i)\,dv$ gets related to the time rate of change of the quadrupole moment of the charge distribution as follows:

$$\int (j_i x_k + j_k x_i)\,dv = \frac{\partial}{\partial t}\int \rho x_i x_k\,dv = \frac{\partial}{\partial t} Q_{ik}$$

We have already seen that the magnetic dipole moment of a current distribution is $\frac{1}{2}\int (\mathbf{r}\times\mathbf{j})\,dv$, when written in tensor notation, it is of the form $\frac{1}{2}\int (j_i x_k - j_k x_i)\,dv$, which is just the antisymmetric tensor corresponding to the symmetric tensor we have linked up with the time rate of change of electric quadrupole moment. We may separate out the symmetric and antisymmetric[1] parts.
Then the first integral in Eq. (15.6) is purely symmetric. The integrand is $n_i n_k j_i x_k'$, which is equal to $n_k n_i j_k x_i'$, hence the symmetry follows, while the integrand in the second integral is identical with the integral in Eq. (8.15). Hence, considering the antisymmetric part only, we get

$$\mathbf{E}_{1\omega}(\mathbf{r}) = \frac{\mu_0}{4\pi}\frac{(i\omega)^2}{c} f(r)\,\mathbf{n}\times\boldsymbol{\mu}_\omega \tag{15.8}$$

We can similarly consider the antisymmetric part of Eq. (15.7) to get the corresponding magnetic field

$$\mathbf{B}_{1\omega}(\mathbf{r}) = \frac{\mu_0}{4\pi}\left(\frac{i\omega}{c}\right)^2 f(r)\,\left((\mathbf{n}\cdot\boldsymbol{\mu}_\omega)\mathbf{n} - \boldsymbol{\mu}_\omega\right) \tag{15.9}$$

Thus, apart from a multiplying factor to which we shall return a little later, we see that Eqs. (15.8) and (15.9) are identical to Eqs. (15.3) and (15.4), with $\boldsymbol{\mu}$ taking the place of $\mathbf{j}$ and $\mathbf{E}$ and $\mathbf{B}$ interchanging their roles. Hence, the angular dependence of this magnetic dipole radiation will be the same as that of electric quadrupole radiation.
So far as the symmetric part (which arises from the electric quadrupole moment) is concerned, the simplest possible source that we may consider is a straight line current. We may choose one of the coordinate axes along the direction of the current. A look at the expression (15.7) for the magnetic field shows that the angular dependence of the amplitude of the magnetic field will be $\sim \sin\theta\cos\theta$, where θ is the angle between $\mathbf{n}$ and $\mathbf{j}$. Hence, the intensity of radiation would vary as $\sin^2\theta\cos^2\theta$, i.e., as $\sin^2 2\theta$. The polar diagram is shown in Fig. 15.2.
One may calculate the fields explicitly. Taking the x-axis in the direction of the current, the integrals in Eq. (15.6) become

[1] Any tensor quantity can be decomposed as a sum of two parts, one which is symmetric and another which is antisymmetric. For example,

$$(\mathbf{n}\cdot\mathbf{r}')\mathbf{j} = \frac{1}{2}\left((\mathbf{n}\cdot\mathbf{r}')\mathbf{j} + (\mathbf{n}\cdot\mathbf{j})\,\mathbf{r}'\right) + \frac{1}{2}(\mathbf{r}'\times\mathbf{j})\times\mathbf{n}$$

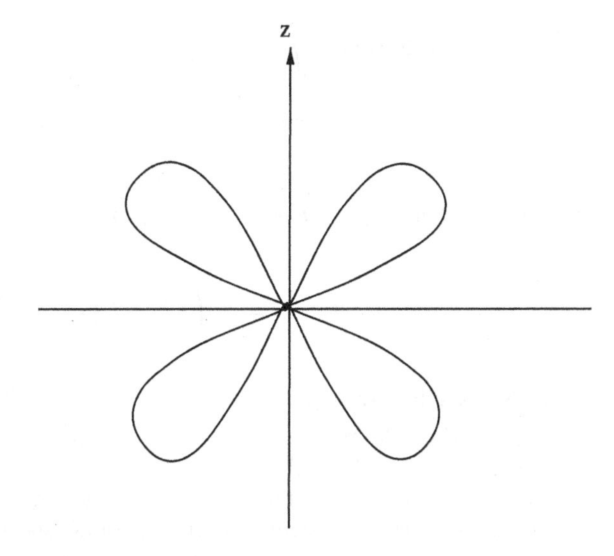

Fig. 15.2 Polar curve showing angular radiation pattern of an electric quadrupole radiation, where the current density **j** is along the z-axis

$$\int \left(n_x^2 jx'\mathbf{n} - \mathbf{j}n_x x'\right) dv' = \int jx' \left(n_x^2\mathbf{n} - n_x\hat{\imath}\right) dv'$$

where $\hat{\imath}$ is the unit vector in the x-direction, so that

$$\begin{aligned}
E_x &= \frac{\mu_0}{4\pi}\frac{(i\omega)^2}{c} f(r)(\cos^3\theta - \cos\theta) \int jx'\, dv' \\
E_y &= \frac{\mu_0}{4\pi}\frac{(i\omega)^2}{c} f(r)\cos^2\theta \sin\theta\cos\phi \int jx'\, dv' \\
E_z &= \frac{\mu_0}{4\pi}\frac{(i\omega)^2}{c} f(r)\cos^2\theta \sin\theta\sin\phi \int jx'\, dv'
\end{aligned}$$

Similarly from Eq. (15.7)

$$\begin{aligned}
B_x &= 0 \\
B_y &= -\frac{\mu_0}{4\pi}\left(\frac{i\omega}{c}\right)^2 f(r)\sin\theta\cos\theta\sin\phi \int jx'\, dv' \\
B_z &= \frac{\mu_0}{4\pi}\left(\frac{i\omega}{c}\right)^2 f(r)\sin\theta\cos\theta\cos\phi \int jx'\, dv'
\end{aligned}$$

A point of some interest is to see the order of time derivative that is effective for the radiation field in different cases. Considering Eq. (15.3)

$$\begin{aligned}\mathbf{E}(\mathbf{r},t) &= \int_{-\infty}^{+\infty} \mathbf{E}_\omega(\mathbf{r})\, e^{i\omega t}\, d\omega \\ &= \frac{\mu_0}{4\pi r}\int\int i\omega e^{i\omega t}\, ((\mathbf{n}\cdot\mathbf{j}_\omega)\,\mathbf{n} - \mathbf{j}_\omega)\, dv'\, d\omega\end{aligned}$$

Now note that $\frac{\partial}{\partial t}\mathbf{j}\left(\mathbf{r}', t - \frac{R}{c}\right) = i\int \mathbf{j}_\omega(\mathbf{r}')\,\omega e^{i\omega(t-R/c)}\, d\omega$. Hence,

$$\mathbf{E} = \frac{\mu_0}{4\pi r}\int\left(\left(\mathbf{n}\cdot\frac{\partial\mathbf{j}}{\partial t}\right)\mathbf{n} - \frac{\partial\mathbf{j}}{\partial t}\right) dv'$$

Thus, the effective source of the field is $\int \frac{\partial\mathbf{j}}{\partial t}\, dv'$. If $\mathbf{j} = \rho\mathbf{v} = \rho\dot{\mathbf{r}}'$, then $\frac{\partial\mathbf{j}}{\partial t} = \frac{\partial}{\partial t}(\rho\dot{\mathbf{r}}')$. Thus, the source is the second derivative of the dipole moment. This result is equivalent to what we shall see in a future chapter that a charge in uniform motion does not radiate, but it does radiate when in accelerated motion. These results are consistent with the idea of the special theory of relativity that a uniform velocity has no absolute significance and can be transformed away. As the second-order time derivative gives rise to the dipole radiation, the fields vary as υ^2, so far as different Fourier components are concerned. Consequently, the intensity of radiation varies inversely as the fourth power of wavelength in the case of dipole radiation.
If we examine the fields in Eqs. (15.6) and (15.7), we see that the effective source involves $\partial^2\mathbf{j}/\partial t^2$, hence considering that $\mathbf{j}$ itself involves the first derivative with respect to time, we see that the field is due to the third-order derivative of the electric quadrupole moment. However, the magnetic dipole moment of a current involves the current itself; therefore, the second-order time derivative of the magnetic dipole moment is involved.

The series expansion in powers of $\frac{\omega}{c}\frac{\mathbf{r}\cdot\mathbf{r}'}{r}$ is meaningful and significant only if this be small compared to unity. This requires that r' is small compared to the relevant wavelength. Again the whole radiation field theory depends on the condition that the distance r is much greater than the wavelength. Hence, this analysis of the radiation field is significant only if the source system is bounded within a region small compared to the wavelength of the radiation concerned; the observation point, however, must be at a distance large compared to the wavelength. This condition is satisfied for atoms and molecules emitting optical radiation (source dimension $\sim 10^{-10}$ m and $\lambda \sim 10^{-7}$ m) or nuclei emitting X-rays or γ-rays (source dimension $\sim 10^{-15}$ m and $\lambda \sim 10^{-10}$ m).

The field expressions involve the factor $(\omega r'/c)^s$ in the s-th term of the power series. Since r'/λ is small, the successive terms will decrease and the energy flux being proportional to the square of the field strengths would decrease quite fast as we go to higher values of s. In practice, one needs to consider only the radiation of the lowest order, i.e., the electric dipole radiation, unless this radiation is specifically forbidden. In fact, the following argument shows that in a *closed system*, in which charge density everywhere bears a constant ratio to the mass density, neither the electric dipole nor the magnetic dipole radiation can appear. For in such a system, electric dipole moment

$$\mathbf{p} = \int \rho_e \mathbf{r}\, dv = \alpha \int \rho_m \mathbf{r}\, dv$$

where ρ_e is the charge density, ρ_m the mass density and $\alpha = \rho_e/\rho_m$ is a constant. Now, because of conservation of linear momentum,

$$\frac{d}{dt} \int \rho_m \mathbf{r}\, dv = \text{constant}$$

Hence,

$$\frac{d^2}{dt^2} \int \rho_e \mathbf{r}\, dv = \alpha \frac{d^2}{dt^2} \int \rho_m \mathbf{r}\, dv = 0$$

Similarly, the magnetic dipole moment

$$\mu = \frac{1}{2} \int \rho_e (\mathbf{r} \times \mathbf{u})\, dv = \frac{\alpha}{2} \int \rho_m (\mathbf{r} \times \mathbf{u})\, dv = \text{constant}$$

because of the conservation of angular momentum. This consideration is, however, not of relevance for the magnetic dipole radiation from atoms as the ratio of the magnetic moment to the angular momentum is different for the orbital motion and the spin of the electron.

A situation, not covered by our method of analysis, is met with when we consider the interaction between the electrons in the atom and the nucleus. Here, if the wavelength considered is comparable with atomic dimensions, our approximation breaks down. However, for particles executing oscillations, if the velocity is not negligible compared to c, the dimension of the system (which is of the order of its velocity times the time period) may not be small compared to the wavelength, which is c times the time period. There a description in terms of the multipole radiations is no longer tenable—this is what happens when an electron is in an orbit or executes a simple harmonic motion with a velocity v that is comparable to c.

15.1 The Hertz Vector and Hertz's Method of Analysis

In the electrostatic case, we arrived at the dipole and higher multipole moments by two methods: (i) by a Taylor series expansion and (ii) by solving Laplace's equation with the help of spherical harmonics. In a similar manner, the radiation field can be analysed into dipole and higher multipole fields either by the series expansion method we have just described or by a method utilizing a 'superpotential' introduced by Hertz, where the spherical harmonics appear directly.

Let us introduce two vectors **p**, called the polarization vector, and **Z**, the Hertz potential, defined by the following equations:

$$\rho = -\nabla \cdot \mathbf{p} \tag{15.10}$$

$$\mathbf{j} = \frac{\partial \mathbf{p}}{\partial t} \tag{15.11}$$

$$\phi = -\nabla \cdot \mathbf{Z} \tag{15.12}$$

$$\mathbf{A} = \mu\epsilon \frac{\partial \mathbf{Z}}{\partial t} \tag{15.13}$$

The beauty of these relations is that the charge conservation relation, as well as the Lorentz gauge condition, are implicit in them and so they may be said to be automatically satisfied. Substituting these relations in the inhomogeneous equations for the scalar and vector potential, we get

$$\Box\phi = -\frac{\rho}{\epsilon_0} \rightarrow \Box(\nabla \cdot \mathbf{Z}) = \frac{1}{\epsilon_0}\nabla \cdot \mathbf{p}$$

$$\text{or} \qquad \nabla \cdot \left(\Box\mathbf{Z} + \frac{1}{\epsilon_0}\mathbf{p}\right) = 0 \tag{15.14}$$

$$\Box\mathbf{A} = -\mu_0\mathbf{j} \rightarrow \Box\left(\frac{\partial \mathbf{Z}}{\partial t}\right) = -\frac{1}{\epsilon_0}\frac{\partial \mathbf{p}}{\partial t}$$

$$\text{or} \qquad \frac{\partial}{\partial t}\left(\Box\mathbf{Z} + \frac{1}{\epsilon_0}\mathbf{p}\right) = 0 \tag{15.15}$$

Both Eqs. (15.14) and (15.15) are satisfied if

$$\Box\mathbf{Z} = -\frac{1}{\epsilon_0}\mathbf{p} \tag{15.16}$$

Conversely, if Eq. (15.16) is satisfied, both the inhomogeneous equations for the potentials will be satisfied. Thus, Eq. (15.16) in a sense unifies the equations for ϕ and **A**. However, the number of equations in Eq. (15.16) is only three, while the equations for the potentials are four in number. The reason is that because of the Lorentz gauge condition and the conservation of charge, the four equations for the potentials are not completely independent—this has been taken advantage of in defining **p** and **Z**.

The solution of Eq. (15.16) is already known in the form

$$\mathbf{Z}(\mathbf{r}, t) = \frac{1}{4\pi\epsilon_0}\int \frac{\mathbf{p}\left(\mathbf{r}', t - \frac{|\mathbf{r}-\mathbf{r}'|}{c}\right)}{|\mathbf{r}-\mathbf{r}'|}\, dv' \tag{15.17}$$

or in terms of its Fourier components

$$\mathbf{Z}_\omega(\mathbf{r}) = \frac{1}{4\pi\epsilon_0}\int \frac{\mathbf{p}_\omega(\mathbf{r}')\, e^{-i\omega|\mathbf{r}-\mathbf{r}'|/c}}{|\mathbf{r}-\mathbf{r}'|}\, dv' \tag{15.18}$$

We shall now introduce the two conditions.

(i) The wavelength is large compared to the dimensions of the source region, i.e., $\frac{\omega r'}{c} \ll 1$.
(ii) The field point is at a distance large compared to the wavelength, i.e., $\frac{\omega r}{c} \gg 1$.

With these conditions satisfied, Eq. (15.18) may be expressed as a series. We have the well-known formula

$$\exp\left(\frac{i\omega r'}{c}\cos\Theta\right) = \sum_{l=0}^{\infty} i^l (2l+1)\, j_l\left(\frac{\omega r'}{c}\right) P_l(\cos\Theta) \tag{15.19}$$

where j_l are the spherical Bessel functions, which may be written in terms of the Bessel functions of half-integral order

$$j_l\left(\frac{\omega r'}{c}\right) = \sqrt{\frac{\pi c}{2\omega r'}}\, J_{l+\frac{1}{2}}\left(\frac{\omega r'}{c}\right) \approx \frac{2^l l!}{(2l+1)!}\left(\frac{\omega r'}{c}\right)^l$$

for $\frac{\omega r'}{c} \ll 1$. Hence, Eq. (15.18) may be written as

$$\mathbf{Z}_\omega(\mathbf{r}) = \sum_l \mathbf{Z}_\omega^{(l)}(\mathbf{r}) \tag{15.20}$$

$$\mathbf{Z}_\omega^{(l)}(\mathbf{r}) = \frac{e^{-i\omega r/c}}{4\pi\epsilon_0 r}\frac{(2i)^l l!}{(2l)!}\int \left(\frac{\omega r'}{c}\right)^l \mathbf{p}_\omega(\mathbf{r}') P_l(\cos\Theta)\, dv' \tag{15.21}$$

The angle Θ is the angle between $\mathbf{r}$ and $\mathbf{r}'$, so that in general if (θ, φ) are the angular coordinates for $\mathbf{r}$ and (θ', φ') are those for $\mathbf{r}'$, the Legendre polynomial $P_l(\cos\Theta)$ is given by

$$P_l(\cos\Theta) = \sum_{m=-l}^{+l} (-1)^m P_l^m(\cos\theta)\, P_l^{-m}(\cos\theta')\, e^{im(\varphi-\varphi')}$$

The different terms in the sum (15.20), i.e., $\mathbf{Z}_\omega^{(l)}$ give the different multipole fields, as may be verified fairly easily. If the source distribution be linear in space as is the case for a simple dipole, we may take the dipole direction as the polar axis, then $\Theta = \theta$. Thus, for $l = 0$, we get

$$\mathbf{Z}_\omega^{(0)} = \frac{e^{-i\omega r/c}}{4\pi\epsilon_0 r}\int \mathbf{p}_\omega(\mathbf{r}')\, dv' \tag{15.22}$$

Note that

$$0 = \int \nabla\cdot(\mathbf{p}x_i')\, dv' = \int \left(x_i'\nabla\cdot\mathbf{p} + (\mathbf{p}\cdot\nabla)x_i'\right) dv' = -\int \left(\rho x_i' - p_i\right) dv'$$

Thus, in Eq. (15.22), the source term is simply the electrical dipole moment and the components of the Hertz vector $\mathbf{Z}_{\omega}^{(0)}$ are

$$Z_{\omega r}^{(0)} = |Z_{\omega}^{(0)}| \cos\theta = \frac{|\boldsymbol{\mu}_{\omega}| e^{-i\omega r/c}}{4\pi\epsilon_0 r} \cos\theta \tag{15.23}$$

$$Z_{\omega\theta}^{(0)} = \frac{|\boldsymbol{\mu}_{\omega}| e^{-i\omega r/c}}{4\pi\epsilon_0 r} \sin\theta, \qquad Z_{\omega\varphi}^{(0)} = 0 \tag{15.24}$$

where $\boldsymbol{\mu}_{\omega} = \int \rho_{\omega} \mathbf{r}' \, dv'$ and $\mu = |\boldsymbol{\mu}_{\omega}|$ is the magnitude of μ_{ω}.

In order to find the explicit expressions for the magnetic and electric fields, define a vector $\boldsymbol{\xi}$ by

$$\boldsymbol{\xi} = \nabla \times \mathbf{Z} \tag{15.25}$$

Now we now have from Eqs. (15.12), (15.13) and (15.16)

$$\mathbf{B} = \mu_0 \epsilon_0 \dot{\boldsymbol{\xi}} \tag{15.26}$$

$$\mathbf{E} = \nabla \times \boldsymbol{\xi} - \frac{1}{\epsilon_0} \mathbf{p} \tag{15.27}$$

Using Eq. (15.24) in Eq. (15.25), we find that the only non-vanishing component of $\boldsymbol{\xi}$ is

$$\xi_{\phi} = \frac{\mu}{4\pi\epsilon_0} \left(\frac{1}{r^2} + \frac{i\omega}{cr} \right) \sin\theta \, e^{i\omega(t - r/c)} \tag{15.28}$$

where we have restored the time dependence. Henceforth, we shall assume that the sources vary in a simple harmonic manner so that only one Fourier component has to be considered. Equations (15.26) and (15.27) now determine the electric and magnetic fields

$$\begin{aligned}
B_r &= B_\theta = E_\varphi = 0 \\
B_\varphi &= \frac{\mu_0}{4\pi} i\omega\mu \sin\theta \left(\frac{1}{r^2} + \frac{i\omega}{cr} \right) e^{i\omega(t - r/c)} \\
E_r &= \frac{1}{2\pi\epsilon_0} \mu \cos\theta \left(\frac{1}{r^3} + \frac{i\omega}{cr^2} \right) e^{i\omega(t - r/c)} \\
E_\theta &= \frac{1}{4\pi\epsilon_0} \mu \sin\theta \left(\frac{1}{r^3} + \frac{i\omega}{cr^2} - \frac{\omega^2}{c^2 r} \right) e^{i\omega(t - r/c)}
\end{aligned}$$

Note that the polarization vector $\mathbf{p}$ vanishes everywhere except at the origin, so that the above expressions give the field at all points except at the origin, where in any case the field blows up. The study of this oscillating dipole field is facilitated by considering three distinct regions according to the values of $r\omega/c$, i.e., of r/λ. (It is permissible to consider small values of $r\omega/c$ even though we have assumed r'/r to be small, because the source is now just a singularity at the origin with $r' \to 0$.)

Very close to the origin $r\omega/c \ll 1$, so if we retain only the highest power of $c/(r\omega)$, the fields become

$$\begin{aligned}
E_r &= \frac{1}{4\pi\epsilon_0}\frac{2\mu\cos\theta}{r^3}e^{-i(\omega(r/c-t)+\phi_{\rm el})} = \frac{1}{4\pi\epsilon_0}\frac{2\mu\cos\theta}{r^3}\cos(\omega t' - \phi_{\rm el}) \\
E_\theta &= \frac{1}{4\pi\epsilon_0}\frac{\mu\sin\theta}{r^3}e^{-i(\omega(r/c-t)+\phi_{\rm el})} = \frac{1}{4\pi\epsilon_0}\frac{\mu\sin\theta}{r^3}\cos(\omega t' - \phi_{\rm el}) \\
E_\phi &= B_r = B_\theta = 0 \\
B_\phi &= -\frac{\mu_0}{4\pi}\frac{\omega\mu\sin\theta}{r^2}\sin(\omega t')
\end{aligned}$$

where $t' = t - \frac{r}{c}$ and $\phi_{\rm el} = \tan^{-1}\left(-\frac{\omega r}{c}\right) \to 0$ as $r \to 0$.

The interpretation is simple. The electric field is the same as the static field due to a dipole whose moment at any instant is $\mu\cos(\omega t - \phi_{\rm el})$ directed along the polar axis, where the phase difference $\phi_{\rm el}$ tends to vanish as $r \to 0$. The magnetic field is due to an infinitesimal current element at the origin directed along the polar axis, while the strength of the current is $\frac{d}{dt}\left(\mu\cos\omega t'\right)$, so that it can be interpreted as linked with the oscillation of the dipole moment. However, the strength of the magnetic field is relatively small. For all these reasons, this region is sometimes referred to as the static zone.

At the other extreme where $r\omega/c \gg 1$, i.e., r is much greater than the wavelength, retaining again only the lowest power of $c/(\omega r)$, the fields assume the form

$$\begin{aligned}
E_\theta &= -\frac{1}{4\pi\epsilon_0}\frac{\omega^2\mu}{c^2 r}\sin\theta\cos\left[\omega\left(\frac{r}{c}-t\right)\right], \quad E_r = E_\phi = 0 \\
B_\varphi &= -\frac{\mu_0}{4\pi}\frac{\omega^2\mu}{cr}\sin\theta\cos\left[\omega\left(\frac{r}{c}-t\right)\right], \quad B_r = B_\theta = 0
\end{aligned}$$

Thus, we have the typical form of the radiation field, with **E**, **B** and **r** orthogonal to each other, **E** and **B** in phase and the magnitude of **E** being c times the magnitude of the magnetic field. One may calculate the rate of radiation of energy by integrating the Poynting vector over the sphere at infinity. This turns out to be

$$\frac{2\mu^2\omega^4}{3c^3}\cos^2\left[\omega\left(\frac{r}{c}-t\right)\right]$$

which is the result for an oscillating dipole. As $\mu\cos\left(\omega(t - r/c)\right)$ is the dipole moment at the retarded time $t - r/c$, we may interpret the result in the following manner. The rate of loss of energy by radiation for an oscillating dipole is

$$\frac{2}{3c^3}\times\left[\frac{d^2}{dt^2}(\text{dipole moment})\right]^2$$

Let us recall our earlier result that the dipole radiation depends on the second-order time derivative of the dipole moment.

In between this radiation zone and the static zone, in the region where r is comparable with the wavelength, one has to consider the complete expressions for the field. In this region, the electric field is not transverse ($E_r \neq 0$); the magnetic field, however, is transverse throughout. The latter fact is easy to understand from the Biot-Savart law). The Poynting vector has both r- and θ-components, and during a complete cycle, the flux through any closed surface must equal the total energy radiated during the cycle. This intermediate zone is called the *induction* or *transition zone*—it extends over a distance several times the wavelength. Therefore, while for atomic systems, the optical wavelength being $\sim 10^{-5}$ cm, all observations are in the radiation zone; for radio waves, observation in the induction zone, which may extend over several hundred metres, is quite easy.

Problems

1. Show that for the advanced field, energy flows inwards from infinity. Also that for half advanced and half retarded field, the energy flux vanishes.
2. Show that the $l = 1$ term in the Hertz potential gives rise to magnetic dipole and electric quadrupole radiation.
3. By direct calculation, verify that the flux of energy through a closed surface in the induction zone of the oscillating dipole field is equal to the energy flux through the sphere at infinity on the average.
4. Obtain the equation of the lines of electric force in the induction zone. Hence or otherwise show that the phase velocity in this region exceeds c.

Chapter 16
Field Due to a Moving Charged Particle

Our problem is to investigate the field due to a moving charged system—the dimensions of the region in which the charge is situated being so small compared to the distance from the field point that the charged system may be considered to be a particle and the source described by a Dirac δ-function. Often the field we are going to find is spoken of as the field due to a moving electron but it should be understood that the theory leaves the electron itself as an unanalysed singularity.

The most direct method of solving our problem would be to put in the δ-function distribution in our formula for the retarded potentials. Thus, in this case,

$$\rho(\mathbf{r}', t') = e\,\delta(\mathbf{r}' - \mathbf{f}(t')) \tag{16.1}$$

where $\mathbf{f}(t')$ is a vector function of time giving the position of the particle whose charge is e. Hence, the required potential is

$$\phi(\mathbf{r}, t) = \frac{e}{4\pi\epsilon_0}\int \frac{\delta(\mathbf{r}' - \mathbf{f}(t'))}{|\mathbf{r} - \mathbf{r}'|}\,dv' \tag{16.2}$$

where t' is the retarded time $t' = t - |\mathbf{r} - \mathbf{r}'|/c$. To evaluate the integral of Eq. (16.2), we transform it to the form $\int \Psi(\mathbf{r})\delta(\mathbf{r})dv$, where dv is the volume element pertaining to $\mathbf{r}$. To do this, we define

$$\mathbf{r}'' = \mathbf{r}' - \mathbf{f}(t') \tag{16.3}$$

and consider it as a transformation of coordinates from $\mathbf{r}' = (x', y', z')$ to $\mathbf{r}'' = (x'', y'', z'')$. When we perform a coordinate transformation from x_i to x'_k, the volume element dv in a volume integral is to be replaced by dv' where $dv' = J\,dv$, with J being the Jacobian of the transformation defined by

A. K. Raychaudhuri, *Classical Theory of Electricity and Magnetism*, Texts and Readings in Physical Sciences 21, https://doi.org/10.1007/978-981-16-8139-4_16

$$J = \begin{vmatrix} \frac{\partial x'_1}{\partial x_1} & \frac{\partial x'_1}{\partial x_2} & \frac{\partial x'_1}{\partial x_3} \\ \frac{\partial x'_2}{\partial x_1} & \frac{\partial x'_2}{\partial x_2} & \frac{\partial x'_2}{\partial x_3} \\ \frac{\partial x'_3}{\partial x_1} & \frac{\partial x'_3}{\partial x_2} & \frac{\partial x'_3}{\partial x_3} \end{vmatrix}$$

(Note that for the simple example of transformation from rectangular cartesian to spherical polar, the Jacobian turns out as $r^2 \sin\theta$, the volume integral $dv = dx\, dy\, dz$ is replaced by $r^2 dr\ \sin\theta\, d\theta\, d\varphi$.) The transformation (16.3) can be written as

$$x''_i = x'_i - f_i(t') \tag{16.4}$$

where

$$\frac{\partial f_i(t')}{\partial x'_k} = \frac{\partial f_i(t')}{\partial t'}\frac{\partial t'}{\partial x'_k} = u_i(t')\frac{x_k - x'_k}{c|\mathbf{r} - \mathbf{r}'|}$$

$u_i(t')$ being the velocity of the particle at time t'. Hence, from (16.4),

$$\frac{\partial x''_i}{\partial x'_k} = \delta_{ik} - \frac{(x_k - x'_k)}{c|\mathbf{r} - \mathbf{r}'|}u_i(t')$$

The Jacobian is, thus,

$$J = \begin{vmatrix} 1 - \frac{u_1}{c}\frac{R_1}{|R|} & -\frac{u_1}{c}\frac{R_2}{|R|} & -\frac{u_1}{c}\frac{R_3}{|R|} \\ -\frac{u_2}{c}\frac{R_1}{|R|} & 1 - \frac{u_2}{c}\frac{R_2}{|R|} & -\frac{u_2}{c}\frac{R_3}{|R|} \\ -\frac{u_3}{c}\frac{R_1}{|R|} & -\frac{u_3}{c}\frac{R_2}{|R|} & 1 - \frac{u_3}{c}\frac{R_3}{|R|} \end{vmatrix}$$

where $\mathbf{R} = \mathbf{r} - \mathbf{r}'$. A direct calculation of the determinant gives

$$J = 1 - \frac{u_i R_i}{cR}$$

where the summation over repeated indices is implied, as usual. Hence, we get the scalar potential from Eq. (16.2), and in an exactly similar manner, the vector potential as

$$\phi(\mathbf{r}, t) = \frac{e}{4\pi\epsilon_0}\int \frac{\delta(\mathbf{r}''_2)\, dv''}{R\left(1 - \frac{u_i R_i}{cR}\right)} = \frac{e}{4\pi\epsilon_0}\left[\frac{1}{R - \frac{u_i R_i}{c}}\right]_{\mathbf{r}'=\mathbf{f}(t')} \tag{16.5}$$

$$\mathbf{A}(\mathbf{r}, t) = \frac{\mu_0}{4\pi}\int \frac{\rho(\mathbf{r}', t')\mathbf{u}(t')\, dv'}{|\mathbf{r} - \mathbf{r}'|} = \frac{\mu_0}{4\pi}\left[\frac{e\mathbf{u}}{R - \frac{u_i R_i}{c}}\right]_{t'} = \frac{\mathbf{u}}{c^2}\phi(\mathbf{r}, t) \tag{16.6}$$

The solutions (16.5) and (16.6) are referred to as the Lienard-Wiechert potentials. Formally, they have a striking resemblance to the Coulomb and Biot-Savart potentials for the stationary case but the time-independent distance $|\mathbf{r} - \mathbf{r}'|$ occurring there is replaced here by $R - \frac{u_i R_i}{c}$, the value of this quantity being taken at the retarded time (i.e., the time of emission of the influence from the source particle). Equations (16.5) and (16.6) are generally true, holding for arbitrary motion of the source. We shall first consider the case of uniform rectilinear motion, when $\mathbf{u}$ is a constant vector.

Before going to the study of that case, we consider the application of our solution to distributions which extend over a finite region of space and thus depart from the assumed δ-function form. Obviously, the condition is that R and $\mathbf{u}$ should not vary appreciably as we move over the source region. This would require that the linear dimension of the source (say l) must be small compared to R and also that the change of velocity over l must be small compared to $\mathbf{u}$, i.e.,

$$\frac{l}{R} \ll 1 \quad \text{and} \quad \frac{l}{c}\dot{\mathbf{u}} \ll \mathbf{u}$$

16.1 Field of a Particle in Uniform Rectilinear Motion

We shall calculate the fields from (16.5) and (16.6) using the formulae

$$\mathbf{E} = -\nabla\phi - \frac{\partial \mathbf{A}}{\partial t}, \qquad \mathbf{B} = \nabla \times \mathbf{A}$$

However, the partial differentiations require some care

$$\frac{\partial t'}{\partial x_i} = \frac{\partial}{\partial x_i}\left(t - \frac{R}{c}\right) = -\frac{R_i}{cR} + \frac{u_k R_k}{cR}\frac{\partial t'}{\partial x_i}$$

$$\text{or} \qquad \frac{\partial t'}{\partial x_i} = -\frac{R_i}{cR}\left(1 - \frac{u_k R_k}{cR}\right)^{-1}$$

since x_i' varies with x_k via the variation of t'. Likewise,

$$\frac{\partial t'}{\partial t} = 1 + \frac{u_i R_i}{cR}\frac{\partial t'}{\partial t}$$

$$\text{or} \qquad \frac{\partial t'}{\partial t} = \left(1 - \frac{u_i R_i}{cR}\right)^{-1}$$

Hence,

$$\frac{\partial R_i}{\partial x_k} = \delta_{ik} - u_i \frac{\partial t'}{\partial x_k} = \delta_{ik} + \frac{u_i R_k}{cR}\left(1 - \frac{u_l R_l}{cR}\right)^{-1}$$
$$\frac{\partial R_i}{\partial t} = \frac{\partial}{\partial t}(x_i - x_i') = -u_i \frac{\partial t'}{\partial t} = -u_i\left(1 - \frac{u_k R_k}{cR}\right)^{-1}$$

Using these

$$\frac{\partial}{\partial x_k}\left(R - \frac{u_i}{c}R_i\right) = \frac{R_i}{R}\frac{\partial R_i}{\partial x_k} - \frac{u_i}{c}\frac{\partial R_i}{\partial x_k}$$
$$= \frac{1}{R}\left(R_i - \frac{u_i}{c}R\right)\left[\delta_{ik} + \frac{u_i R_k}{cR}\left(1 - \frac{u_l R_l}{cR}\right)^{-1}\right]$$
$$\frac{\partial}{\partial t}\left(R - \frac{u_i}{c}R_i\right) = -\frac{u_i}{R}\left(R_i - \frac{u_i}{c}R\right)\left(1 - \frac{u_l R_l}{cR}\right)^{-1}$$

we finally express the components of the electric field

$$E_k = \frac{e}{4\pi\epsilon_0 R}\frac{\left(R_i - \frac{u_i}{c}R\right)}{\left(R - \frac{u_l}{c}R_l\right)^2}\left[\delta_{ik} + \left(\frac{u_i R_k}{cR} - \frac{u_i u_k}{c^2}\right)\left(1 - \frac{u_l R_l}{cR}\right)^{-1}\right]$$
$$= \frac{e}{4\pi\epsilon_0}\left(1 - \frac{u^2}{c^2}\right)\frac{\left(R_k - \frac{u_k}{c}R\right)}{\left(R - \frac{u_l}{c}R_l\right)^3} \tag{16.7}$$

where we have used the expression Eq. (16.6) for $\dot{\mathbf{A}}$, simplified using the algebraic relations above and substituted $\mu_0\epsilon_0 = 1/c^2$. The vector $\mathbf{R} - (\mathbf{u}/c)R$ has a simple meaning, the shift $\mathbf{u}R/c$ being the displacement of the source in time R/c, i.e., from the position of the source at the instant of emission to its position at the instant of reception. Therefore, it is the vector joining the field point to the source point at the instant of reception. This seems rather strange and would have been more apparent had there been instantaneous action at a distance. However, that may be, we would do well to remember that the electric field is now directed away from the position of the source at the instant of reception. This differs from central fields in that the intensity would be angle dependent, as we shall see soon afterwards.

The magnetic field is easy to calculate

$$\mathbf{B} = \nabla \times \mathbf{A} = \nabla \times \frac{\mathbf{u}\phi}{c^2} = -\frac{\mathbf{u}}{c^2} \times \nabla\phi = \frac{\mathbf{u}}{c^2} \times \mathbf{E} \tag{16.8}$$

where, in simplying the above, we have used $\mathbf{u} \times \dot{\mathbf{A}} = \mathbf{u} \times \frac{\mathbf{u}\dot{\phi}}{c^2} = 0$. The fact that $\mathbf{B}$ is orthogonal to $\mathbf{E}$ means that the Poynting vector does not vanish. In fact, the Poynting vector **S** is given by

$$\mathbf{S} = \frac{1}{\mu_0}\mathbf{E} \times \left(\frac{\mathbf{u}}{c^2} \times \mathbf{E}\right) = \epsilon_0 \left(E^2 \mathbf{u} - (\mathbf{u} \cdot \mathbf{E})\,\mathbf{E}\right)$$

However, the following argument is usually advanced to conclude that the charge in uniform motion does not radiate energy. The fields **E** and **B** both show a $1/R^2$ dependence; hence, the Poynting vector would fall off as $1/R^4$. If we consider the flux through a sphere at infinity, the surface integral $\int \mathbf{S} \cdot \mathbf{ds}$ would go to zero as $1/R^2$. The rather shaky point in the argument is that if one is to consider the surface at infinity, one may also be tempted to consider the integral

$$\frac{1}{2}\int \left(\epsilon_0 E^2 + \frac{B^2}{\mu_0}\right) R^2 dR\, d\Omega$$

as denoting the electromagnetic energy of the entire system—this clearly diverges in case of radiation fields for which $E \sim R^{-1}$ and $B \sim R^{-1}$. Therefore, it seems appropriate that one should not bring in infinity in discussing radiation problems.

Consider the element of volume dv and the charge moving with velocity **u** along the line ABC. In the time the charge moves from A, the values of the field intensities within the element dv continually increases, and consequently, the energy within the element also increases. This means that there must have been an influx of energy into the volume element. As the charge moves further away, the energy within the element decreases, indicating an outward flux of energy. Thus, the non-vanishing Poynting vector does represent a flux of energy. However, this indicates only a redistribution of energy; whether the charge in the process loses energy is a different question. We may now consider an observer at dv moving with the velocity u. At each stage, the relative position of the source and the observer remains unchanged, so the observer would not perceive any change in the energy density distribution, so he concludes that there is no radiation loss of the charge. In view of this situation, we would take as the criterion of radiation the non-existence of *any* physical observer for whom the Poynting vector vanishes. In this particular case, there does exist such an observer and so there is no radiation.

The reasoning can be made at once clearer and more rigorous by an appeal to the special theory of relativity. According to it, all frames in uniform relative motion are equivalent so far as the laws of physics are concerned. We can, in order to decide whether a charge in uniform motion radiate energy or not, consider a frame in which the charge is at rest. In this frame, the field is static and we have only the electrostatic field so that there is no radiation of energy. In this frame, therefore, the particle continues to be at rest. In the other frame, if the particle radiates, a force is required to keep the velocity uniform. Thus, there is an intrinsic difference between the two frames in that one has to posit a force in one frame while there is none in the other frame. The resolution of the difficulty lies in recognizing that a charge in uniform motion does not radiate.

We shall express the field in a slightly different form to facilitate the discussion of the direction dependence of the field. Writing $\mathbf{R}_0 = \mathbf{R} - \frac{\mathbf{u}}{c}R$, we have

$$\begin{aligned}
\mathbf{R} \times \mathbf{u} &= \mathbf{R}_0 \times \mathbf{u} \\
R_0^2 &= R^2 + \frac{u^2}{c^2}R^2 - \frac{2R}{c}\mathbf{u}\cdot\mathbf{R} \\
&= \left(R - \frac{\mathbf{u}\cdot\mathbf{R}}{c}\right)^2 + \frac{R^2u^2}{c^2} - \frac{(\mathbf{u}\cdot\mathbf{R})^2}{c^2} \\
&= \left(R - \frac{\mathbf{u}\cdot\mathbf{R}}{c}\right)^2 + \left(\frac{\mathbf{u}\times\mathbf{R}}{c}\right)^2
\end{aligned}$$

Therefore, $\left(R - \frac{\mathbf{u}\cdot\mathbf{R}}{c}\right)^2 = R_0^2 - \left(\frac{\mathbf{u}\times\mathbf{R}}{c}\right)^2$, thus

$$\mathbf{E} = \frac{1}{4\pi\epsilon_0}\left(1 - \frac{u^2}{c^2}\right)\frac{c\mathbf{R}_0}{R_0^3\left(1 - \frac{u^2}{c^2}\sin^2\theta\right)^{3/2}} \tag{16.9}$$

where θ is the angle between $\mathbf{u}$ and $\mathbf{R}_0$. In the direction of motion, i.e., for $\theta = 0$, $E \propto (1 - u^2/c^2)$ whereas for $\theta = \pi/2$, $E \propto 1/\sqrt{1 - u^2/c^2}$.

Thus, while for $u = 0$ (the point charge at rest), we have the radially symmetric Coulomb field, as $|u|$ increases, the field becomes relatively stronger and stronger in the transverse direction (Fig. 16.1). Thus, the lines of force become concentrated in the transverse direction at the cost of the longitudinal direction. Indeed as $|u| \to c$, the field approaches a purely transverse form $\mathbf{E}\cdot\mathbf{r} \to 0$. Moreover, the magnetic field, which is weak for low velocities, approaches the following conditions as $|u| \to c$

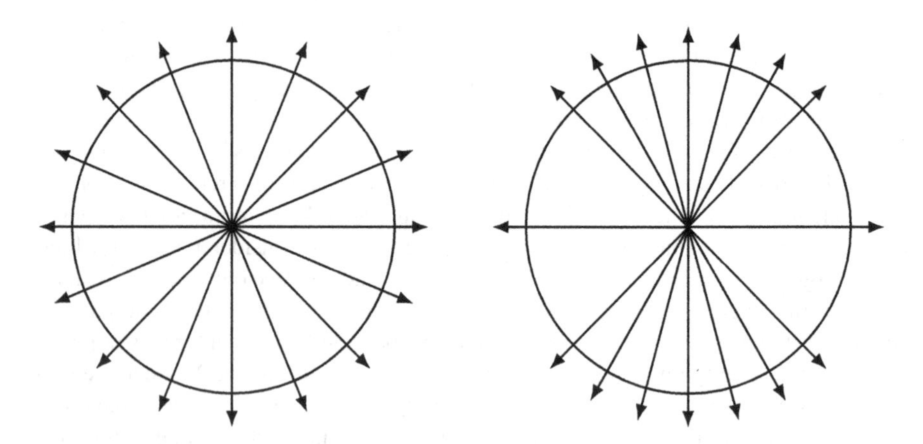

Fig. 16.1 Electric field lines for a charge at rest ($u = 0$) and those for a charge moving with relativistic speed ($u \to c$)

$$c^2B^2 - E^2 \to 0, \qquad \mathbf{B}\cdot\mathbf{E} = \mathbf{B}\cdot\mathbf{r} = 0$$

These are exactly the conditions that we found for the radiation field. Hence, the result that a charged particle moving with a velocity close to that of light generates a field that simulates a pure radiation field. We shall see later that this result has been utilized to simplify some calculations. Note that if the velocity $\mathbf{u}$ be along the x-axis,

$$\left(R - \frac{\mathbf{u}\cdot\mathbf{R}}{c}\right)^2 = R_0^2 - \left(\frac{\mathbf{u}}{c}\times\mathbf{R}_0\right)^2 = (x-ut)^2 + \left(1 - \frac{u^2}{c^2}\right)(y^2+z^2) \quad (16.10)$$

so that

$$\phi = \frac{e}{4\pi\epsilon_0}\left[(x-ut)^2 + \left(1-\frac{u^2}{c^2}\right)(y^2+x^2)\right]^{-1/2} \quad (16.11)$$

$$\mathbf{A} = \frac{\phi\,\mathbf{u}}{c^2} = \frac{\mu_0 e\mathbf{u}}{4\pi}\left[(x-ut)^2 + \left(1-\frac{u^2}{c^2}\right)(y^2+x^2)\right]^{-1/2} \quad (16.12)$$

16.2 The Method of Virtual Quanta

We have seen that as $|u| \to c$, the field of the electron simulates a radiation field. Hence, in its interaction with other charged particles, a sufficiently high energy electron (or proton or α-particle) may be regarded as equivalent to a pulse of radiation. This analogy was first pointed out by Fermi and exploited in numerous cases by Weizsäcker and Williams. To see how the method works, we perform a Fourier analysis of the particle's field (or rather the transverse field, which alone is the effective field at these high velocities). Writing the transverse field as

$$E_{\perp\omega}(\mathbf{r}) = \frac{1}{2\pi}\int_{-\infty}^{\infty} E_\perp(\mathbf{r},t)e^{-i\omega t}dt$$

and using Eqs. (16.9) and (16.10), we get

$$E_{\perp\omega}(\mathbf{r}) = \frac{eb(1-\beta^2)}{8\pi^2\epsilon_0}\int_{-\infty}^{\infty}\frac{e^{-i\omega t}}{\left((x-ut)^2 + b^2(1-\beta^2)\right)^{3/2}}dt \quad (16.13)$$

where $\beta = u/c$ and $b = \sqrt{y^2+z^2}$ is the distance of the field point from the line of motion of the charged particle; see Fig. 16.2. Introducing a new variable $\xi = t - \frac{x}{u}$, we rewrite the above[1] as

[1] The electric charge e and the exponential e should not be confused.

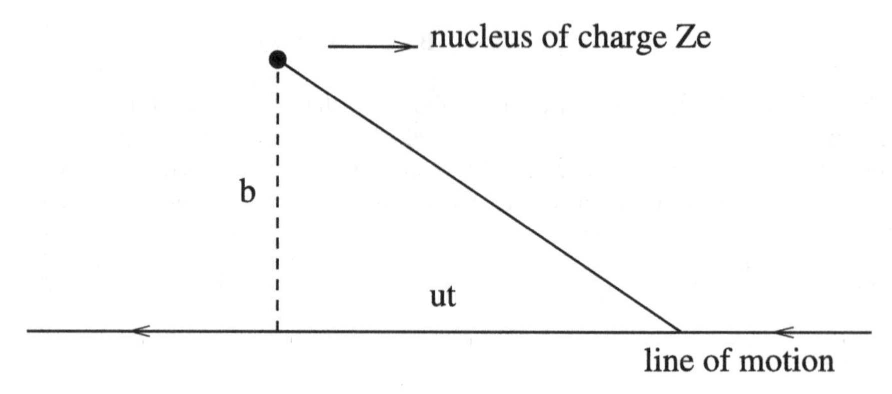

Fig. 16.2 Geometry of an interaction of a charged particle in the field of a fast-moving electron

$$
\begin{aligned}
E_{\perp\omega}(\mathbf{r}) &= \frac{eb(1-\beta^2)}{8\pi^2\epsilon_0} e^{-i\omega x/u} \int_{-\infty}^{\infty} \frac{e^{-i\omega\xi}}{\left(u^2\xi^2 + b^2(1-\beta^2)\right)^{3/2}} d\xi \\
&= -\frac{e}{8\pi^2\epsilon_0} e^{-i\omega x/u} \frac{\partial}{\partial b} \int_{-\infty}^{\infty} \frac{e^{-i\omega\xi}}{\sqrt{u^2\xi^2 + b^2(1-\beta^2)}} d\xi \\
&= -\frac{e}{8\pi^2 u\epsilon_0} e^{-i\omega x/u} \frac{\partial}{\partial b} \int_{-\infty}^{\infty} \frac{\cos\omega\xi}{\sqrt{\xi^2 + \frac{b^2}{u^2}(1-\beta^2)}} d\xi \\
&= -\frac{e}{2\pi^2 u\epsilon_0} e^{-i\omega x/u} \frac{\partial}{\partial b} \int_{0}^{\infty} \frac{\cos\omega\xi}{\sqrt{\xi^2 + \frac{b^2}{u^2}(1-\beta^2)}} d\xi \\
&= -\frac{e}{2\pi^2 u\epsilon_0} e^{-i\omega x/u} \frac{\partial}{\partial b} K_0\left(\sqrt{1-\beta^2}\,\frac{\omega b}{u}\right)
\end{aligned}
\tag{16.14}
$$

where we have used the fact that the sine term from $e^{-i\omega\xi}$, being an odd function, of ξ vanishes on integration. The final expression is in terms of the Hankel function of order zero. In going from the first to the second line above, we have traded the exponent in the denominator of the integrand from 3/2 to 1/2 with a differentiation with respect to b. Had we not done it, the resulting function would have been the Hankel function of order 1. Thus, the solution is presented as either the first derivative of the Hankel function of order zero or, equivalently, as the Hankel function of order 1.

The Hankel function is a combination of the Bessel functions. However, the only property of the Hankel function that will be utilized in our present discussion is their asymptotic behaviour. We have

$$
K_0(x) \sim \begin{cases} -\ln(x/2) & \text{for } x \ll 1 \\ \sqrt{\dfrac{\pi}{2x}}\, e^{-x} & \text{for } x \gg 1 \end{cases}
\tag{16.15}
$$

Thus, it has appreciable value only in the region $x < 1$. We now estimate the range up to which an incident energetic particle will be effective as an ionizing agent. For ionization, the relevant frequencies of electromagnetic disturbances are about 10^{15} Hz, to be in near resonance with the optical electrons. Thus, the range for ionization is determined by the condition

$$\sqrt{1-\beta^2}\,\frac{\omega b}{u} < 1 \tag{16.16}$$

where $\omega \sim 10^{15}$ rad/s. The value of b up to which ionization occurs depends crucially on $\sqrt{1-\beta^2}$, which varies sensitively with u, for $|u|$ approaching c. Now from the total energy $E = mc^2 = \dfrac{m_0c^2}{\sqrt{1-\beta^2}}$, we see that $\sqrt{1-\beta^2} = \dfrac{m_0c^2}{E}$, i.e., it is proportional to m_0 for given total energy E. Hence, from (16.16), the smaller the mass of the ionizing particle, the larger the distance up to which it ionizes.

As a realistic example, we may consider the case of α-particles. Taking the velocity to be 10^9 cm/s, $1-\beta^2 \sim 1$, consequently, Eq. (16.14) gives

$$b < \frac{10^9}{10^{15}} \sim 10^{-6}\ \text{cm}$$

The non-appearance of the charge of the particle in (16.16) may appear strange—it means that the range of ionization is determined directly by the velocity of the ionizing particle and indirectly by its mass and total energy, the charge having no influence on the range. The charge, however, determines the field intensity, hence the number of photons in the virtual radiation field.

16.3 Number of Equivalent Photons in the Virtual Radiation Field

We now calculate the energy in the range of frequency ν to $\nu + d\nu$ in the virtual radiation field. For the convenience of calculation, we take note of the behaviour of the Hankel function $K_0(x)$ in Eq. (16.15) and make the somewhat drastic approximation

$$K_0(x) \sim 0 \quad \text{for } x > 1$$

Then

$$\begin{aligned} E_{\perp\omega}(\mathbf{r}) &= \frac{e}{4\pi^2 u\epsilon_0} e^{-i\omega x/u} \frac{\partial}{\partial b} \ln\left(\sqrt{1-\beta^2}\,\frac{b\omega}{2u}\right) \\ &= \frac{e}{4\pi^2 ub\epsilon_0} e^{-i\omega x/u} \qquad \text{for } b \le \frac{2u}{\omega\sqrt{1-\beta^2}} \end{aligned}$$

We see that $E_\perp$ has a singularity at $b = 0$; therefore, the field and the energy diverge at this point, i.e., on the line of motion. The energy flux per unit area for the frequency range between ω and $\omega + d\omega$ for a particle passing at distance b is

$$J_\omega(b)\, d\omega = 2\epsilon_0 c\ |2\pi E_{\perp\omega}(b)|^2\, d\omega$$

where the factor 2 has been introduced to take care of the fact that $E_{\perp\omega}$ and $E_{\perp(-\omega)}$ contribute equally to this energy. To obtain the total energy, we integrate over b

$$2\pi\, d\omega \int J_\omega(b)\, b\, db$$

The limits of b are apparently 0 and ∞. However, owing to the assumed approximate form of the Hankel function, the upper limit is effectively $2u/\left(\omega\sqrt{1-\beta^2}\right)$. Mathematically, the zero limit would make the integral diverge—physically, one gets rid of the zero limit by introducing the idea of finite sizes of the reacting systems. Unfortunately, this procedure does not give any well-defined value of the lower limit. Let us just put it as $2b_0$. We, thus, get for the total energy flux

$$\frac{e^2 c}{\pi \epsilon_0 u^2} \ln\left(\frac{u}{b_0 \omega \sqrt{1-\beta^2}}\right) d\omega$$

Equating this to $N_\nu h\nu\, d\nu$, where N_ν is the number of equivalent photons, we get

$$N_\nu d\nu = \frac{2e^2 c}{u^2\, h\nu \epsilon_0} \ln\left(\frac{u}{2\pi \nu b_0 \sqrt{1-\beta^2}}\right) d\nu$$

In the range of validity of our approximations, $|u| \approx c$, hence

$$N_\nu d\nu = \frac{2e^2}{hc\nu\epsilon_0} \ln\left(\frac{\lambda}{2\pi b_0 \sqrt{1-\beta^2}}\right) d\nu = 4\alpha \ln\left(\frac{\lambda}{2\pi b_0\sqrt{1-\beta^2}}\right) \frac{d\nu}{\nu} \quad (16.17)$$

In case of collision between electrons, b_0 may be taken to be of the order of the Compton length, namely $h/(m_0 c)$, whereas for ionization the value of b_0 is expected to be of the order of atomic dimensions. As the fine structure constant $\alpha = \dfrac{e^2}{4\pi\epsilon_0 \hbar c} \approx \frac{1}{137}$, the factor $\frac{2\pi}{\alpha} \sim \frac{1}{200}$, hence the number of equivalent photons per electron will be small unless the logarithmic terms be sufficiently large. (In the relevant region, the value of ν is high. This is because the collisions are effective only when there is some sort of resonance—for atomic or nuclear phenomena, this requires $\nu \geq 10^{15}$ Hz.)

We now mention some cases where this method of virtual quanta may be used.

1. The disintegration of Beryllium nuclei by electronic bombardment.

The threshold energy for photo-disintegration of Be nuclei is 1.5 MeV and the cross-section for photo-disintegration as 3×10^{-28} cm^2 for photon energies above the threshold value. Now consider a beam of electrons of energy 2 MeV on Be. The virtual quanta which will be effective for the disintegration of Be nuclei are those having energy between 1.5 and 2 MeV, the cut off at 2 MeV being the maximum energy that the electrons can give to the virtual photon. Taking the closest distance of approach as the Compton length, namely $b_0 = \frac{h}{mc} \sim 10^{-10}$ cm, and $1 - \beta^2 \approx 0.04$ for an electron of this energy, we get from (16.16)

$$N_\nu d\nu \approx 10^{-2} \frac{d\nu}{\nu}$$

Considering the entire relevant range of frequency, the total number of equivalent photons is

$$N \sim 10^{-2} \int_{\nu=1.5\,\mathrm{MeV}/h} \frac{d\nu}{\nu} \sim 10^{-2} \ln(1.33)$$

Thus, the cross-section for disintegration by electrons is N times the cross-section for photo-disintegration which is approximately 10^{-30} cm^2. Of course, the crudeness of approximations is apparent and one can expect at most a very rough order of magnitude agreement with observations.

2. Bremsstrahlung from very high energy electrons passing near nuclei.
A proper calculation of Bremsstrahlung will involve the consideration of the field of an accelerated electron. However, if the velocity be high enough, the change of velocity may be quite small in the short time the electron is in the field of a nucleus. The velocity of the electron is then uniform(!) and always close to c. Now let us make a transformation to a relatively moving frame, in which the electron is brought to rest and the nucleus moves with a velocity of the order of c. The electromagnetic field of the nucleus is now analysed into the Fourier components and the number of equivalent photons of different frequencies obtained from (16.17). (Of course, the source charge is now the nuclear charge Ze.) These virtual photons now suffer the Compton scattering by the electron. As the cross-section of the Compton scattering is known, one finally obtains the spectrum of the scattered radiation from the electron. Thus, the following statements are equivalent.

Bremsstrahlung due to acceleration of high energy electrons in the field of a nucleus	$\equiv$	Quasi-radiation pulse due to the nucleus moving in with high velocity in rest frame of electrons and the Compton scattering of this pulse by the electrons
Acceleration and change to in the path of electrons due to the field of a nucleus	$\equiv$	Deceleration and the change of path of the electrons due to Compton recoil in scattering the pulse from the nucleus

3. Pair formation due to the passage of a high energy photon near a nucleus.
   Two photons may produce an electron-positron pair provided the product of photon energies exceeds $(mc^2)^2$. The cross-section for the process is fairly constant from energy values close to the threshold up to about 15 times that value, the cross-section being 10^{-25} cm^2. When a photon interacts with a nucleus, we make a transformation to a frame in which the photon energy becomes mc^2. This makes the nucleus move with a high velocity and we get a distribution of photons in the quasi radiation pulse due to the nucleus. These photons are now supposed to interact with the original photon producing electron-positron pairs. As the cross-section for this process is already known, we get finally the cross-section for the production of pairs by a photon in the field of a nucleus.

Problems

1. Obtain the Lienard-Wiechert potential without using the delta function, i.e., consider a small charged sphere and proceed to the limit of a point charge.
2. Use the idea of a photon and the principles of conservation of energy and momentum to justify the conclusion that a charge in uniform motion cannot radiate.
3. Assuming that the force on a charged particle moving with velocity $\mathbf{v}$ is given by the Lorentz equation $\mathbf{F} = e(\mathbf{E} + \mathbf{v} \times \mathbf{B})$, show that the force between two equal point charges e in relative rest, as calculated in a moving frame (velocity u in the line joining the charges), is given by

$$\mathbf{F} = -\nabla \left(\frac{e^2 \left(1 - \frac{u^2}{c^2}\right)}{r - \dfrac{\mathbf{r} \cdot \mathbf{u}}{c}} \right)$$

   Is this force central?
4. Show that the interaction between two charges at rest in a moving frame (as in Problem 3) may be described in terms of an anisotropic medium and find the apparent dielectric constant.

Chapter 17
Field of a Particle in Non-Uniform Motion

We find the field from the Liénard-Wiechert potential, this time taking into account that **u** is no longer independent of time. This brings in some additional terms. Thus,

$$\frac{\partial}{\partial t}\left(R-\frac{u_i R_i}{c}\right)=-\frac{\left(R_i-\frac{Ru_i}{c}\right)u_i}{R\left(1-\frac{R_k u_k}{cR}\right)}-\frac{R_i\dot{u}_i}{c\left(1-\frac{R_k u_k}{cR}\right)}$$

$$\frac{\partial}{\partial x_k}\left(R-\frac{u_i R_i}{c}\right)=\frac{1}{R}\left(R_i-\frac{Ru_i}{c}\right)\left[\delta_{ik}+\frac{u_i R_k}{cR\left(1-\frac{u_l R_l}{cR}\right)}\right]+\frac{\dot{u}_i R_i R_k}{c^2 R\left(1-\frac{u_l R_l}{cR}\right)}$$

The overhead dots signify differentiation with respect to the *retarded time* t', and all variables pertaining to the source are to be given their retarded values. A straightforward calculation now gives the components of the electric field

$$E_k=\frac{e}{4\pi\epsilon_0}\left[\frac{\left(R_k-\frac{Ru_k}{c}\right)\left(1-\frac{u^2}{c^2}\right)}{\left(R-\frac{R_l u_l}{c}\right)^3}+\frac{\dot{u}_i R_i\left(R_k-\frac{Ru_k}{c}\right)-\dot{u}_k R\left(R-\frac{R-R_l u_l}{c}\right)}{c^2\left(R-\frac{R_l u_l}{c}\right)^3}\right]$$

or in vector notation

$$\mathbf{E}=\frac{e}{4\pi\epsilon_0}\left[\frac{\left(1-\frac{u^2}{c^2}\right)}{\left(R-\frac{\mathbf{u}\cdot\mathbf{R}}{c}\right)^3}\left(\mathbf{R}-\frac{R\mathbf{u}}{c}\right)+\frac{\mathbf{R}\times\left[\left(\mathbf{R}-\frac{R\mathbf{u}}{c}\right)\times\dot{\mathbf{u}}\right]}{c^2\left(R-\frac{\mathbf{u}\cdot\mathbf{R}}{c}\right)^3}\right] \tag{17.1}$$

The first term in (17.1) is just the field we obtain for the uniformly moving charged particle. The second term is specifically due to acceleration of the charged particle. The magnetic field may also be calculated—it turns out to be

$$\mathbf{B}=\frac{\mathbf{R}\times\mathbf{E}}{cR}=\frac{1}{c}\mathbf{n}\times\mathbf{E} \tag{17.2}$$

where **n** is the unit vector in the direction of **R**.

A. K. Raychaudhuri, *Classical Theory of Electricity and Magnetism*, Texts and Readings in Physical Sciences 21, https://doi.org/10.1007/978-981-16-8139-4_17

The relation (17.2) holds good in the case of uniform velocity as well. It may be noted that the part of **E** which arises due to acceleration (called the acceleration field in short) as well as **B** are orthogonal to **R**. (In the case of **B**, the orthogonality holds for both the velocity and acceleration fields.) The vector

$$\mathbf{R} - \frac{\mathbf{u}}{c} R$$

is no longer the vector joining the position of the source at time t to the field point, as $\mathbf{u}$ is no longer a constant. This would have been the case if the velocity of the source were frozen at the value it had at the retarded time t'.

If $u \ll c$ (a condition which does not mean any loss of generality from the point of view of the special theory of relativity since a Lorentz transformation may reduce the velocity at any instant to zero) terms in u/c may be neglected, the first term in (17.1) reduces to the static Coulomb field with $1/R^2$ dependence. (This is a non-radiative field.) The second part

$$\frac{e}{4\pi\epsilon_0 c^2 R^3} (\mathbf{R} \times (\mathbf{R} \times \dot{\mathbf{u}})) = \frac{e}{4\pi\epsilon_0 c^2 R^3} \left((\mathbf{R} \cdot \dot{\mathbf{u}})\mathbf{R} - R^2 \dot{\mathbf{u}} \right)$$

apparently falls off as $1/R$. Hence, at large enough distances, this part (called the radiation field) will be dominant. We also have

$$\mathbf{B} \cdot \mathbf{E} = 0, \qquad c|\mathbf{B}| = |\mathbf{E}|, \qquad \mathbf{B} \cdot \mathbf{R} = 0 = \mathbf{E} \cdot \mathbf{R}$$

Thus, we have a radially outgoing radiation field. The Poynting vector would fall off as $1/R^2$, hence its integral over the 'sphere at infinity' would be non-vanishing. This is taken as indicating that energy is being radiated out. However, as we have already pointed out, it is rather awkward to proceed to infinity and call $\frac{1}{2}\epsilon_0(E^2 + c^2B^2)$ the energy density, for this quantity behaves as $1/R^2$ and its volume integral diverges.

The splitting up of the field into $1/R$ and $1/R^2$ terms has been done by taking $u \ll c$. While, in general, such a process may not raise any difficulty, there are exceptional situations like the case of hyperbolic motion $x^2 = b^2 + c^2t^2$ (which is the case of uniform acceleration according to the special theory of relativity) where the apparent dependence on R is quite different. However, this does not invalidate our earlier analysis leading to a $1/R$ dependence of the radiation field when the source is confined to a finite bounded region—in the case of uniform acceleration, the source extends from $x = -\infty$ to $x = +\infty$.

Let us return to the radiation field

$$\mathbf{E}_{\text{rad}} = \frac{e}{4\pi\epsilon_0 c^2 R} (\dot{u} \cos\theta \, \mathbf{n} - \dot{\mathbf{u}})$$

$$\mathbf{B}_{\text{rad}} = \frac{e}{4\pi\epsilon_0 c^3 R} \dot{\mathbf{u}} \times \mathbf{n}$$

where θ is the angle between $\mathbf{u}$ and $\mathbf{R}$. Note that these agree with the far field of a dipole found with the help of the Hertz vectors, if we put $e\dot{\mathbf{u}} = \boldsymbol{\mu}\cos\omega t'$ and $t' = t - \frac{r}{c}$. These correspond to dipole radiation with the energy flux $\mathbf{E}\times\mathbf{H}$ having the typical $\sin^2\theta$ dependence. The second time derivative of the time-dependent dipole moment is $e\dot{\mathbf{u}}$. The rate of energy loss

$$\begin{aligned}\frac{1}{\mu_0}\int(\mathbf{E}\times\mathbf{B})\cdot d\mathbf{S} &= \frac{1}{\mu_0}\frac{e^2}{16\pi^2\epsilon_0^2c^5R^2}\int \dot{u}^2\sin^2\theta\; 2\pi R^2\sin\theta\, d\theta \\ &= \frac{e^2\dot{u}^2}{8\pi\epsilon_0 c^3}\int_0^\pi \sin^3\theta d\theta \;=\; \frac{e^2}{6\pi\epsilon_0}\frac{\dot{u}^2}{c^3}\end{aligned} \tag{17.3}$$

is in agreement with our previous result.

The calculation of the rate of radiation needs a little care in case the velocity u is comparable with c. (From the point of view of the special theory of relativity, the difference arises because acceleration is not a Lorentz invariant quantity.) The rate of radiation of energy can no longer be obtained by simply integrating over the surface of a distant sphere. To see this, consider two successive positions of the source P_1 and P_2 separated by a time interval dt, so that the infinitesimal vector $\overrightarrow{P_1P_2} = \mathbf{u}dt$. (After time t, the radiation emitted at P_1 reaches the sphere S_1 of radius ct. Similarly, at that instant ,the radiation emitted at P_2 has reached the sphere S_2 centred at P_2 of radius $c(t-dt)$ (see Fig. 17.1)). If the source was stationary, the region between the spheres would have been a shell of uniform width $c\,dt$. However, now the width is non-uniform and is equal to $(c - u\cos\theta)dt$, where θ is the angle between the particular direction and the velocity vector u. The energy radiated in

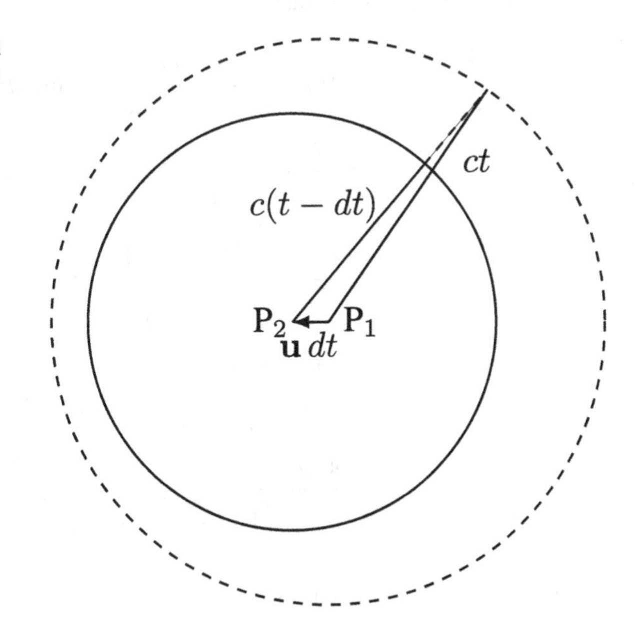

Fig. 17.1 The circle, centred at P_1 of radius ct denotes the location of the wavefront of energy radiated by an electron when it is at P_1. As it moves from to P_2, with a velocity $\mathbf{u}$ as indicated, the circle centred at P_2 of radius $c(t-dt)$ denotes the corresponding wavefront

time dt is simply the energy contained within this region, so that the rate of radiation is

$$\begin{aligned}\frac{\epsilon_0}{2}\int \left(E^2+c^2B^2\right)\, R^2(c-u\cos\theta)\, d\omega &= \epsilon_0\int E^2\, R^2(c-u\cos\theta)\, d\Omega \\ &= \frac{e^2}{16\pi^2\epsilon_0 c^3}\int \left[\dot{u}^2\left(1-\frac{\mathbf{u}\cdot\mathbf{n}}{c}\right)^2+\frac{2(\dot{\mathbf{u}}\cdot\mathbf{n})(\dot{\mathbf{u}}\cdot\mathbf{u})}{c}\left(1-\frac{\mathbf{u}\cdot\mathbf{n}}{c}\right)\right. \\ &\qquad \left. -(\dot{\mathbf{u}}\cdot\mathbf{n})^2\left(1-\frac{u^2}{c^2}\right)\right]\left(1-\frac{\mathbf{u}\cdot\mathbf{n}}{c}\right)^{-5} d\Omega \qquad (17.4)\end{aligned}$$

where $\mathbf{n}$ is the unit vector in the direction of $\mathbf{R}$ and we have considered only the acceleration field as the velocity part of the field does not contribute to the radiation.

To see the difference between the general formula valid for arbitrary u and that for $u \ll c$, consider the case of a linear harmonic oscillator, for which $\mathbf{u}$ and $\dot{\mathbf{u}}$ are always in the same direction. The numerator in the integrand in the above simplifies to

$$\dot{u}^2\left(1-\frac{\mathbf{u}\cdot\mathbf{n}}{c}\right)^2+\frac{2(\dot{\mathbf{u}}\cdot\mathbf{n})(\dot{\mathbf{u}}\cdot\mathbf{u})}{c}\left(1-\frac{\mathbf{u}\cdot\mathbf{n}}{c}\right)-(\dot{\mathbf{u}}\cdot\mathbf{n})^2\left(1-\frac{u^2}{c^2}\right)=\dot{u}^2\sin^2\theta$$

One can make repeated use of

$$\frac{d}{d\theta}\frac{1}{\left(1-\frac{u}{c}\cos\theta\right)^k}=-\frac{ku}{c}\frac{\sin\theta}{\left(1-\frac{u}{c}\cos\theta\right)^{k+1}}$$

to integrate by parts to show that

$$\begin{aligned}\int\frac{\dot{u}^2\sin^2\theta}{\left(1-\frac{u}{c}\cos\theta\right)^5}\, d\Omega &= 2\pi\dot{u}^2\int_0^\pi\frac{\sin^2\theta}{\left(1-\frac{u}{c}\cos\theta\right)^5}\sin\theta\, d\theta \\ &= \frac{\pi\dot{u}^2 c}{u}\int_0^\pi\frac{\cos\theta}{\left(1-\frac{u}{c}\cos\theta\right)^4}\sin\theta\, d\theta \\ &= \frac{\pi\dot{u}^2c^2}{3u^2}\left[\frac{2\left(1+\frac{3u^2}{c^2}\right)}{\left(1-\frac{u^2}{c^2}\right)^3}-\int_0^\pi\frac{1}{\left(1-\frac{u}{c}\cos\theta\right)^3}\sin\theta\, d\theta\right] \\ &= \frac{2\pi\dot{u}^2c^2}{3u^2}\frac{1+\frac{3u^2}{c^2}}{\left(1-\frac{u^2}{c^2}\right)^3}+\frac{\pi\dot{u}^2c^3}{6u^3}\int_{\theta=0}^{\theta=\pi} d\left(\frac{1}{\left(1-\frac{u}{c}\cos\theta\right)^2}\right) \\ &= \frac{8\pi\dot{u}^2}{3}\frac{1}{\left(1-\frac{u^2}{c^2}\right)^3}\end{aligned}$$

Hence, the radiation rate in this case is

$$\frac{e^2\dot{u}^2}{6\pi\epsilon_0 c^3}\left(1-\frac{u^2}{c^2}\right)^{-3} \qquad \boxed{\frac{2e^2\dot{u}^2}{3c^3}\left(1-\frac{u^2}{c^2}\right)^{-3} \quad \text{(Gaussian)}} \tag{17.5}$$

Thus, the radiation rate is enhanced by the factor $1/(1-u^2/c^2)^3$. For a harmonic oscillator,

$$x = a\cos\omega t, \qquad u = -a\omega\sin\omega t, \qquad \dot{u} = -a\omega^2\cos\omega t$$

Hence, the radiation rate is

$$\frac{e^2}{6\pi\epsilon_0 c^3}\frac{a^2\omega^4\cos^2\omega t}{\left(1-\frac{a^2\omega^2}{c^2}\sin^2\omega t\right)^3} \tag{17.6}$$

In the limiting case of $\frac{a\omega}{c}\to 1$, i.e., the maximum velocity approaching c, the radiation rate at $t=0$ is $\frac{e^2a^2\omega^4}{6\pi\epsilon_0 c^3}$, which is just double of the average radiation rate for $a\omega \ll c$. In the latter case, the average radiation rate is $\frac{e^2a^2\omega^4}{12\pi\epsilon_0 c^3}$, because $\langle\cos^2\omega t\rangle = \frac{1}{2}$.

Notes

1. The analysis of the radiation field utilized a power series expansion that is valid only if the dimensions of the radiating system are small compared to the relevant wavelength. When $u\sim c$, the dimensions of the system are of the order of $cT\sim\lambda$, hence the analysis breaks down. With u/c not negligible, the angular dependence is no longer of the form $\sin^2\theta$, which we have found to be a characteristic of dipole radiation.
2. In general, for $\mathbf{u}$ and $\dot{\mathbf{u}}$ not in the same direction, the radiation rate may be calculated to be
$$\frac{e^2\dot{u}^2}{6\pi\epsilon_0 c^3}\left(1-\frac{u^2}{c^2}\right)^{-3}\left[\dot{u}^2-\left(\frac{\mathbf{u}}{c}\times\dot{\mathbf{u}}\right)^2\right]$$
Thus, any inclination between the two decreases the radiation rate.

17.1 Radiation from a Particle Describing a Circular Orbit with Uniform Velocity—Acceleration Always Orthogonal to the Velocity

If $\mathbf{u}$, $\dot{\mathbf{u}}$ and $\mathbf{n}$ are in the same plane, Eq. (17.4) gives for the radiation rate at an angle θ to be

$$\frac{e^2\dot{u}^2}{16\pi^2\epsilon_0 c^3}\frac{\left(\frac{u}{c}-\cos\theta\right)^2}{\left(1-\frac{u}{c}\cos\theta\right)^5} \tag{17.7}$$

so that the intensity is concentrated in the forward direction, the ratio between the forward and backward intensities being

$$\left(\frac{1+u/c}{1-u/c}\right)^3$$

The discussion is somewhat complicated if the three are not in the same plane. Take the instantaneous directions of $\mathbf{u}$ and $\mathbf{n}$ as the z- and x-axis, respectively, so that we have

$$\mathbf{u}\cdot\mathbf{n} = u\cos\theta, \qquad \mathbf{u}\cdot\dot{\mathbf{u}} = 0, \qquad \dot{\mathbf{u}}\cdot\mathbf{n} = \dot{u}\sin\theta\cos\varphi$$

Substituting in Eq. (17.4), we get for the radiation rate

$$\begin{aligned}
&\frac{e^2}{16\pi^2\epsilon_0 c^3}\int_0^{2\pi} d\varphi \int_0^{\pi} d\theta \sin\theta \, \frac{\dot{u}^2\left(1-\frac{u}{c}\cos\theta\right)^2 - \dot{u}^2\sin^2\theta\cos^2\varphi\left(1-\frac{u^2}{c^2}\right)}{\left(1-\frac{u}{c}\cos\theta\right)^5} \\
&= \frac{e^2}{16\pi\epsilon_0 c^3}\int_0^{\pi} d\theta \sin\theta \, \frac{2\dot{u}^2\left(1-\frac{u}{c}\cos\theta\right)^2 - \dot{u}^2\sin^2\theta\left(1-\frac{u^2}{c^2}\right)}{\left(1-\frac{u}{c}\cos\theta\right)^5} \\
&= \frac{e^2\dot{u}^2}{8\pi\epsilon_0 c^3}\int_0^{\pi}\frac{\sin\theta d\theta}{\left(1-\frac{u}{c}\cos\theta\right)^3} - \frac{e^2\dot{u}^2\left(1-\frac{u^2}{c^2}\right)}{16\pi\epsilon_0 c^3}\int_0^{\pi}\frac{\sin^3\theta d\theta}{\left(1-\frac{u}{c}\cos\theta\right)^5} \\
&= \frac{e^2\dot{u}^2}{4\pi\epsilon_0 c^3}\frac{1}{\left(1-\frac{u^2}{c^2}\right)^2} - \frac{e^2\dot{u}^2\left(1-\frac{u^2}{c^2}\right)}{16\pi\epsilon_0 c^3}\frac{4}{3\left(1-\dfrac{u^2}{c^2}\right)^3} \\
&= \frac{e^2\dot{u}^2}{6\pi\epsilon_0 c^3}\left(1-\frac{u^2}{c^2}\right)^{-2}
\end{aligned} \tag{17.8}$$

For the particle in a circular orbit, if ω be the angular velocity and a the radius of the circle $u = a\omega$ and $\dot{u} = a\omega^2$, hence the rate of radiation is

$$\frac{e^2}{6\pi\epsilon_0 c^3}\frac{a^2\omega^4}{\left(1-\frac{a^2\omega^2}{c^2}\right)}.$$

The intensity has strong directional properties. A look at Eq. (17.8) shows that in the plane of motion (i.e., for $\phi = 0$), the intensity of radiation vanishes at an angle $\theta = \cos^{-1}\frac{u}{c}$. It has a fairly sharp maximum in the direction normal to the plane, where it has the value

$$\frac{e^2\dot{u}^2}{16\pi^2 c^3}$$

It is of some interest to calculate the ratio of the energy loss (over a full period of oscillation) to the kinetic energy of the particle. It is

$$\approx \left(\frac{e^2a^2\omega^4}{16\pi^2\epsilon_0c^3} \times \frac{2\pi}{\omega}\right) \Big/ \left(\frac{1}{2}ma^2\omega^2\right) = \frac{2e^2\omega}{3\epsilon_0c^3m}$$

provided $u \ll c$. If the radiating particle be an electron, the ratio is $\sim r_0\omega/c$ where $r_0 = e^2/(4\pi\epsilon_0mc^2)$ is of the order of magnitude of the classical radius of the electron. Substituting values appropriate for the electron, the ratio comes out as $10^{-23}\omega$, hence unless the frequency is exceptionally high, the radiation loss does not effectively decrease the energy of a radiating system.

17.2 Classical Theory of Bremsstrahlung

We are now going to study the application of our formulae to cases which occur in nature—where there is an emission of electromagnetic radiation due to an acceleration of an electron. Familiar examples are the continuous radiation from an X-ray tube where electrons are stopped in the target, or the radiation accompanying the emission of a β-particle, which may be looked upon as an increase of velocity from zero to a finite value. Mathematically, the simplest case is that of an acceleration in the form of a δ-function, i.e., when the velocity remains constant throughout, except for a discontinuous jump at a certain instant. Such a situation is realized if the event bringing about the jump in velocity is a collision between two ideally 'hard' spheres. However, more realistic models must consider a finite interaction time with a potential like say the Coulomb potential of nuclear charge.

The total radiated energy is

$$\frac{1}{c\mu_0}\int E_{\text{rad}}^2 R^2\, d\Omega\, dt$$

With the δ-function acceleration $\dot{u} \sim \delta(t)$, consequently $E_{\text{rad}} \sim \delta(t)$ and the above integral will obviously diverge. This is a basic defect of the classical theory and has to be rectified (admittedly artificially) by grafting to the classical theory the quantum consideration that the emitted radiation frequency spectrum has a cut-off as the photon energy cannot exceed the kinetic energy of the decelerating particle (i.e., $h\nu \leq mu^2/2$).

Let us instead calculate the radiated energy in a particular frequency range. As usual, we decompose the radiation field into its Fourier components

$$\mathbf{E} = \int_{-\infty}^{\infty} E_\nu\, e^{2\pi i \nu t}\, d\nu$$

$$\mathbf{E}_\nu = \int_{-\infty}^{\infty} \mathbf{E}\, e^{-2\pi i \nu t}\, dt$$

$$= \frac{e}{4\pi \epsilon_0 c^2 R} \int_{-\infty}^{\infty} (\dot{u} \cos\theta \mathbf{n} - \dot{\mathbf{u}})\; e^{-2\pi i \nu t}\, dt$$

where we have taken $u \ll c$. With $\dot{\mathbf{u}} = (\mathbf{u}_2 - \mathbf{u}_1)\delta(t)$, where $\mathbf{u}_2$ and $\mathbf{u}_1$ are the final and initial velocities, respectively, and the interaction takes place at $t = 0$, we get

$$\mathbf{E}_\nu = \frac{e}{4\pi \epsilon_0 c^2 R} [(u_2 - u_1) \cos\theta\, \mathbf{n} - (\mathbf{u}_2 - \mathbf{u}_1)]$$

$$|\mathbf{E}_\nu| = \frac{e}{4\pi \epsilon_0 c^2 R} |\mathbf{u}_2 - \mathbf{u}_1| \sin\theta$$

The radiated energy of frequency in the range ν to $\nu + d\nu$ is, therefore,

$$R_\nu\, d\nu = \frac{2}{c\mu_0} \int_\Omega |E_\nu|^2 R^2\, d\Omega\, d\nu$$

(where the factor 2 comes from the fact that the contribution of $\mathbf{E}_{-\nu}$ is equal to that of $\mathbf{E}_\nu$). Integrating over all angles, we have

$$R_\nu\, d\nu = \frac{e^2}{4\pi \epsilon_0^2 c^5 \mu_0} \int_{-1}^{1} |\mathbf{u}_2 - \mathbf{u}_1|^2 \sin^2\theta\, d(\cos\theta)\, d\nu$$

$$= \frac{e^2}{3\pi \epsilon_0^2 c^5 \mu_0} |\mathbf{u}_2 - \mathbf{u}_1|^2\, d\nu$$

The above result is independent of the frequency. Note that the radiation has the $\sin^2\theta$ dependence characteristic of dipole radiation. As we have already noted, to have a finite value of total energy radiated, one has to cut off at frequency $\frac{mu^2}{2h}$.

Consider the scatterer to be a hard sphere of radius a (see Fig. 17.2). If N be the incident flux (i.e., the number of particles passing through the unit area normal to the direction of the velocity per unit time), then the number incident on the element of surface of the sphere is $N \cos\theta\, a^2\, d\Omega = Na^2 \cos\theta \sin\theta\, d\theta\, d\phi$, θ being the angle between the normal to the element of area and the velocity of the incoming particles. For an elastic collision, the outgoing particles will also make the same angle with the radius, so that the angle of scattering is $\gamma = \pi - 2\theta$. Then

$$(\mathbf{u}_2 - \mathbf{u}_1)^2 = u_1^2 + u_2^2 - 2u_1 u_2 \cos\gamma$$

$$= 2u^2(1 - \cos\gamma) \; = \; \left(2u \sin\frac{\gamma}{2}\right)^2$$

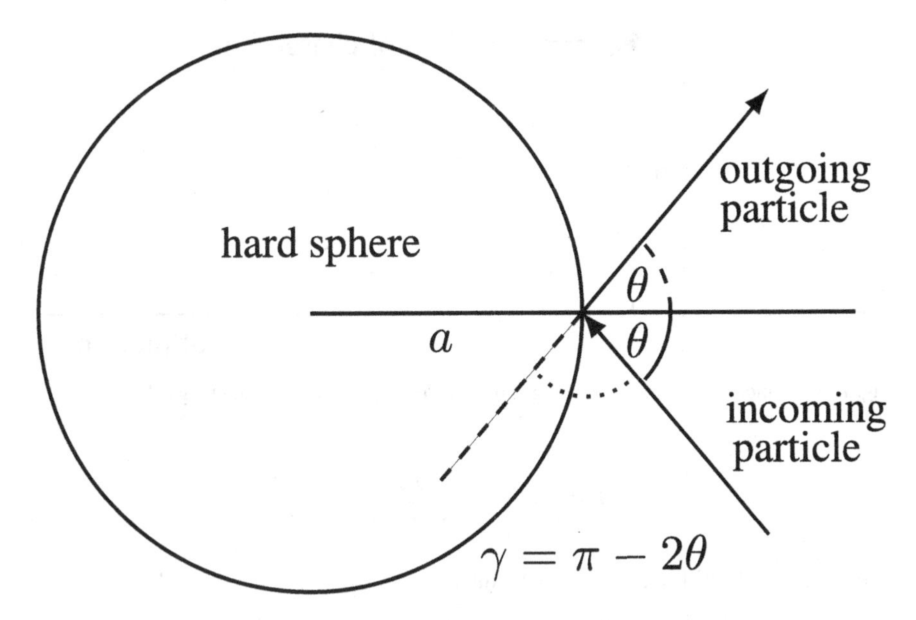

Fig. 17.2 Scattering of charged particles off a hard sphere of radius a

where we have used the fact that $|\mathbf{u}_2| = |\mathbf{u}_1| = u$ (say) for elastic scattering. Therefore, the number of particles scattered at an angle γ is

$$\frac{1}{2}Na^2 \sin 2\theta \, d\theta \, d\varphi = \frac{1}{4}Na^2 \sin\gamma \, d\gamma \, d\varphi$$

Hence, the differential cross-section $d\sigma$ (defined as the fraction of the incident flux scattered into the infinitesimal solid angle $d\Omega$) is

$$d\sigma = \frac{a^2}{4} \sin\gamma \, d\gamma \, d\varphi$$

The total cross-section is

$$\sigma = \int_0^{\pi} \int_0^{2\pi} \frac{a^2}{4} \sin\gamma \, d\gamma \, d\varphi = \pi a^2$$

which is just the geometrical cross-section of the scatterer. Also, the product $d\sigma \; R_\nu d\nu$ is

$$\frac{a^2}{4} \sin\gamma \, d\gamma \, d\varphi \times \frac{e^2}{3\pi\epsilon_0^2 c^5 \mu_0} (\mathbf{u}_2 - \mathbf{u}_1)^2 d\nu$$

Integrating over all value of $0 \le \varphi \le 2\pi$ and $0 \le \gamma \le \pi$, we get for the energy radiated in the frequency interval $[\nu, \nu + d\nu]$ as

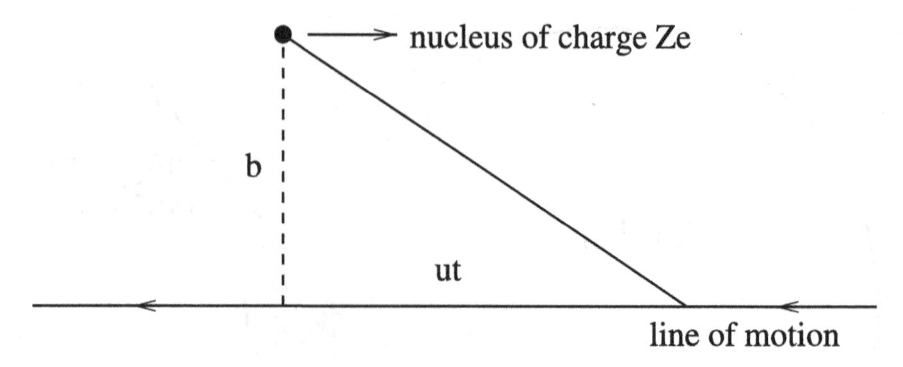

Fig. 17.3 Coulomb scattering of a charged particle in the field of a nuclear charge Ze

$$\int d\sigma \; R_\nu d\nu = \frac{2e^2a^2}{3\epsilon_0^2 c^5 \mu_0} u^2 d\nu$$

The 'cross-section' for energy loss is, thus,

$$\left(\frac{2e^2a^2u^2}{3\epsilon_0^2 c^5 \mu_0} \int_0^{\nu_{\max}} d\nu \right) \Big/ \left(\frac{1}{2} mu^2 \right) = \frac{4e^2a^2}{3\epsilon_0 c^3 \, m} \nu_{\max}$$

with $\nu_{\max} = \frac{mu^2}{2h}$. So finally, the above becomes

$$\frac{2e^2a^2u^2}{3\epsilon_0 c^3 \, h} = \frac{4}{3\pi} \frac{\pi a^2}{\alpha} \left(\frac{u}{c} \right)^2 \approx \frac{\pi a^2}{300} \beta^2$$

where we have used the value of the fine structure constant $\alpha = \frac{e^2}{4\pi\epsilon_0 c\hbar} \simeq \frac{1}{137}$.

The discussion above suffers from the main defect that the radiation comes out independent of the frequency. If we consider more realistic potentials, so that the collision time is finite there is a decrease of radiated energy at high frequencies. Indeed, the decrease may be quite significant as we shall find presently for the Coulomb potential but there is no sharp cut-off frequency. For that, one must appeal to ideas from quantum physics.

Next, we consider the scattering field to be Coulomb, due to a nucleus of charge Ze (see Fig. 17.3). The magnitude of the acceleration, for an electron of charge e, is

$$\dot{u} = \frac{Ze^2}{4\pi\epsilon_0 m} \frac{1}{(p^2 + u^2t^2)}$$

The transverse component of the acceleration and the corresponding radiation field are

$$\dot{u}_\perp = \frac{Ze^2}{4\pi\epsilon_0 m}\frac{p}{(p^2+u^2t^2)^{3/2}}$$
$$|E| = \frac{e}{4\pi\epsilon_0 c^2 r}\,\dot{u}_\perp \sin\theta$$

The Fourier component of frequency ν of the latter is

$$\begin{aligned}
|E_\nu| &= \int |E|\, e^{-2\pi i\nu t}\, dt \\
&= \frac{Ze^2}{4\pi\epsilon_0 m}\frac{e}{4\pi\epsilon_0 c^2 r}\int_{-\infty}^{\infty}\frac{p\sin\theta}{(p^2+u^2t^2)^{3/2}}\, e^{-2\pi i\nu t}\, dt \\
&= \frac{1}{(4\pi\epsilon_0)^2}\frac{Ze^2}{m}\frac{2e}{c^2 r}\int_0^{\infty}\frac{p\sin\theta\cos 2\pi\nu t}{(p^2+u^2t^2)^{3/2}}\, dt \\
&= -\frac{1}{(4\pi\epsilon_0)^2}\frac{Ze^2}{m}\frac{e}{c^2 r}\frac{2\sin\theta}{u}\frac{\partial}{\partial p}\int_0^{\infty}\frac{cos\xi}{\left(\xi^2+\frac{4\pi^2p^2\nu^2}{u^2}\right)}\, d\xi \\
&= -\frac{1}{(4\pi\epsilon_0)^2}\frac{Ze^2}{m}\frac{e}{c^2 r}\frac{2\sin\theta}{u}\frac{\partial}{\partial p}\left[K_0\left(\frac{2\pi p\nu}{u}\right)\right]
\end{aligned}$$

where $K_0(x)$ is the Hankel function of order zero—we have already come across this function in the last chapter. We recall that

$$K_0(x) = \begin{cases} -\ln(x/2) & \text{for } x < 1 \\ \sim \exp(-x) & \text{for } x > 1 \end{cases}$$

Hence, E_ν decreases sharply for $\nu > u/(2\pi p)$. For frequencies low enough where the Hankel function may be approximated by $-\ln(x/2)$, we have

$$E_\nu = \frac{1}{(4\pi\epsilon_0)^2}\frac{Ze^2}{m}\frac{e}{c^2 r}\frac{2\sin\theta}{u}\frac{1}{p}$$

Hence, the radiation is

$$\begin{aligned}
\frac{2}{c\mu_0}\int |E_\nu|^2 r^2\, d\Omega\, d\nu &= \frac{Z^2e^6}{32\pi^4\epsilon_0^4 m^2 c^5\mu_0 u^2 p^2}\, 2\pi \int \sin^2\theta\, d(\cos\theta) d\nu \\
&= \frac{Z^2e^6}{12\pi^3\epsilon_0^2 m^2 c^3 u^2 p^2}\, d\nu
\end{aligned}$$

To obtain the cross-section for the process, one has to integrate the above expression over all possible values of p. The upper limit may be infinite, although in realistic cases there is a cut-off at a finite value as the nuclear field becomes screened by the electrons outside the nucleus. The lower limit cannot be pushed to zero—that would make the integral diverge. For the lower limit, one may set a value corresponding to the Compton length (the classical electron radius).

Problems

1. Examine the following argument: Maxwell's equations are linear so the sum of two solutions is also a solution. Two simple harmonic motions may combine to form an elliptic motion. Hence, the radiation due to a charge in such a motion will be the same as the sum of the radiations due to the constituent simple harmonic motion of the same charge.
2. Argue on the basis of Gauss' flux theorem that the $1/r$ radiation field (for both the electric and magnetic fields) must be transverse.

Chapter 18
Electrons in Material Media

18.1 Cerenkov Radiation

An electron moving with uniform velocity in a dielectric medium can emit electromagnetic radiation if its speed exceeds the speed of electromagnetic waves in the medium. This radiation is called Cerenkov radiation. It may appear somewhat paradoxical in view of our earlier conclusion that for a charged panicle in unaccelerated motion *in vacuo*, there is no radiation. We proceed to show that a variety of arguments that may be used to arrive at the no-radiation conclusion *in vacuo* may not hold in case of motion in a material medium.

The first argument is an appeal to the special theory of relativity according to which all frames in uniform relative motion are equivalent. Hence, to decide whether there is any radiation, we may go over to the frame in which the electron is at rest. In this frame the field is simply electrostatic, hence obviously, there is no radiation. In case the electron is moving in a material medium, the relevant factor is the relative velocity between the electron and the medium. In the rest frame of the electron, the charges in the medium (originating from polarization of the dielectric due to the field of the electron) are in motion and these may be looked upon as the source of the radiation.

A second argument is that the fields of the uniformly moving charge fall off as $1/r^2$ for large r, hence, the energy flux is given by the integral of the Poynting vector (which falls off at least as fast as $1/r^4$) over the sphere at infinity vanishes. However, the field expression

$$\mathbf{E} = \frac{e}{4\pi\epsilon_0} \frac{\mathbf{R} - \frac{\mathbf{u}}{c}R}{\left(R - \frac{\mathbf{u}\cdot\mathbf{R}}{c}\right)^3} \left(1 - \frac{u^2}{c^2}\right)$$

has a singularity at $u = c$. While this velocity is unattainable for an electron, the corresponding singularity occurs at $u = c/n$ (n being the refractive index of the medium, so c/n is the velocity of electromagnetic waves in the medium). We may, therefore, expect that the argument about the field falling off too fast may not hold good if $u \geq c/n$.

A. K. Raychaudhuri, *Classical Theory of Electricity and Magnetism*, Texts and Readings in Physical Sciences 21, https://doi.org/10.1007/978-981-16-8139-4_18

Lastly, we may look at the problem from the point of view of conservation of energy and momentum. A photon of energy $h\nu$ has a momentum $h\nu/c$. If E_i and E_f be the initial and final energy and p_i, p_f the corresponding momenta of the electron, then for the photon emission to be possible

$$\begin{aligned} \Delta E &= E_i - E_f, \\ \Delta p &= p_i - p_f = \frac{E_i - E_f}{c} = \frac{\Delta E}{c} \end{aligned} \tag{18.1}$$

But for an electron $E^2 = p^2c^2 + m_0^2c^4$, so that

$$E > pc \tag{18.2a}$$

$$E\,\Delta E = pc^2\Delta p \tag{18.2b}$$

From Eqs. (18.2a) and (18.2b)

$$\Delta E < c\Delta p \tag{18.3}$$

which contradicts Eq. (18.1), hence there cannot be any photon emission. We have used formulas from special relativity and the idea of photon from quantum theory, however, the argument holds with Newtonian formulas and field energy and momentum. Again this argument does not hold good in the dielectric medium, for Eq. (18.1) is changed to

$$\Delta E = \Delta p\frac{c}{n} < c\Delta p$$

which is consistent with Eq. (18.3). (Note that the above argument is not valid for an accelerated electron *in vacuo*, for the agency which accelerates the electron can compensate for the energy momentum imbalance. In other words, the electron and the field do not constitute a closed system in case of the accelerated electron.)

To make a mathematical calculation we recall that the radiation flux is still given by $\mathbf{E} \times \mathbf{H}$. In the absence of ferromagnetics, we may take $\mathbf{B} = \mu_0\mathbf{H}$ but now instead of $|\mathbf{E}| = c\mu_0|\mathbf{H}|$ as *in vacuo*, we have $|\mathbf{E}| = c|\mathbf{B}|/n$ in the dielectric. Thus, we have

$$S = \frac{cB^2}{\mu_0 n} \tag{18.4}$$

for the energy flux per unit area per unit time. Now we use Fourier transform to write

$$\int_{-\infty}^{\infty} |\mathbf{B}|^2\,dt = 2\pi\int_{-\infty}^{\infty} \mathbf{B}_\omega \cdot \mathbf{B}_{-\omega}\,d\omega = 4\pi\int_0^{\infty} |\mathbf{B}_\omega|^2\,d\omega \tag{18.5}$$

In the above, the factor 2 comes from the change of limits of integration and we have used the relation $\mathbf{B}_{-\omega} = \mathbf{B}_\omega^*$. The inhomogeneous wave equation for the vector potential now reads

$$\nabla^2 \mathbf{A} - \frac{n^2}{c^2}\frac{\partial \mathbf{A}}{\partial t^2} = -\mu_0 \mathbf{j}, \qquad \mu = \mu_0, \quad \epsilon = n^2$$

so that in the far field region

$$\mathbf{A}_\omega = \frac{\mu_0}{4\pi r}\int \mathbf{j}_\omega(\mathbf{r}')e^{-i\omega n|\mathbf{r}-\mathbf{r}'|/c}dv' \tag{18.6}$$

$$\mathbf{B}_\omega(\mathbf{r}) = \nabla \times \mathbf{A}_\omega(\mathbf{r}) = -\frac{i\omega n}{c}\mathbf{n} \times \mathbf{A}_\omega(\mathbf{r}) \tag{18.7}$$

(In the above, n, the scalar, is the refractive index and $\mathbf{n}$ is the unit vector in the direction of r, there should not be any confusion between the two.)

Using Eqs. (18.4)–(18.7), the energy crossing the surface element subtending a solid angle $d\Omega$ at a distance r may now be written as

$$\begin{aligned}\frac{c}{\mu_0 n}\int B^2 r^2 d\Omega dt &= \frac{\mu_0 n}{4\pi c}\int_0^\infty d\omega\,\omega^2 \left|\int_{v'} \left(\mathbf{n} \times \mathbf{j}_\omega(\mathbf{r}')\right) e^{\frac{i\omega n}{c}\mathbf{n}\cdot\mathbf{r}'}dv'\right|^2 d\Omega \qquad (18.8)\\ &= \frac{\mu_0 n}{16\pi^3 c}\int_0^\infty d\omega\,\omega^2 \left|\int_{v'}\int_{t'} \left(\mathbf{n} \times \mathbf{j}(\mathbf{r}', t')\right) e^{\frac{i\omega n}{c}(t'-\mathbf{n}\cdot\mathbf{r}')}dv'\right|^2 d\Omega\end{aligned}$$

In the last step above we have replaced $\mathbf{j}_\omega$ by its inverse Fourier transform $\frac{1}{2\pi}\int \mathbf{j}(\mathbf{r}', t)e^{-i\omega t}dt$ (cf. Eq. (14.35)).

Now for a point charge e moving with uniform velocity u along the x-axis, if the origin be taken at the position of the charge at time $t' = 0$, we have for the current density vector

$$\mathbf{j}(\mathbf{r}', t') = e\mathbf{u}\,\delta(x' - ut')\,\delta(y')\,\delta(z') \tag{18.9}$$

Using this in (18.8) one gets the energy emitted in the range $d\omega$ and solid angle $d\Omega$ (taking the x-axis as the polar axis)

$$\begin{aligned}&\frac{\mu_0 n e^2 u^2}{16\pi^3 c}\omega^2 d\omega \sin^2\theta d\Omega \left|\int dv' \int dt' e^{-i\omega(t'-\frac{n}{c}x'\cos\theta)}\delta(x'-ut')\delta(y')\delta(z')\right|^2\\ &= \frac{\mu_0 n e^2 u^2}{16\pi^3 c}\omega^2 d\omega\, \sin^3\theta d\theta d\varphi \left|\int dt' e^{-i\omega t'(1-\frac{n}{c}u\cos\theta)}\right|^2 \qquad (18.10)\end{aligned}$$

If the limits of t' are taken as $-\infty$ and $+\infty$, the integral diverges (it is a δ-function)—this is to be expected as it involves radiation over an infinite time span. In actuality, the radiation will occur over a finite time interval and we need to calculate the energy radiated over a finite path length of the particle.

Thus taking the integral between the limits $t' = -l/2u$ to $+l/2u$ (l being the length), the total radiation emitted in the angular frequency range ω to $\omega + d\omega$ is

$$\frac{\mu_0 n e^2 u^2}{8\pi^2 c}\omega^2 d\omega \int_0^{\pi} d\theta \, \sin^3\theta \left| \frac{2\sin\left(\frac{\omega l}{2u}(1-\frac{nu}{c}\cos\theta)\right)}{\omega\left(1-\frac{nu}{c}\cos\theta\right)} \right|^2 \tag{18.11}$$

The angular frequency in the optical region is large ($\sim 10^{15}\ \mathrm{s}^{-1}$), hence if $1-\frac{nu}{c}\cos\theta$ departs even slightly from zero, the denominator becomes a large quantity while the numerator, the sine function being bounded, remains below 4—thus the integrand then practically vanishes. We therefore get a finite radiation only over a cone with semi-vertical angle $\cos^{-1}\frac{c}{nu}$. Physically, the underlying situation is that the waves radiated by the electron fluctuate in phase very quickly and the resultant vanishes except for that particular θ where the phase differences vanish.

Returning to Eq. (18.11), as the contribution to the integral is significant only for $\cos\theta = \frac{c}{nu}$, we replace $\sin^2\theta$ by $1-\frac{c^2}{n^2u^2}$ and extend the limits of the integral to $\cos\theta \equiv \xi = \pm\infty$. We thus obtain the following expression from Eq. (18.11)

$$\frac{\mu_0 n e^2 u^2}{8\pi^2 c}\omega^2 d\omega \left(1-\frac{c^2}{n^2u^2}\right)\int_{-\infty}^{+\infty} d\xi \, \frac{\sin^2\frac{\omega l}{2u}\left(1-\frac{nu}{c}\xi\right)}{\omega^2\left(1-\frac{nu\xi}{c}\right)^2}$$

Using the result

$$\int_{-\infty}^{+\infty}\frac{\sin^2 x}{x^2}dx = \pi$$

finally we get the radiated energy in the range ω to $\omega + d\omega$ as

$$\frac{\mu_0 e^2 l}{16\pi}\left(1-\frac{c^2}{n^2u^2}\right)\omega d\omega$$

The corresponding value for the frequency range ν to $\nu + d\nu$

$$\frac{\mu_0 \pi e^2 l}{4}\left(1-\frac{c^2}{n^2u^2}\right)\nu d\nu$$

is obtained using $\nu = \omega/2\pi$.

Thus the energy radiated is proportional to the path length, and exists only if $u \geq c/n$. The simple theory shows a divergence as ν goes to infinity. This is due to our erroneous assumption that the refractive index n is independent of the frequency. In reality, the molecular polarizability depends on the frequency of the incident electric field—in particular if the frequency be high enough, the molecules cannot respond to the electric field and thus the dielectric constant and consequently the refractive index tends to the value unity. Hence for such frequencies, the relation $u \geq c/n$ may not be satisfied and there would thus be no Cerenkov radiation. (As the semi-vertical angle of the cone of Cerenkov radiation depends on n which in turn depends on the frequency, the Cerenkov radiation shows a dispersion.) The number of quanta emitted in the frequency range ν to $\nu + d\nu$ (per unit path length) is

$$N_\nu d\nu = \frac{\mu_0 e^2 \pi}{4h}\left(1 - \frac{c^2}{n^2u^2}\right) = \frac{\pi\alpha}{2c}\left(1 - \frac{c^2}{n^2u^2}\right) \tag{18.12}$$

where $\alpha = \frac{\mu_0}{4\pi}\frac{e^2 c}{\hbar}$ is the fine structure constant which is approximately $1/137$. So long as n can be regarded as independent of frequency, $N(\nu)d\nu$ can be taken to be the same for all ν. Taking $d\nu = 6 \times 10^{14}$ Hz, the frequency interval for the optical region, we have

$$N_\nu d\nu = \frac{\pi}{2 \times 137}\frac{6 \times 10^{14}}{3 \times 10^8}\left(1 - \frac{c^2}{n^2u^2}\right) < 10^5 \text{ quanta/m}$$

It may be verified that the energy loss of the electron due to Cerenkov radiation is negligible compared to the electron's kinetic energy for path length of a few centimetres and this justifies our taking u as a constant[1] in the calculation. Besides, in the ultra-relativistic region, even significant decrease in energy causes only a small change in velocity.

The theory presented above gives the Cerenkov radiation due to a charged particle. However, the neutron which has no electrical charge does give rise to this type of radiation. This can be ascribed to the magnetic moment of the neutron and the number of quanta emitted may be calculated by writing for the current density vector $\mathbf{j} = \nabla \times \mathbf{M}$, where $\mathbf{M}$ is the magnetic moment of the neutron.

18.2 Scattering of Electromagnetic Waves by Electrons

The electric vector in an incident radiation sets the electrons into oscillations and the electrons, in their turn, being accelerated begin radiating. This is the basic idea of scattering of electromagnetic radiation. Ordinarily under the action of the wave field the electron never attains a velocity comparable with c, so one need not take into account relativistic effects. Further as in the radiation $|\mathbf{B}| = |\mathbf{E}|/c$, the magnetic force $e\mathbf{u} \times \mathbf{B}$ is consequently much smaller than $e\mathbf{E}$ and can be neglected.

We shall consider the electron to be elastically bound to its equilibrium position so that there will be a restoring force varying directly as the displacement from the equilibrium position. The motion of the electron is also subject to damping which may arise either from collision or due to the radiation of energy. The former is proportional to the velocity while the radiation reaction is $\mu_0 e^2 \ddot{\mathbf{u}}/(6\pi c)$ (proportional to the rate of change of acceleration). Although formally different, the effect of both types of damping is very similar in the problem we are now considering. Further, as our interest is more in cases where collisions are rather rare, we shall take the damping to be due to radiation reaction.

[1] The energy lost per cm is less than 1000×10 eV, i.e., $\lesssim 10^4$ eV/cm. Hence even for a path length of 1 m, the energy loss is less than 1 MeV. An electron whose energy is several MeV (say ~ 10 MeV) would suffer a negligible change in velocity by losing 1 MeV.

Assuming the incident radiation to be monochromatic, and using complex variables, the equation of motion of the electrons are

$$m\ddot{x} + \gamma x - \frac{\mu_0 e^2}{6\pi c}\dddot{x} = eE_0 e^{ipt} \tag{18.13}$$

The resulting motion will be a superposition of the forced and natural vibration of which only the forced part survives. Hence substituting $x = Ae^{ipt}$

$$\left(-mp^2 + \gamma + \frac{i\mu_0 e^2}{6\pi c}p^3\right)A = eE_0$$

Writing

$$\frac{\gamma}{m} \equiv p_0^2 \quad \text{and} \quad \frac{\mu_0 e^2}{6\pi mc} \equiv \frac{1}{\omega}$$

We get $A = \dfrac{eE_0}{m\left(p_0^2 - p^2 + \frac{ip^3}{\omega}\right)}$. The induced dipole moment and its second time derivative are therefore

$$ex = \frac{e^2 E_0 e^{ipt}}{m\left(p_0^2 - p^2 + \frac{ip^3}{\omega}\right)} \tag{18.14}$$

$$e\ddot{x} = -\frac{e^2 E_0 p^2 e^{ipt}}{m\left(p_0^2 - p^2 + \frac{ip^3}{\omega}\right)} \tag{18.15}$$

The complex form indicates a phase difference between the oscillating dipole moment and the incident electric field. The real part gives

$$e\ddot{x} = -\frac{e^2 E_0}{m}\frac{\left(\frac{p_0^2}{p^2} - 1\right)\cos pt + \frac{p}{\omega}\sin pt}{\left(\frac{p_0^2}{p^2} - 1\right)^2 + \frac{p^2}{\omega^2}} \tag{18.16}$$

As we have already seen the radiation field due to such a dipole is

$$cB_\phi = E_\theta = \frac{e\ddot{x}\sin\theta}{4\pi\epsilon_0 rc^2}$$

$$B_r = B_\theta = E_r = E_\phi = 0$$

where we are using spherical polar coordinates with the x-axis as the polar axis. The energy radiated in time dt in solid angle $d\Omega$ is therefore

$$\frac{1}{\mu_0 c} E_\theta^2 r^2 d\Omega dt = \frac{1}{\mu_0 c^5}\left(\frac{e\ddot{x}}{4\pi\epsilon_0}\right)^2 \sin^2\theta d\Omega dt$$

$$= \frac{1}{\mu_0 c^5}\left(\frac{e^2 E_0}{4\pi\epsilon_0 m} \frac{\left(\frac{p_0^2}{p^2}-1\right)\cos pt + \frac{p}{\omega}\sin pt}{\left(\frac{p_0^2}{p^2}-1\right)^2 + \frac{p^2}{\omega^2}}\right)^2 \sin^2\theta d\Omega dt$$

If we take the time average, i.e., integrate over a time period T $(= 2\pi/p)$ and divide by T, the terms with $\cos pt$ and $\sin pt$ yield zero and $\langle\cos^2 pt\rangle = \langle\sin^2 pt\rangle = \frac{1}{2}$. Thus the time averaged radiation rate is

$$\frac{1}{2\mu_0 c^5}\left(\frac{e^2 E_0}{4\pi\epsilon_0 m}\right)^2 \frac{\sin^2\theta}{\left(\frac{p_0^2}{p^2}-1\right)^2 + \frac{p^2}{\omega^2}} d\Omega \tag{18.17}$$

In deducing the above, we have considered the forced oscillation of the electron along the x-axis only. If the incident wave be unpolarized, then for direction of propagation along the z-axis, there would be electric field components along both x- and y-axes. Consequently in the electronic motion there will be a superposition of vibrations in these directions, which are uncorrelated but with equal amplitudes on average. Then, if we consider the scattered beam in a direction ϕ to the incident beam, the total scattered energy per unit time in the solid angle $d\Omega$ will be (note that θ for the x-axis is $\frac{\pi}{2} - \phi$ and for y-axis is $\pi/2$, see Fig. 18.1)

$$\frac{1}{4\mu_0 c^5}\left(\frac{e^2 E_0}{4\pi\epsilon_0 m}\right)^2 \frac{(1+\cos^2\phi)}{\left(\frac{p_0^2}{p^2}-1\right)^2 + \frac{p^2}{\omega^2}} d\Omega \tag{18.18}$$

as the amplitudes of the electric vector in x- and y-directions are

$$\langle E_x^2\rangle = \langle E_y^2\rangle = \frac{1}{2}E_0^2$$

The factor $(1+\cos^2\phi)$ in the numerator is called the polarization factor.

We consider first the limiting case of free electrons, i.e., $p_0^2 \ll p^2$. The p^2/ω^2 term is in any case usually small ($\frac{1}{\omega} = \frac{\mu_0 e^2}{6\pi mc} \approx 6.3\times 10^{-24}$ s, while p for the X-ray region $10^{18}\,\text{s}^{-1}$), hence even for X- or γ-rays, the expression (18.18) reduces to

$$\frac{1}{4\mu_0 c^5}\left(\frac{e^2 E_0}{4\pi\epsilon_0 m}\right)^2 (1+\cos^2\phi)\, d\Omega$$

Defining the differential scattering cross-section $d\sigma$ as follows:

$$\frac{d\sigma}{d\Omega} = \frac{\text{energy radiated / unit time / unit solid angle}}{\text{energy incident / unit time / unit area } (\perp \text{ to the direction of propagation})}$$

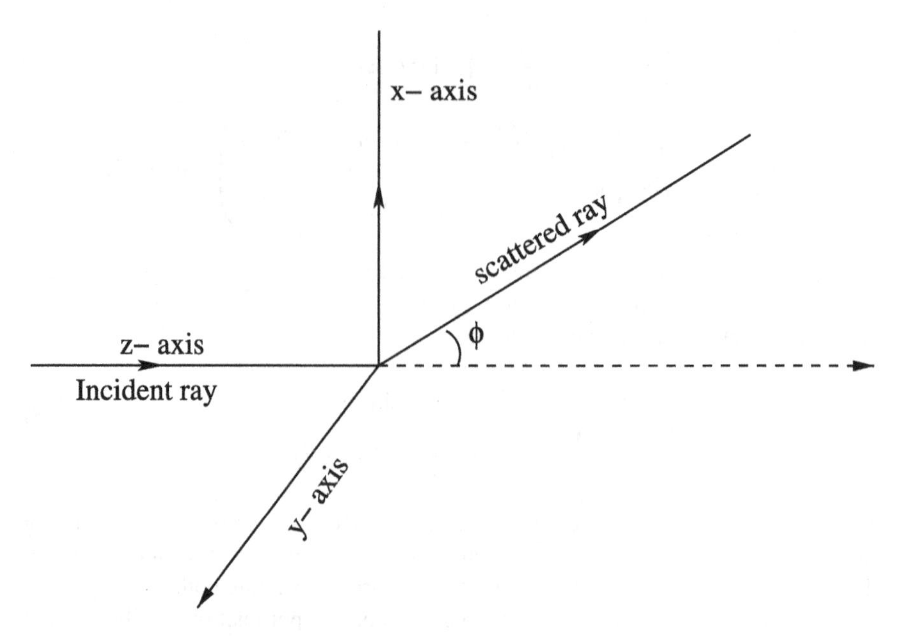

Fig. 18.1 Geometry of scattering of electromagnetic wave scattered from a bound electron at the origin

we get

$$\frac{d\sigma}{d\Omega} = \frac{1}{2}\left(\frac{e^2}{4\pi\epsilon_0 mc^2}\right)^2 (1+\cos^2\phi) \tag{18.19}$$

since the denominator is $E_0^2/(2\mu_0 c)$. To obtain the total scattering cross-section, we integrate over all angles

$$\begin{aligned}\frac{d\sigma}{d\Omega} = \int \frac{d\sigma}{d\Omega} d\Omega &= \frac{1}{2}\left(\frac{e^2}{4\pi\epsilon_0 mc^2}\right)^2 \int_0^{\pi} (1+\cos^2\phi)\, 2\pi \sin\phi d\phi \\ &= \frac{8\pi}{3}\frac{1}{(4\pi\epsilon_0)^2}\frac{e^4}{m^2c^4} \qquad \boxed{\frac{8\pi}{3}\frac{e^4}{m^2c^4}\ \text{(Gaussian)}}\end{aligned} \tag{18.20}$$

The type of scattering that we have just considered is known as Thomson scattering and (18.19) and (18.20) are called Thomson scattering cross-sections. Observationally, this type of scattering occurs if the frequency of the incident radiation is high as in X- or γ-rays. The notable feature of this scattering is the independence of the scattering cross-section on wavelength.

In the opposite limit, when the electron is so strongly bound that $p_0^2 \gg p^2$, we get

$$\sigma = \frac{8\pi}{3}\frac{1}{(4\pi\epsilon_0)^2}\frac{e^4}{m^2c^4}\frac{p^4}{p_0^4} \propto \frac{1}{\lambda^4} \tag{18.21}$$

for the scattering cross-section. The strong dependence on wavelength is characteristic of Rayleigh scattering which is held responsible for the blue of the sky and the redness of the rising and setting sun.

Thus on the one hand we expect Thomson scattering for X- and γ-rays and on the other Rayleigh type scattering for comparatively low frequencies, the cross-section in either case being small. In between these two extremes lies the condition of resonance. For exact resonance $p_0 = p$ and

$$\sigma = \frac{8\pi}{3}\frac{1}{(4\pi\epsilon_0)^2}\frac{e^4}{m^2c^4}\frac{\omega^2}{p^2} \tag{18.22}$$

which can have quite large values.

18.3 Dispersion and Absorption

The above analysis has shown that the induced dipole moments depend on the frequency of the incident electric field. Consequently, the dielectric constant and the refractive index will also depend on the frequency. This will account for the dispersion phenomenon. We have seen that the radiation damping term makes the induced dipole moment complex (this will be the case for collision damping as well—indeed any force term involving an odd order derivative of the displacement will bring in an imaginary part). This leads to a complex dielectric constant which can be interpreted as an absorption of the radiation. (Cf. the discussion of wave propagation in conducting media in Chap. 11.)

We shall consider the force of restitution to be isotropic so that the polarization (recall the definition—it is the dipole moment per unit volume) is therefore given by

$$\mathbf{P} = \frac{Ne^2e^{ipt}\mathbf{E_0}}{m\left(p_0^2 - p^2 + \frac{ip^3}{\omega}\right)}$$

where N is the number of electrons per unit volume, each having the natural angular frequency p_0. (We have used N to represent the number density, rather than n used in Chap. 5, so as not to confuse with the notation for the refractive index.) The volume polarizability is thus

$$\alpha = \frac{Ne^2}{m(p_0^2 - p^2 + ip^3/\omega)}$$

Hence the Clausius–Mossotti relation (5.23) reads

$$\frac{\epsilon_r - 1}{\epsilon_r + 2} = \frac{Ne^2}{3\epsilon_0 m(p_0^2 - p^2 + ip^3/\omega)} \tag{18.23}$$

The above relation needs to be modified when there are different natural frequencies. Let f_i be the fraction of the total number of electrons with a natural angular frequency p_i. We have to add up the contributions due to all these in the volume polarizability, which consequently modifies (18.23) to

$$\frac{\epsilon_r - 1}{\epsilon_r + 2} = \frac{Ne^2}{3\epsilon_0 m} \sum_i \frac{f_i}{(p_i^2 - p^2 + ip^3/\omega)} \tag{18.24}$$

Note that ω is a characteristic constant for electrons independent of p or p_i. If $p_i^2 - p^2$ be not small for any i (i.e., if the incident frequency be not close to any resonance frequency) the imaginary part may be neglected, hence one may write

$$\frac{\epsilon_r - 1}{\epsilon_r + 2} = \frac{Ne^2}{3m\epsilon_0} \sum_i \frac{f_i}{p_i^2 - p^2}$$

$$\text{or,} \qquad \epsilon_r = \frac{1 + \dfrac{2Ne^2}{3m\epsilon_0} \displaystyle\sum_i \frac{f_i}{p_i^2 - p^2}}{1 - \dfrac{Ne^2}{3m\epsilon_0} \displaystyle\sum_i \frac{f_i}{p_i^2 - p^2}} \tag{18.25}$$

Now the refractive index $n = \sqrt{\epsilon_r}$ and if n and ϵ_r be close to unity, Eq. (18.25) may be transformed to the form

$$n = 1 + \sum_i \frac{\alpha_i \lambda^2}{\lambda^2 - \lambda_i^2} \tag{18.26}$$

where λ's are the wavelength *in vacuo* corresponding to the frequencies p. The above equation is known as Sellmeier's formula. It agrees well with observations except near resonance frequencies.

Near any resonance, the imaginary part must be taken into account and we can usually retain only the term corresponding to the resonance in the summation in (18.24). (This is not true if there be two close values of p_i, e. g., for the D lines of sodium.) The dielectric constant is now complex and writing $\sqrt{\epsilon_r} = n(1 - ik)$, we have from (18.24)

$$n = 1 + \frac{Ne^2}{2m\epsilon_0} \frac{f_i(p_i^2 - p^2)}{\left(p_i^2 - p^2\right)^2 + \frac{p^6}{\omega^2}} \tag{18.27}$$

$$nk = \frac{Ne^2}{2m\epsilon_0} \frac{f_i p^3}{\omega\left((p_i^2 - p^2)^2 + \frac{p^6}{\omega^2}\right)} \tag{18.28}$$

where we have taken $\mathrm{Re}(\epsilon_r) = 1$, so that $\mathrm{Re}(\epsilon_r) + 2 = 3$. The plane wave equation is now of the form

$$\sim e^{ip(t - x\sqrt{\epsilon_r}/c)} = e^{-pnkx/c}\, e^{ip(t - xn/c)}$$

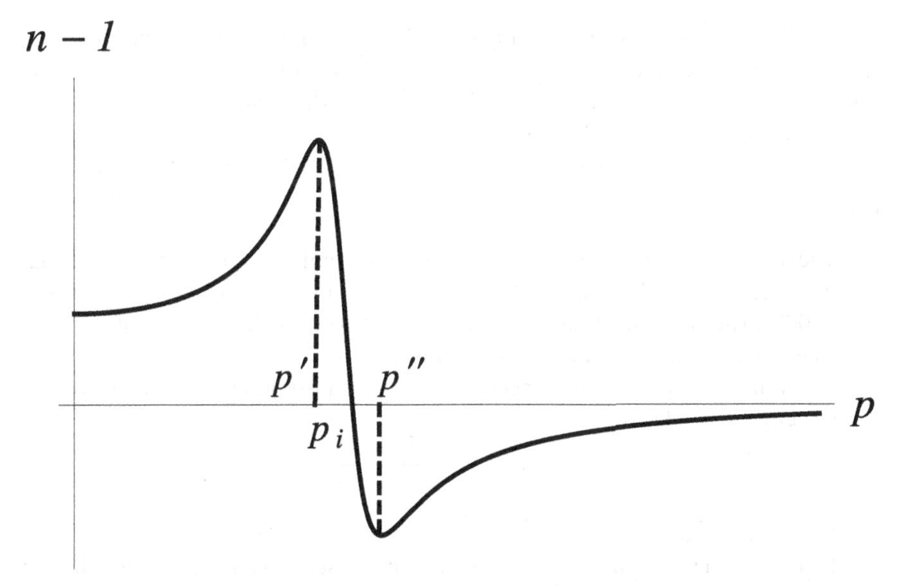

Fig. 18.2 Dispersion of electromagnetic radiation by dielectric

so that n gives the refractive index and $\exp(-nkpx/c)$ shows an absorption of the wave. Figure 18.2 shows the variation of n with p as given by (18.27).

We see that n increases with p as we approach p_i, attaining a maximum at p' (where $p' < p_i$), then decreases with increasing p. It reaches a minimum at p'' (where $p'' > p_i$) and after that again shows an increase with increasing p. The region $p' < p_i < p''$ may be identified with the absorption band and the phenomenon of n decreasing with increasing frequency (decreasing wavelength) is referred to as anomalous dispersion.

For a range of frequencies $p > p_i$, the refractive index is less than unity, hence no Cerenkov radiation of these frequencies can be observed in the medium under consideration, however high the electron velocity may be. A refractive index less than unity is also obtained if the incident frequency be sufficiently high so that the right-hand side in (18.24) becomes negative. This is the case for X-rays as well as in plasma and the ionosphere. The phase velocities are now greater than the speed of light *in vacuo* but this does not contradict the special theory of relativity as the velocity with which energy (i.e., a signal) travels remains below c.

In case the electrons are really free (fraction $f_i = 1$ and $p_i = 0$), one may have

$$n^2 = 1 - \frac{Ne^2}{m\epsilon_0 p^2} = 1 - \frac{Ne^2\lambda^2}{4\pi^2 m\epsilon_0 c^2}$$

so that there exists a value of wavelength above which the dielectric constant is negative and the refractive index is purely imaginary. There cannot then be any propagated wave motion and one gets a 'total reflection' at all angles of incidence.

Quantum mechanics gives formally very similar formulae for the dielectric constant and refractive index, but f_i's are then the oscillator strength for the transition giving rise to radiation of frequency $p_i/2\pi$. Further, while classically the characteristic frequencies are obtained empirically, quantum mechanics allows one, at least in principle, to calculate both the transition frequencies and the transition probabilities.

Problems

1. The atomic number of carbon (C) is 6 and the atomic weight 12. Assuming that the electrons in $^{12}_{6}C$ scatter incoherently, show that its mass scattering coefficient is 0.2. (The mass scattering coefficient is defined as the fraction of incident energy scattered per unit mass of the substance.)
2. Assuming that in the region concerned there is no absorption frequency p_i, obtain Cauchy's formula
$$n = A + \frac{B}{\lambda^2} + \frac{C}{\lambda^4}$$
from Sellmeier's formula.
3. Find the value of the differential scattering cross-section if the incident radiation be elliptically polarized.
4. The following table gives the values of the refractive index of sodium vapour for different wavelengths as observed by Wood.

Wavelength (in A.U.)	$(n - 1) \times 10^5$
5400	−15
5700	−40
5750	−50
5807	−91
5843	−150
5850	−180
5867	−310
5877	−460
5882	−920
5886	−2600
5888.4	−5600
5889.6	−38600
5896.4	+38000
5897	+9000
5899	+2000
5904	+1000

Plot a graph to locate p_i, p' and p'', and discuss the wavelengths of Cerenkov radiations if the electron velocity in sodium vapour be $0.95 \times c$.

Chapter 19
Motion of Charged Particles in Electromagnetic Fields

The programme of our present study is to investigate the solutions of the equation of motion of a charged particle

$$\frac{d\mathbf{p}}{dt} = e\,(\mathbf{E} + \mathbf{v} \times \mathbf{B}) \tag{19.1}$$

where $\mathbf{p}$ is the momentum vector of the particle of charge e and the right-hand side, known as the Lorentz expression for the force, gives the force on the particle due to the electric field $\mathbf{E}$ and magnetic field $\mathbf{B}$, $\mathbf{v}$ being the velocity of the particle. The Lorentz force expression is sometimes considered as a definition of electric and magnetic intensities, the Maxwell equations giving the laws governing these fields. Alternatively, one may regard the Maxwell equations as the description of two fields, whose interaction with matter is given by the Lorentz force expression. Equation (19.1) does not contain any radiation reaction term—the implication being that these forces are negligible compared to the Lorentz force. Further, the fields $\mathbf{E}$ and $\mathbf{B}$ will be regarded as independent of the position or velocity of the charged particle whose motion is being studied. Thus rigorously speaking, our considerations will apply only to the motion of test charges.

Case I. *Motion in a spatially uniform and temporally constant magnetic field* $\mathbf{B}$.
In this case, Eq. (19.1) reads

$$\frac{d\mathbf{p}}{dt} = e\mathbf{v} \times \mathbf{B} \tag{19.2}$$

By taking the scalar product of Eq. (19.2) with $\mathbf{p}$ we find that p^2 is a constant of motion, so that even in the relativistic case, the mass remains constant during motion. Also taking the component of (19.2) in the direction of $\mathbf{B}$, we get $p_{\parallel} = \text{constant}$, where $p_{\parallel}$ represents the component of $\mathbf{p}$ in the direction of $\mathbf{B}$. Thus, the particle

A. K. Raychaudhuri, *Classical Theory of Electricity and Magnetism*, Texts and Readings in Physical Sciences 21, https://doi.org/10.1007/978-981-16-8139-4_19

drifts with a constant velocity in the direction of the applied field. As both p^2 and $p_\parallel$ are constants, the magnitude of $p_\perp$, (the component of $\mathbf{p}$ normal to $\mathbf{B}$) is also constant. Using these, we get

$$\begin{aligned}\frac{d\mathbf{v}}{dt} &= \frac{e}{m}\mathbf{v}\times\mathbf{B}\\ \frac{d^2\mathbf{v}_\perp}{dt^2} &= \frac{d^2\mathbf{v}}{dt^2} = \frac{e}{m}\dot{\mathbf{v}}\times\mathbf{B} = \frac{e^2}{m^2}(\mathbf{v}\times\mathbf{B})\times\mathbf{B} = -\frac{e^2}{m^2}\left(B^2\mathbf{v} - (\mathbf{v}\cdot\mathbf{B})\mathbf{B}\right)\\ &= -\frac{e^2B^2}{m^2}(\mathbf{v}-\mathbf{v}_\parallel) = -\frac{e^2B^2}{m^2}\mathbf{v}_\perp\end{aligned}$$

the solution of which is

$$\mathbf{v}_\perp = \boldsymbol{\alpha}\cos\frac{eBt}{m} + \boldsymbol{\beta}\sin\frac{eBt}{m} \tag{19.3}$$

where $\boldsymbol{\alpha}$ and $\boldsymbol{\beta}$ are constant vectors. The condition $\mathbf{v}_\perp^2$ = constant gives

$$\alpha^2 = \beta^2 \quad \text{and} \quad \boldsymbol{\alpha}\cdot\boldsymbol{\beta} = 0 \tag{19.4}$$

Also, integrating Eq. (19.3) with a suitable choice of origin

$$\mathbf{r}_\perp = \frac{m}{Be}\left(\boldsymbol{\alpha}\sin\frac{eBt}{m} - \boldsymbol{\beta}\cos\frac{eBt}{m}\right) \tag{19.5}$$

Hence from Eqs. (19.3)–(19.5)

$$r_\perp^2 = \frac{m^2v_\perp^2}{B^2e^2} \tag{19.6}$$

showing that the projection of the path of the particle in the plane normal to $\mathbf{B}$ is a circle of radius

$$r_\perp = \frac{mv_\perp}{Be} \tag{19.7}$$

Note that the angular velocity and the time to complete one circle in the xy-plane are independent of the velocity of the particle. The angular velocity Be/m is called the cyclotron frequency. Hence the path, in general, is a helix with its axis parallel to the direction of the magnetic field. Equation (19.7) has played a very important role in the experimental study of microscopic particles—an observation of the curvature of the path in a known uniform magnetic field enables one to determine the momentum of the particle, if its charge is known. Alternatively, the formula may also be used to determine the value of e/m. To conclude the discussion of this case, we note that in view of Eqs. (19.3) and (19.4), if we choose the coordinate axes x and y suitably in the plane perpendicular to $\mathbf{B}$, we may write

$$v_x = A\cos\frac{eBt}{m} \quad \text{and} \quad v_y = -A\sin\frac{eBt}{m} \tag{19.8}$$

Case II. **E** *and* **B** *both spatially uniform and constant in time and perpendicular to one another.*

There are two possibilities here.

(i) First, we shall consider the case $|\mathbf{E}| < c|\mathbf{B}|$. We can then introduce a velocity **u** satisfying the relation

$$\mathbf{E} + \mathbf{u} \times \mathbf{B} = 0 \tag{19.9}$$

However **u** is not uniquely determined by the above relation, so we introduce the further condition that **u** is perpendicular to **B** as well. Then Eq. (19.9) gives

$$\mathbf{u} = \frac{(\mathbf{E} \times \mathbf{B})}{B^2} \tag{19.10}$$

From Eq. (19.1) we find

$$\frac{d\mathbf{p}}{dt} = e\,(\mathbf{E} + \mathbf{v} \times \mathbf{B}) = e\,(\mathbf{E} + \mathbf{u} \times \mathbf{B}) + e\,((\mathbf{v} - \mathbf{u}) \times \mathbf{B})$$

$$\frac{d\mathbf{p}}{dt} - m\frac{d\mathbf{u}}{dt} = e(\mathbf{v} - \mathbf{u}) \times \mathbf{B} \tag{19.11}$$

If the velocity of the electrically charged particle be small enough to justify taking the mass to be constant, the above equation reduces to

$$\frac{d}{dt}(\mathbf{v} - \mathbf{u}) = \frac{e}{m}(\mathbf{v} - \mathbf{u}) \times \mathbf{B} \tag{19.12}$$

This is effectively Eq. (19.2) of the previous case and thus represents the motion of a particle of velocity $(\mathbf{v} - \mathbf{u})$ in a uniform magnetic field **B**. We can therefore just take over the results deduced previously. If the z-axis be taken in the direction of the magnetic field **B**, y-axis in the direction of **E**, then **u** will be in the x-direction and we have

$$v_z - u_z = \text{constant} = v_z \tag{19.13}$$

$$v_y^2 + (v_x - u)^2 = \text{constant} \tag{19.14}$$

$$v_y = -A\sin\left(\frac{eBt}{m}\right) \quad \text{and} \quad v_x - u = A\cos\left(\frac{eBt}{m}\right) \tag{19.15}$$

Thus, besides the sinusoidal velocities in the x- and y-directions (with phase difference $\pi/2$) there is a constant drift velocity in the direction orthogonal to **E** and **B** given by Eq. (19.10). This drift is in the direction of the Poynting vector and is independent of the charge or mass of the particle (of course the drift does not exist if the charge vanishes). Thus this drift can be made to vanish by going over to a relatively moving frame—in that frame the electric field vanishes and the motion will be a simple helical one under the action of the uniform magnetic field. Integrating Eqs. (19.14) and (19.15), we get

$$
\begin{aligned}
y - y_0 &= r\cos\left(\frac{eBt}{m}\right) \\
x - x_0 &= ut + r\sin\left(\frac{eBt}{m}\right)
\end{aligned}
\tag{19.16}
$$

with

$$r = \frac{mA}{Be} = \frac{m}{Be}\sqrt{v_y^2 + (v_x - u)^2}$$

The curve represented by Eq. (19.16) is called a trochoid. In case v_x vanishes at $t = 0$, from Eq. (19.15) we have

$$
\begin{aligned}
u &= -A = -\frac{eBr}{m} \\
r &= -\frac{mu}{eB} = -\frac{mE}{eB^2} \\
y &= y_0 + r\cos\frac{eBt}{m} \\
x &= x_0 - r\left(\frac{eBt}{m} - \sin\frac{eBt}{m}\right)
\end{aligned}
\tag{19.17}
$$

The curve represented by Eq. (19.17) is called a cycloid.

The above theory has been applied to an experimental determination of e/m for electrons. Suppose AB is a plate emitting electrons (say under the action of ultraviolet light), CD another plate at a higher potential than AB so that the electrons experience a force towards CD due to the action of the electric field. However, there is also a uniform magnetic field in a direction perpendicular to the plane of the paper. The paths of the electrons are cycloids as shown in Fig. 19.1. Obviously if the plate CD is beyond the line EFGH, it will not receive them. A detecting arrangement thus enables us to determine the distance between AB and EFGH and from the theory just developed this distance is $|r| = mE/(eB^2)$. Hence knowing E and B, the ratio e/m is determined.

In case $\mathbf{E}$ is not perpendicular to $\mathbf{B}$ but makes some angle, the analysis needs only a little modification. The component of $\mathbf{E}$ parallel to $\mathbf{B}$, i.e., in our notation E_z, will give a uniform acceleration in the z-direction while for motion in the xy-plane, we are simply to replace E by E_y in our analysis.

(ii) Now let us analyze the case $|\mathbf{E}| \geq c|\mathbf{B}|$. Our earlier arguments based as it is on Eq. (19.10) breaks down in this case. We first show that with $E \geq cB$, the motion will necessarily be relativistic, i.e., the particle velocity will be comparable with c.

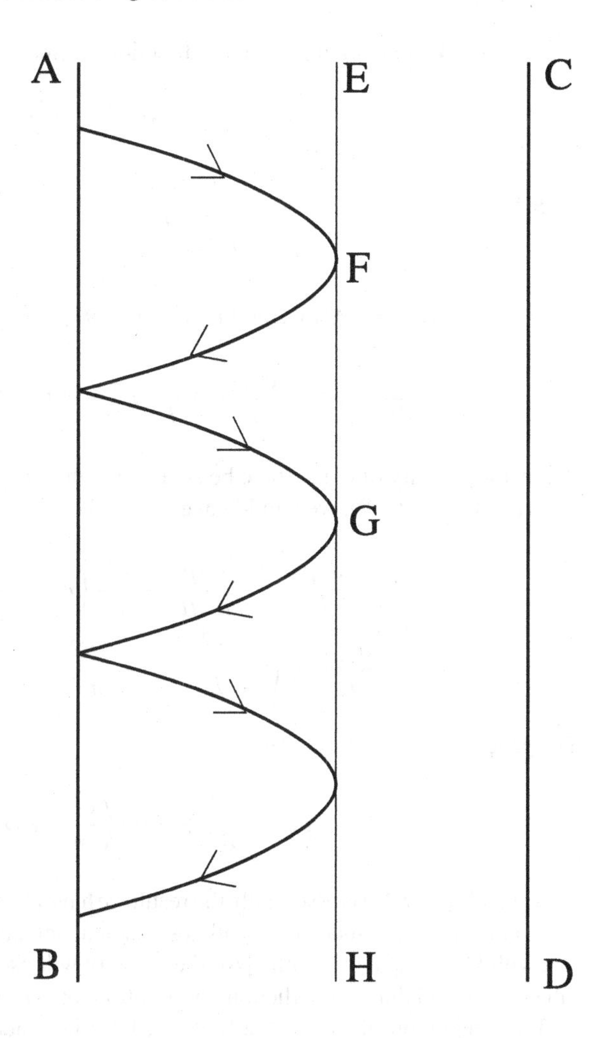

Fig. 19.1 Motion of an electron in crossed electric field **E** and magnetic field **B**, with **B** being perpendicular to the plane of the paper and the electric field directed towards left from the positively charged plate CD

If m be the mass at velocity v (momentum $p = mv$) we have from the special theory of relativity

$$m^2c^4 = p^2c^2 + m_0^2c^4 \tag{19.18}$$

where m_0 is the rest mass. As mc^2 is the total energy, we have

$$c^2\frac{dm}{dt} = e\mathbf{v}\cdot(\mathbf{E}+\mathbf{v}\times\mathbf{B}) = e\mathbf{v}\cdot\mathbf{E} = \frac{e}{m}\mathbf{p}\cdot\mathbf{E}$$

$$c^2\frac{dm}{dt} = \frac{eEp_y}{m}$$

since **E** and **B** are along the y- and z-directions, respectively. Integrating

$$\frac{dp_x}{dt} = ev_y B = \frac{ep_y B}{m} = \frac{c^2 B}{E}\frac{dm}{dt} \tag{19.19}$$

we get

$$p_x = \frac{B}{E}mc^2 - \alpha \tag{19.20}$$

where α is a constant. Also, using Eqs. (19.1) and (19.20)

$$\frac{dp_y}{dt} = eE - \frac{ep_x B}{m} = \frac{e}{E}\left(E^2 - c^2B^2\right) + \frac{\alpha c^2 B}{Ep_y}\frac{dm}{dt} \tag{19.21}$$

If now the velocity of the particle be small enough as to justify taking $m = m_0$, then in place of Eq. (19.19), we would have

$$\frac{d^2 p_x}{dt^2} = \frac{eB}{m}\frac{dp_y}{dt} = \frac{eB}{m}\left(eE - \frac{eB}{m}p_x\right)$$

$$\frac{d^2 p_x}{dt^2} + \left(\frac{eB}{m}\right)^2 p_x = \frac{e^2 BE}{m} \tag{19.22}$$

The general integral of Eq. (19.22) is

$$p_x = \frac{mE}{B} + A\sin\left(\frac{eBt}{m} + \phi\right) \tag{19.23}$$

(Note that Eq. (19.23) is essentially the result we have already obtained—a drift velocity of constant magnitude along with a sinusoidal velocity.) Equation (19.23) shows that with $|E|$ comparable with $c|B|$, the velocity will oscillate about a value comparable with c and thus contradict our assumption that m may be regarded as constant.

The integration of Eqs. (19.20) and (19.21) in general does not lead to simple results. However, one may go over to a relatively uniformly moving frame in which the magnetic field vanishes and solves the problem of motion in the transformed electric field (which is also uniform). Transforming back to the original frame, the motion in this frame is obtained. (Note that this transformation to a simple electric field is not possible in case $E^2 - c^2B^2 = 0$. As, $E^2 - c^2B^2$ and $\mathbf{E}\cdot\mathbf{B}$ are Lorentz invariants, if these vanish in one frame, they will vanish in all frames.)

Before going to the case $E^2 - c^2B^2 = 0$, we shall consider a mathematically simple but rather artificial situation. Suppose the constant α in Eq. (19.20) vanishes, then

$$p_x = \frac{mc^2 B}{E}, \qquad v_x = \frac{c^2 B}{E} \tag{19.24}$$

so that

$$\frac{dp_y}{dt} = eE - eBv_x = \frac{e}{E}(E^2 - c^2B^2)$$

$$p_y = \frac{e}{E}(E^2 - c^2B^2)t - \beta \tag{19.25}$$

$$p_z = \gamma \tag{19.26}$$

where β and γ are constants. Using Eqs. (19.24)–(19.26) in Eq. (19.18), we get on choosing $t = 0$ when $p_y = 0$

$$m^2c^2 = e^2\left(E^2 - c^2B^2\right)t^2 + \mu^2 \tag{19.27}$$

$$\text{where} \quad \mu^2 = \frac{E^2(\gamma^2 + m_0^2c^2)}{E^2 - c^2B^2} \tag{19.28}$$

We now have

$$\frac{dy}{dt} = \frac{ect(E^2 - c^2B^2)}{E\sqrt{e^2(E^2 - c^2B^2)t^2 + \mu^2}}$$

$$x - x_0 = \frac{c^2Bt}{E} \tag{19.29}$$

$$y - y_0 = \frac{c}{E}\sqrt{(E^2 - c^2B^2)\left(t^2 + \frac{\mu^2}{e^2(E^2 - c^2B^2)}\right)} \tag{19.30}$$

$$z = \frac{c\gamma}{e\sqrt{E^2 - c^2B^2}} \tag{19.31}$$

The projection of the path in the xy-plane (i.e., the plane orthogonal to **B**) is a hyperbola.

We next consider the case $E^2 = c^2B^2$. On integration Eq. (19.21) gives

$$p_y^2 = 2\alpha c(m - \mu) \tag{19.32}$$

where μ is a constant of integration. Hence $\frac{mc^2}{eE}\frac{dm}{dt} = \sqrt{2\alpha c(m - \mu)}$. Introducing the variable $\xi \equiv m - \mu$, we can integrate and write

$$\frac{2}{3}\sqrt{\xi^3} + 2\mu\sqrt{\xi} = \sqrt{2\alpha c}\,\frac{eE}{c^2}(t - t_0) \tag{19.33}$$

From Eq. (19.20) we now have

$$\frac{dx}{d\xi} = \frac{\dot{x}}{\dot{\xi}} = \frac{c\sqrt{\xi} + (c\mu - \alpha)/\sqrt{\xi}}{\sqrt{2\alpha c}(Ee/c^2)} \tag{19.34}$$

Integrating again

$$x - x_0 = \frac{1}{\sqrt{2\alpha c}\,(eE/c^2)} \left(\frac{2c}{3} \sqrt{\xi^3} + 2(c\mu - \alpha)\sqrt{\xi} \right) \tag{19.35}$$

Moreover

$$\frac{dy}{d\xi} = \frac{\dot{y}}{\dot{\xi}} = \frac{c^2}{eE}$$
$$\frac{dz}{d\xi} = \frac{\dot{z}}{\dot{\xi}} = \frac{\gamma c^2}{eE\sqrt{2\alpha c \xi}} \quad (y = \text{constant})$$

The integration of which gives

$$y - y_0 = \frac{c^2 \xi}{eE} \tag{19.36}$$
$$z - z_0 = \frac{2\gamma c^2 \sqrt{\xi}}{eE\sqrt{2\alpha c}} \tag{19.37}$$

Thus the parameter ξ is just the y-coordinate, scaled and translated. We note that the projection of the path on the plane normal to **B** is a curve with a cubic equation of the form

$$(x - x_0)^2 = ay^3 + by^2 + cy + d$$

while that on the plane containing **E** and **B** is a parabola.

19.1 Weak Perturbations and Drift of the Guiding Centre

We have just seen that in a spatially uniform time-independent magnetic field, a charged particle describes a helix which can be considered to be a combination of uniform circular motion in the plane perpendicular to **B** with a drift of the centre of this circular motion (called the guiding centre) parallel to **B**. We shall now consider small departures from this situation of the following types.

(a) *An Additional Force Field* **F** *Other Than the Magnetic Field*
We shall take this force field to be uniform and constant and further that it is normal to **B**. (Any component in the direction of **B** merely gives a uniform acceleration in that direction which does not affect the circular motion in the perpendicular plane, in the non-relativistic case. In the relativistic case, the variation of mass would indirectly influence the circular motion.) The equation of motion with this additional field **F** is

$$\frac{d\mathbf{v}}{dt} = \frac{e}{m}(\mathbf{v} \times \mathbf{B}) + \frac{\mathbf{F}}{m}$$

$$\frac{d}{dt}(\mathbf{v} - \mathbf{u}) = \frac{e}{m}(\mathbf{v} - \mathbf{u}) \times \mathbf{B} \tag{19.38}$$

$$\text{where,} \qquad \mathbf{u} = \frac{\mathbf{F} \times \mathbf{B}}{eB^2} \tag{19.39}$$

The equation implies a drift of the guiding centre with velocity $\mathbf{u}$. In case $\mathbf{F}$ is due to an electrostatic field, we get back the drift caused by $\mathbf{E} \times \mathbf{B}$.

(b) *A Magnetic Field That Is Weakly Time Dependent*
The electric field is considered to be spatially uniform and perpendicular to $\mathbf{B}$ and the variation with time is slow, by which we mean that the characteristic time of variation of $\mathbf{E}$ is large compared to the time taken by the particle to go once round the circle. In other words, if we consider the Fourier integral

$$\mathbf{E}(t) = \int \mathbf{E}_\omega e^{i\omega t} d\omega \tag{19.40}$$

then E_ω has an appreciable value only for ω small compared to the cyclotron frequency $\omega_0 = Be/m$. On differentiation with respect to time, the equation of motion (19.1) gives

$$\frac{d^2\mathbf{v}}{dt^2} + \omega_0^2\mathbf{v} = \frac{e}{m}\left(\frac{\partial \mathbf{E}}{\partial t} + \frac{e}{m}\mathbf{E} \times \mathbf{B}\right) \tag{19.41}$$

The approximate solution of the above, valid for slow variation of E, is

$$\mathbf{v} = \mathbf{v}_0 + \frac{e}{m\omega_0^2}\frac{\partial \mathbf{E}}{\partial t} + \frac{\mathbf{E} \times \mathbf{B}}{B^2} \tag{19.42}$$

where $\mathbf{v}_0$ is the velocity in absence of $\mathbf{E}$ (i.e., the velocity giving the simple helical motion). The last term on the right is the usual drift due to electric field (irrespective of time dependence) while the second term arises due to time dependence of $\mathbf{E}$ and is known as the polarization drift of the guiding centre. This drift

$$\mathbf{v}_p = \frac{e}{m\omega_0^2}\frac{\partial \mathbf{E}}{\partial t} = \frac{m}{eB^2}\frac{\partial \mathbf{E}}{\partial t} \tag{19.43}$$

is in opposite directions for charges of opposite sign—hence it will cause a net current in a neutral plasma and may also bring about a separation of the charges,

The reason, why the $\mathbf{E} \times \mathbf{B}$ drift is independent of the characteristics of the particles, while the polarization drift is not, may be explained as follows. For a uniform time-independent electric field $\mathbf{E}$ (which is perpendicular to $\mathbf{B}$ and of magnitude less than $\mathbf{B}$), there exists a Lorentz transformation which reduces the electromagnetic field to a simple magnetic field, i.e., in the transformed frame the electric field vanishes. However, in case the field is time dependent, no such transformation exists, hence the responses of the particles depend on e/m of the particles.

Equation (19.43) allows us to deduce a simple expression for the current due to the polarization drift. If n_+ and n_- be the number densities of positively and negatively charged particles carrying charges e_+ and e_-, respectively, the current density $\mathbf{j}$ is

$$\mathbf{j} = n_+ e_+ \mathbf{v}_{p_+} + n_- e_- \mathbf{v}_{p_-} = \frac{(n_+ m_+ + n_- m_-)}{B^2} \frac{\partial \mathbf{E}}{\partial t} = \frac{\rho}{B^2} \frac{\partial \mathbf{E}}{\partial t} \tag{19.44}$$

where m_+ and m_- are the masses of the two types and ρ is the combined mass density. Since this current is proportional to $\partial \mathbf{E}/\partial t$, it simulates a displacement current and gives rise to an effective (relative) dielectric constant

$$\epsilon_{\text{eff}} = 1 + \frac{\rho}{\epsilon_0 B^2}$$

Hence, unless the density be very low, a plasma will have a very high effective dielectric constant.

(c) *A Non-uniform Magnetic Field*
Like other perturbations, the non-uniformity will again assumed to be small, i.e., the variation of the magnetic field over a length of the order of the Larmor radius r_0 is small compared to the magnetic field itself. In effect this means that the nine components of the tensor $\alpha_{ik} = \frac{\partial B_i}{\partial x_k}$ may be considered to be constants. We shall consider the basic field to be in the z-direction, so that $|\mathbf{B}| \approx B_z$, and divide the α_{ik}'s into four types.

(I) The diagonal terms of α_{ik}, namely, $\partial B_x/\partial x$, $\partial B_y/\partial y$ and $\partial B_z/\partial z$ are not independent because of the relation $\nabla \cdot \mathbf{B} = 0$. Further, we shall assume a rotational symmetry about the z-axis, so that, in cylindrical polar coordinates, $\frac{\partial B_\varphi}{\partial \varphi} = 0$ and

$$\frac{1}{r}\frac{\partial}{\partial r}(r B_r) + \frac{\partial B_z}{\partial z} = 0 \tag{19.45}$$

Thus, there is only one independent component.

(II) For the terms giving variation of B_z in orthogonal directions, namely, $\partial B_z/\partial x$ and $\partial B_z/\partial y$, there is again only one independent component with rotational symmetry.

(III) The terms giving variation of B_x and B_y in z-direction, namely, $\partial B_x/\partial z$ and $\partial B_y/\partial z$ are related to terms in (II) if the particles are test particles moving in a region where there is no source current, because of $\nabla \times \mathbf{B} = 0$.

(IV) The terms $\partial B_x/\partial y$ and $\partial B_y/\partial x$ are related by $\nabla \times \mathbf{B} = 0$, under conditions stated in (III).

The small non-uniformities introduce a perturbation, while the basic motion is still circular in the xy-plane. We consider the four cases outlined above separately. For non-uniformity of type (I), from Eq. (19.45) we have

$$r_0 B_{r_0} = -\int_0^{r_0} r\frac{\partial B_z}{\partial z}\,dr = -\frac{r_0^2}{2}\frac{\partial B_z}{\partial z}$$

$$\text{or,} \qquad B_{r_0} = -\frac{r_0}{2}\frac{\partial B_z}{\partial z} \tag{19.46}$$

This radial magnetic field gives rise to a force in the z-direction of magnitude

$$F_z = eB_{r_0}\frac{d\varphi}{dt} = -\frac{1}{2}r_0^2\omega_0\frac{\partial B_z}{\partial z} = -\frac{mv_\perp^2}{2B}\frac{\partial B_z}{\partial z} \tag{19.47}$$

where we have used Eq. (19.7) for the Larmor radius. This force will cause a drift of the guiding centre along z-axis with a non-uniform velocity. Writing $v_\parallel$ for the velocity in z-direction, we have from (19.47)

$$\frac{d}{dt}\left(\frac{1}{2}mv_\parallel^2\right) = F_z v_\parallel = -\frac{mv_\perp^2}{2B}\frac{dB}{dt} \tag{19.48}$$

As the particle is moving in a simple magnetic field, its kinetic energy remains constant, so that $\frac{m}{2}\left(v_\parallel^2 + v_\perp^2\right)$ is a constant. Then Eq. (19.48) may be written as $\frac{d}{dt}\left(\frac{1}{2}mv_\perp^2\right) = \frac{mv_\perp^2}{2B}\frac{dB}{dt}$, which integrates to

$$\frac{mv_\perp^2}{B} = \text{constant} \tag{19.49}$$

The above result is sometimes expressed in terms of the magnetic moment of the orbital motion of the particle

$$\mu = \frac{e}{T}\pi r_0^2 = \frac{mv_\perp^2}{2B}$$

The magnetic moment is thus a constant of motion. Equation (19.49) plays an important role in the understanding of some natural phenomena and also in some plasma confining laboratory devices. We shall come to their discussion a little later.

Meanwhile, we consider the effect of non-uniformities of type II. In this case B_z varies, so the circular orbits will be perturbed, but there would be no drift in the z-direction. Considering that only $\partial B_z/\partial x$ is non-vanishing, we have the equations of motion

$$m\dot{v}_x = ev_y B_z = ev_y\left(B_0 + x\frac{\partial B_0}{\partial x}\right) \tag{19.50}$$

$$m\dot{v}_y = -ev_x B_z = -ev_x\left(B_0 + x\frac{\partial B_0}{\partial x}\right) \tag{19.51}$$

The particle settles down to a motion in which the time average of the accelerations $\langle\dot{v}_x\rangle$ and $\langle\dot{v}_y\rangle$ vanish, so that

$$B_0\langle v_y\rangle = -\frac{\partial B_0}{\partial x}\langle x v_y\rangle = \frac{\partial B_0}{\partial x}\langle \omega x_0^2\rangle = \frac{1}{2}\omega r_0^2\frac{\partial B_0}{\partial x} \tag{19.52}$$

$$B_0\langle v_x\rangle = -\frac{\partial B_0}{\partial x}\langle x v_x\rangle = -\frac{\partial B_0}{\partial x}\langle \omega x_0 y_0\rangle = 0 \tag{19.53}$$

where in calculating $\langle x v_x\rangle$ and $\langle x v_y\rangle$ we have used the unperturbed values, as our calculation is restricted to quantities of the first order of smallness. Thus, from Eqs. (19.52) and (19.53), we have finally

$$\langle v_y\rangle = \frac{\omega r_0^2}{2B_0}\frac{\partial B_0}{\partial x}, \qquad \langle v_x\rangle = 0 \tag{19.54}$$

Thus the drift velocity of the guiding centre may be expressed by the vector equation

$$\mathbf{v}_d = \frac{\omega r^2}{2B_0^2}\left(\mathbf{B}\times\nabla B\right) \tag{19.55}$$

This drift, known as gradient drift, depends, through ω, on the sign of charge like the polarization drift, and hence may lead to a net current and a separation of charges in a neutral plasma.

For a type III non-uniformity, we may study the effect by taking $\frac{\partial B_x}{\partial z} \neq 0$, but $\frac{\partial B_y}{\partial z} = 0$. The equations of motion are then

$$\ddot{z} = -\frac{ez\dot{y}}{m}\frac{\partial B_x}{\partial z} \tag{19.56}$$

$$\ddot{x} = \frac{eB_z\dot{y}}{m} \tag{19.57}$$

$$\ddot{y} = -\frac{eB_z\dot{x}}{m} + \frac{ez\dot{z}}{m}\frac{\partial B_x}{\partial z} \tag{19.58}$$

Equation (19.56) on integration shows that $\dot{z}$ is a constant plus a term involving the product of y and $z\frac{\partial B_x}{\partial z}$. At our level of approximation, the time average of this vanishes and z is just a constant, independent of the magnetic field. In other words, there may be a constant drift velocity $v_\parallel$, but it is not related to the characteristic of the magnetic field. Equation (19.57) on integration gives

$$\dot{x} = \frac{eB_0 y}{m} + u_0$$

Substituting this result in Eq. (19.58), we get

$$\ddot{y} + \left(\frac{eB_0}{m}\right)^2 y = -\frac{eB_0}{m}u_0 + \left(\frac{1}{B_0}\frac{\partial B_x}{\partial z}\right)\left(\frac{eB_0}{m}\right)v_\parallel^2 t$$

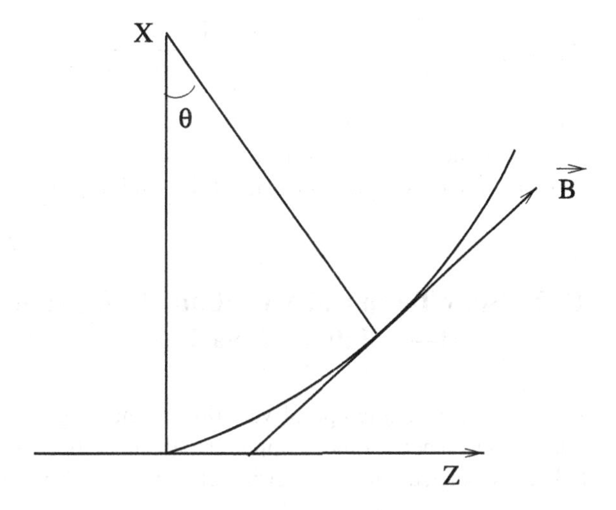

Fig. 19.2 Curvature drift: if the magnetic lines of force are curved, a charged particle experiences a drift due to the centrifugal force

The solution of this equation is

$$y = y_0 \sin\left(\frac{eB_0 t}{m} + \alpha\right) - \frac{mu_0}{eB_0} + \frac{1}{B_0}\frac{\partial B_x}{\partial z}\frac{v_\parallel^2 t}{(eB_0 e/m)}$$

The first term on the right-hand side is simply the velocity for the Larmor motion, the second constant term fixes the origin of the coordinates, while the third term shows a constant drift velocity of magnitude

$$\frac{v_\parallel^2}{\omega_0 B_0}\frac{\partial B_x}{\partial z}$$

This expression is sometimes written in terms of the curvature of the lines of magnetic force.

From Fig. 19.2

$$B_z = |B| \cos\theta = B_0$$

$$B_x = |B| \sin\theta = B_0\theta = z\frac{\partial B_x}{\partial z}$$

The radius of curvature of the lines is

$$\frac{1}{R} = \frac{\theta}{z} = \frac{1}{B_0}\frac{\partial B_x}{\partial z}$$

so that the drift velocity is

$$\frac{\mathbf{R} \times \mathbf{B}}{R^2 B} \frac{v_{\parallel}^2}{\omega_0}$$

This drift is known as the curvature drift.

We shall not go into a formal discussion of non-uniformities of type IV but merely note that it does not give any drift of the guiding centre.

19.2 Slow Temporal Variation of Magnetic Field—Adiabatic Invariants

We now consider a temporal variation of the magnetic field, the variation over a Larmor period being small compared to the field itself. The variation of the magnetic field is associated with an electric field given by Maxwell's equation

$$\nabla \times \mathbf{E} = -\frac{\partial \mathbf{B}}{\partial t}$$

This electric field, having components in directions normal to the z-direction (the direction of the magnetic field), causes an increase of kinetic energy of the particles. The increase, as the particle goes once round the Larmor orbit, being

$$\int_0^{2\pi/\omega} e\mathbf{E} \cdot \mathbf{v}\, dt = e \oint \mathbf{E} \cdot d\mathbf{l} = -e \int \dot{\mathbf{B}} \cdot \mathbf{ds}$$

where the surface integral is over the area of the Larmor orbit, and since

$$\omega = -\frac{e}{m}\mathbf{B}$$

determines the sign, for the increase of kinetic energy we get

$$e\pi r_0^2 \dot{B} = \frac{1}{2} e r_0^2 \omega_0 \Delta B = \frac{1}{2} m v_\perp^2 \frac{\Delta B}{B}$$

so that finally $\Delta\left(\frac{1}{2} m v_\perp^2\right) = \frac{1}{2} m v_\perp^2 \frac{\Delta B}{B}$ leading to

$$\frac{m v_\perp^2}{B} = \text{constant} \tag{19.59}$$

This constant of motion may be expressed differently as (i) the flux through the Larmor orbit remains constant, thus if B goes on increasing, the radius of the orbit decreases progressively, or (ii) the magnetic moment corresponding to the circular motion of the particle remains constant. We may note that we observed the constancy of the same quantities for the case where $\frac{\partial B_z}{\partial z} \neq 0$.

When in case of a periodic motion, some parameter controlling the motion changes slowly,[1] then some constants of motion remain invariant—these are named adiabatic invariants. Thus in the case just studied, B_z is the slowly varying parameter and the adiabatic invariants are the following.

(i) The ratio of the kinetic energy due to transverse motion to B_z, namely, $\frac{mv_\perp^2}{2B}$.
(ii) The magnetic moment of the Larmor motion, i.e., the flux through the Larmor orbit.

There is a theorem in classical mechanics that for slow variations, $J = \oint p\,dq$ is an adiabatic invariant, where q is a coordinate undergoing periodic motion and p the corresponding generalized momentum, and the integral is taken over a complete cycle of q. We can deduce our previous results directly from this formula.

For a charged particle in an electromagnetic field, the generalized momenta p_i are

$$p_i = \frac{\partial L}{\partial \dot{q}_i} = p_i^{(0)} + eA_i$$

where $\mathbf{A}$ is the vector potential and $p_i^{(0)}$ is the corresponding momentum in the absence of electromagnetic fields. Thus, the adiabatic invariant is

$$J = \oint p\,dq = \oint \mathbf{p}^{(0)} \cdot d\mathbf{l} + e \oint \mathbf{A} \cdot d\mathbf{l}$$

where the line integral is over the circular orbit. The line integrals may be converted to surface integrals to give

$$\begin{aligned} J &= m \int (\nabla \times \mathbf{v}) \cdot \mathbf{ds} + e \int (\nabla \times \mathbf{A}) \cdot \mathbf{ds} \\ &= 2m \int \omega \cdot \mathbf{ds} + e \int \mathbf{B} \cdot \mathbf{d}s \;=\; -e \int \mathbf{B} \cdot \mathbf{ds} \end{aligned} \tag{19.60}$$

Thus, the flux through the Larmor orbits is proved to be an adiabatic invariant. It is now easy to deduce the invariance of the magnetic moment and $mv_\perp^2/B$.

19.3 Confinement of Charged Particles in Non-homogeneous Magnetic Fields—The Magnetic Bottle and Magnetic Mirror

We consider the case $\frac{\partial B_z}{\partial z} \neq 0$ that we have studied already. Suppose that in an arrangement we have a minimum of B_z at the centre and it increases symmetrically on either side. In the laboratory, one may have such a situation by winding the turns

[1] By slow change we mean that if λ be the parameter and T the time period, then $T\dot{\lambda} \ll \lambda$.

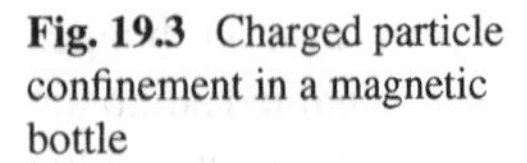
Fig. 19.3 Charged particle confinement in a magnetic bottle

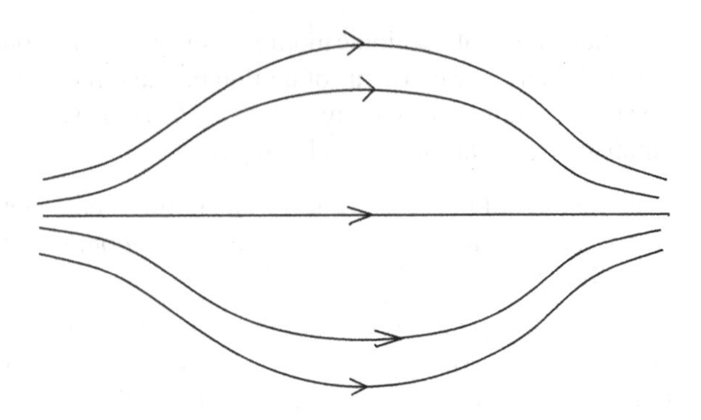

of a solenoid in such a way that the number of turns per unit length is a minimum at the central region and increase outwards. In nature we can often approximate the fields by a dipole field, e.g., in case of the earth's field the dipole axis coincides (approximately) with the axis of the earth, thus the field is a maximum at the poles and has a minimum at the equator. A schematic diagram of the lines of force in the cases envisaged is shown in Fig. 19.3.

As we have already seen (cf. Eq. (19.49)), $v_\perp^2/B$ remains constant during the motion. Moreover, the motion, being in a simple magnetic field, the kinetic energy remains a constant. Thus, we get

$$v_\parallel^2 = v^2 - v_\perp^2 = v_0^2 - B\left(\frac{v_\perp}{B}\right)_0^2 \tag{19.61}$$

where the subscript zero indicates constant values. Differentiating with respect to t, we get the equation of motion of the guiding centre

$$\dot{v}_\parallel = -\frac{\partial}{\partial z}\left[\frac{B}{2}\left(\frac{v_\perp^2}{B}\right)_0\right] \tag{19.62}$$

From Eq. (19.61) we see that as the guiding centre moves along z, if B increases sufficiently, to be exact, it attains the value $B = \left(\frac{B}{\dot{v}_\perp}\right)_0 v^2$, the component $v_\parallel$ will vanish, and then because of Eq. (19.62), the guiding centre will turn back. The reverse motion will also similarly be brought to a halt and reversed again. Thus the guiding centre will execute a to-and-fro motion, remaining confined within the region. Whereas if the field were uniform, a particle with any value of $v_\parallel$ other than zero, would have escaped from the region. The arrangement is sometimes referred to as a magnetic bottle and the turning back of the particle near the two ends may be looked upon as a reflection, hence the name magnetic mirror.

A particle, for which even the maximum value of B, say B_m is less than $\left(\frac{B}{\dot{v}_\perp}\right)_0 v^2$, would not turn back, it would, therefore, escape. This may be expressed in terms of the *pitch angle* α, which the direction of the resultant velocity of the particle makes with the direction of the magnetic field at the centre of the system. Thus

$\left(\frac{B}{v_\perp^2}\right)_0 v_0^2 = \frac{B_c}{\sin^2\alpha}$, where B_c is the magnetic field at the centre. For reflection to occur, $\sin^2\alpha > \frac{B_c}{B_m}$. Particles will readily escape if their pitch angles are smaller. Even for other particles, confinement cannot last indefinitely—interaction between the particles themselves would occasionally change the pitch angles of some particles to low enough values.

Considering the earth's field to be a dipole field and using spherical polar coordinates with origin at the centre of the earth (the position of the hypothetical dipole) and the axis $\theta = 0$ in the direction of the dipole moment, the field at a point (r, θ, ϕ) is

$$B = \frac{\sqrt{1+3\cos^2\theta}}{\sin^6\theta} B_0$$

where B_0 is the field at the equatorial point with same r. For the particle with pitch angle α, the reflection will occur at θ_0 given by

$$\frac{\sin\theta_0}{\sqrt{1+3\cos^2\theta_0}} = \sin^2\alpha$$

Such a confinement of charged particle in the magnetic field of the earth is held to account for the Van Allen radiation belts.

Problems

1. Show that if the relativistic variation of mass be taken into account, the path of a particle in a uniform electrostatic field is a catenary whose equation is of the form $x = a\cosh(by)$ where a and b are constants and the electric field is in the direction of the axis of x. Show further that the path reduces to a parabola if the variation of mass be neglected.
2. A particle of charge e experiences a force proportional to the displacement from a fixed point and directed towards that point so that it executes a simple harmonic motion when displaced. Find the change in motion when there is an additional uniform magnetic field.
3. Study the equation of motion for a charged particle in the field of a magnetic dipole and obtain the two constants of motion and interpret them physically.

Chapter 20
Magnetohydrodynamics—Conducting Fluids and Magnetic Fields

In the last chapter, we have studied the motion of individual charged particles under the action of external impressed fields, i.e., fields to which the particles themselves do not contribute to any appreciable extent. This is a valid approximation when the particle density is sufficiently small. In another extreme situation we have a very high density of particles, so that instead of discrete particles, namely, the electrons and the positive ions, we may consider a continuous distribution of fluid. If during the motion, there is a separation of charges of opposite sign, on a macroscopic scale, electrostatic fields are called into play which tend to restore electrical neutrality. Our analysis in the next chapter will show that this gives rise to oscillations with a frequency $\sqrt{\frac{n_0 e^2}{\epsilon_0 m}} \sim 56 \times \sqrt{n_0}$, which is called the plasma frequency. It has quite a high value even for moderate values of the electron number density n_0 (measured in m^{-3}). Hence in situations where the conditions are static, or the temporal variations occur with a comparatively low frequency, the charge separations will be ineffective and we need not consider the relative motion between the positive ions and the electrons. We can thus consider the motion to be that of a single continuous fluid, which, however, has a very high conductivity. Thus the current is simply an Ohmic current, moreover, in view of the low frequency, the displacement current term may be neglected. This is the situation studied in magneto-hydrodynamics and the basic equations are

$$\rho\left(\frac{\partial u_i}{\partial t} + u_k \frac{\partial u_i}{\partial x_k}\right) = -\frac{\partial p}{\partial x_i} + (\mathbf{j} \times \mathbf{B})_i + F_i \tag{20.1}$$

$$\frac{\partial \rho}{\partial t} + \frac{\partial (\rho u_i)}{\partial x_i} = 0 \tag{20.2}$$

$$\nabla \cdot \mathbf{E} = 0 \tag{20.3}$$

$$\nabla \times \mathbf{E} = -\dot{\mathbf{B}} \tag{20.4}$$

A. K. Raychaudhuri, *Classical Theory of Electricity and Magnetism*, Texts and Readings in Physical Sciences 21, https://doi.org/10.1007/978-981-16-8139-4_20

$$\nabla \cdot \mathbf{B} = 0 \tag{20.5}$$

$$\nabla \times \mathbf{B} = \mu_0 \mathbf{j} \tag{20.6}$$

$$\mathbf{j} = \sigma \left(\mathbf{E} + \mathbf{u} \times \mathbf{B} \right) \tag{20.7}$$

Equation (20.1) is Euler's equation of hydrodynamics, the second term on the right (which is the Lorentz force in the absence of charge separation) couples the fluid motion with the electromagnetic field. The subscript i runs from 1 to 3 and indicates the x, y and z components, respectively. Further a summation over a repeated index is to be understood. As usual, ρ, p and u_i denote the density, pressure and the fluid velocity vector, respectively, while F_i represents the non-electromagnetic forces, which may, for example, be a force due to gravity or viscosity. Equation (20.2) is the equation of continuity, which ensures the conservation of mass. Equations (20.3)–(20.6) are Maxwell's equations with neglect of displacement current and charge separation while Eq. (20.7) is the expression of Ohm's law taking into account the electric intensity as observed in the rest frame of the fluid.

We may first count the number of equations and variables. Equations (20.1), (20.4), (20.6) and (20.7) are vector equations and thus contribute a total number of 12 equations. Along with Eqs. (20.2), (20.3) and (20.5), the total number of equations becomes 15. However, Eqs. (20.3) and (20.5) are sort of initial conditions, which, if they are satisfied at some instant of time, will remain satisfied at all times, because of the vanishing of the displacement current and Eq. (20.4). Hence omitting these two from the count, the effective number of independent equations is 13, while the number of variables is 14 (the four vectors **E**, **B**, **j** and **u** and the scalars p and ρ) disregarding for the moment the force vector F_i. Thus there is one equation too few and the problem becomes completely soluble, in the usual sense, only if one has an additional equation. Such an equation is provided by an equation of state, which may be introduced because of the assumed low frequency of changes. Thus, the equation of state of the fluid

$$p = p(\rho) \tag{20.8}$$

is the supplementary equation.

Assuming the conductivity σ to be constant, from Eqs. (20.4) and (20.7) we get

$$\frac{\partial}{\partial t}\mathbf{B} = -\frac{1}{\sigma}\nabla \times \mathbf{j} + \nabla \times (\mathbf{u} \times \mathbf{B})$$

Eliminating **j** with the help of Eq. (20.6) and using the vector identity $\nabla \times (\nabla \times \mathbf{A}) = \nabla \cdot (\nabla \cdot \mathbf{A}) - \nabla^2 \mathbf{A}$, as well as Eq. (20.5), we get an equation for **B**

$$\frac{\partial}{\partial t}\mathbf{B} = \frac{1}{\mu_0 \sigma}\nabla^2 \mathbf{B} + \nabla \times (\mathbf{u} \times \mathbf{B}) \tag{20.9}$$

We consider Eq. (20.9) under two limiting circumstances according to the relative magnitudes of the two terms on the right in Eq. (20.9).

1. The magnitude of the velocity $|u|$ is small enough to justify neglecting the second term on the right in Eq. (20.9). Then

$$\frac{\partial \mathbf{B}}{\partial t} = \frac{1}{\mu_0 \sigma} \nabla^2 \mathbf{B}$$

If the linear dimension of the system be of the order of l, the above equation may be written as

$$\frac{\partial}{\partial t}|\mathbf{B}| \sim \frac{1}{\mu_0 \sigma} \frac{|\mathbf{B}|}{l^2} \sim \frac{|\mathbf{B}|}{\tau} \tag{20.10}$$

where $\tau \approx \mu_0 \sigma l^2$ is called the diffusion time. It is a measure of the time in which the magnetic field dies out.
For copper, the conductivity at low frequencies is approximately $5 \times 10^6\ \Omega^{-1}\text{m}^{-1}$ so that the diffusion time is less than 100 s for a copper sphere of 1 m radius. However, for the core of the earth (which presumably consists of molten iron) the diffusion time is $\sim 10^4$ years, while for the field of the sun, it may be high as 10^{10} years (which by the way is of the order of the 'age' of the universe).
2. The opposite limit of the first term in the right of Eq. (20.9) being negligible occurs if the conductivity and the linear dimensions are large enough. In this case, Eq. (20.9) reduces to

$$\frac{\partial B}{\partial t} = \nabla \times (\mathbf{u} \times \mathbf{B}) \tag{20.11}$$

which shows that

$$\frac{d}{dt} \int \mathbf{B} \cdot \mathbf{ds} = 0 \tag{20.12}$$

where the symbol d/dt stands for $\frac{\partial}{\partial t} + (\mathbf{u} \cdot \nabla)$ and is the rate of temporal change referred to a fixed element of the fluid. Equation (20.12) indicates that the flux linked with any area bounded by the same particles of the fluid remains constant as the fluid moves.

To see a little more clearly the question of relative magnitude of the two terms in Eq. (20.9), we note that the ratio of their may be taken to be

$$\left| \frac{1}{\mu_0 \sigma} \nabla^2 \mathbf{B} \right| : |\nabla \times (\mathbf{u} \times \mathbf{B})| = \frac{B}{\mu_0 \sigma l^2} : \frac{uB}{l} = 1 : \mu_0 \sigma l u$$

Thus, the deciding factor is the dimensionless number $R_M = \mu_0 \sigma l u = \tau u / l$, which is called the magnetic Reynold's number. From the values of diffusion time given above, it is clear that in ordinary laboratory conditions R_M is likely to be less than unity, while for astronomical bodies there are often cases of $R_M \gg 1$.

In the limit of $\sigma \to \infty$, Eq. (20.7) gives $\mathbf{E} = -\mathbf{u} \times \mathbf{B}$, hence the component of $\mathbf{u}$ orthogonal to $\mathbf{B}$ is

$$\mathbf{u}_\perp = \frac{\mathbf{E} \times \mathbf{B}}{B^2}$$

Thus, the fluid velocity normal to $\mathbf{B}$ is the same as the $\mathbf{E} \times \mathbf{B}$ drift we found for particle motion. When the conductivity is finite, $u_\perp$ will not be equal to the free particle drift velocity, but the magnetic field tends to wipe off the difference between the two. This can be seen from the expression for the electromagnetic interaction

$$\begin{aligned} \mathbf{F}_{\text{em}} &= \mathbf{j} \times \mathbf{B} = \sigma(\mathbf{E} \times \mathbf{B}) + \sigma(\mathbf{u} \times \mathbf{B}) \times \mathbf{B} \\ &= \sigma B^2(\boldsymbol{\omega} - \mathbf{u}_\perp) \end{aligned} \tag{20.13}$$

where $\boldsymbol{\omega} = (\mathbf{E} \times \mathbf{B})/B^2$ is the drift velocity. In the absence of electric fields, $\boldsymbol{\omega}$ vanishes, therefore, the tendency of the magnetic field then will be to reduce $\mathbf{u}$ and make the fluid move along the lines of magnetic force. This phenomenon is sometimes referred to as magnetic viscosity and the corresponding magnetic viscosity coefficient is σB^2.

One can use the vector identity $\nabla(\mathbf{A} \cdot \mathbf{B}) = \mathbf{A} \times (\nabla \times \mathbf{B}) + \mathbf{B} \times (\nabla \times \mathbf{A}) + (\mathbf{B} \cdot \nabla)\mathbf{A} + (\mathbf{A} \cdot \nabla)\mathbf{B}$, to transform the electromagnetic interaction term into a more expressive form

$$\mathbf{j} \times \mathbf{B} = \frac{1}{\mu_0}(\nabla \times \mathbf{B}) \times \mathbf{B} = -\frac{1}{2\mu_0}\nabla B^2 + \frac{1}{\mu_0}(\mathbf{B} \cdot \nabla)\mathbf{B} \tag{20.14}$$

The first term on the right simulates a pressure gradient force with magnetic pressure $B^2/(2\mu_0)$. The second term is a tension along the lines of magnetic force of equal magnitude to the pressure. (To see this suppose $\mathbf{B} = B_x\hat{\mathbf{x}}$, then $\frac{1}{\mu_0}(\mathbf{B} \cdot \nabla)\mathbf{B} = \frac{1}{\mu_0}\frac{\partial B^2}{\partial x}\hat{\mathbf{x}}$, where $\hat{\mathbf{x}}$ denotes a unit vector in the direction of the x-axis, i.e., the direction of the magnetic field.) It is thus said that the lines of force may be considered to behave like stretched strings, therefore, one may expect some transverse waves (distinct from the longitudinal compressional waves) in a fluid when there is a magnetic field. Such waves do indeed exist and are called Alfvén waves.

20.1 Stationary Solutions of the Magneto-Hydrodynamic Equations

A steady or stationary flow is defined as one in which all partial derivatives with respect to time vanish, however, $\mathbf{u}$ and $\mathbf{j}$ do not vanish and these distinguish the flow from a static situation. As a first example of such flow, we consider the Hartmann problem.

A conducting viscous fluid flows subject to the following conditions.

(i) The steady flow velocity is everywhere in the x-direction.
(ii) The velocity has a gradient in the z-direction alone, with the boundary conditions $\mathbf{u} = 0$ at $z = -L$ and $|u| = u_0$ at $z = L$.
(iii) There is a uniform electric field in the y-direction, i.e., perpendicular to both $\mathbf{u}$ and $\nabla|u|$.
(iv) There is a uniform magnetic field B_0 in the z-direction. However, there is a conduction current in the y-direction, consequently, from Eq. (20.6), we have $\partial B_x/\partial z \neq 0$. Thus, there must be a non-uniform magnetic field in the direction of the flow.
(v) There are translation symmetries along x- and y-axes, so that the partial derivatives with respect to x and y vanish.

With these conditions Eq. (20.2) gives

$$\frac{\partial u}{\partial x} = 0 \tag{20.15}$$

which could follow directly from the condition (v). In view of Eq. (20.7)

$$j_y = \sigma\,(E - u B_0) \tag{20.16}$$

The viscous force in this case is $\eta\nabla^2\mathbf{u} = \eta\partial^2 u/\partial z^2$ in the x-direction, where η is the coefficient of fluid viscosity. Hence Eq. (20.1), written out explicitly in term of components, yields two non-trivial relations

$$\sigma B_0\,(E - u B_0) + \eta\frac{\partial^2 u}{\partial z^2} = 0 \tag{20.17}$$

$$\frac{\partial p}{\partial z} + \sigma\,(E - u B_0)\,B_x = 0 \tag{20.18}$$

The first of these is a linear equation for u

$$\frac{d^2 u}{dz^2} - \frac{\sigma B_0^2}{\eta}\left(u - \frac{E}{B_0}\right) = 0 \tag{20.19}$$

which on integration, together with the boundary condition in (ii) (see Fig. 20.1), gives

$$u = u_0\frac{\sinh\left(M\left(\frac{z}{L}+1\right)\right)}{\sinh 2M} + \frac{E}{B_0}\left(1 - \frac{\cosh(Mz/L)}{\cosh M}\right)$$

where

$$M = B_0 L\sqrt{\frac{\sigma}{\eta}}$$

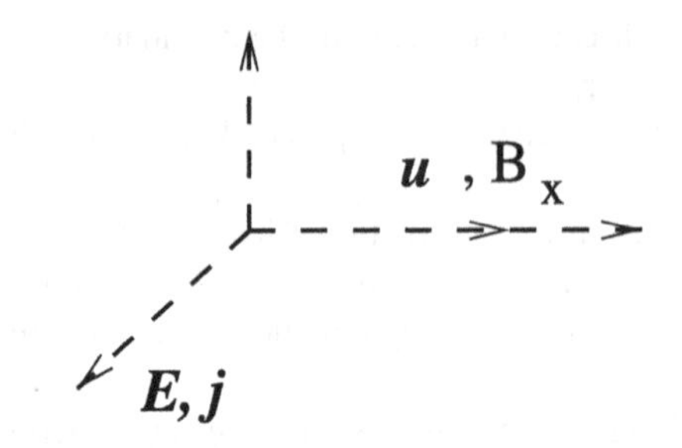

Fig. 20.1 Motion of a viscous magnetic fluid between parallel plates. Velocity **u** has a gradient perpendicular to the plates

is known as the Hartmann number, which depends on the relative magnitude of the magnetic and fluid viscosity. The effect of magnetic viscosity is emphasized in case M is much greater than unity.[1] We then get the following approximate form for the fluid velocity

$$u = \begin{cases} \dfrac{E}{B_0} + \left(u_0 - \dfrac{E}{B_0}\right) e^{M\left(\frac{z}{L}-1\right)} & \text{if } z > 0 \\ \dfrac{E}{B_0}\left(1 - e^{-M\left(\frac{z}{L}-1\right)}\right) & \text{if } z < 0 \end{cases}$$

The above expressions show that while the constraints $u = 0$ at $z = -L$ and $u = u_0$ at $z = L$ are obeyed, the value of u rapidly approaches the standard drift velocity $(\mathbf{E} \times \mathbf{B})/B^2$ as one goes away from the two planes $z = \pm L$. This is to be expected because of the large magnetic viscosity. In this circumstance, Eq. (20.6) reads

$$\frac{\partial B_x}{\partial z} = \mu_0 \sigma \left(E - u B_0\right) \tag{20.20}$$

in which B_x is most easily expressed in terms of u. Substituting from Eq. (20.19) in the above and integrating we get

$$B_x = \frac{\mu_0 \eta}{B_0}\left(b - \frac{du}{dz}\right) \tag{20.21}$$

The constant of integration b remains arbitrary and is usually fixed by imposing the condition that B_x vanishes at both $z = \pm L$. From Eqs. (20.16) and (20.20), it then

[1] In the original text AKR had used *'brought into relief'* to explain that it is emphasized.

follows that the current also vanishes at these planes and the constants u_0, E and B_0 satisfy the relation

$$E - u_0 B_0 = 0 \tag{20.22}$$

Substituting from Eq. (20.21) in Eq. (20.18), we get the pressure as a function of z (again there is a constant of integration). The current density is determined in terms of u by Eq. (20.16).

20.2 The Pinch Effect

As a second example of steady flow we consider the pinch effect—indeed in this case the fluid may not be flowing at all, although there is a non-vanishing current **j**. The effect is of great practical importance as it is of use in the confinement of plasma for controlled thermonuclear fusion.

Consider an infinite cylindrical column of fluid carrying current in the axial direction z (see Fig. 20.2). The magnetic field lines are circles in the plane perpendicular to the axis. The equilibrium is maintained by a balance between pressure gradient force and the electromagnetic interaction, i.e., we must have

$$\nabla p = \mathbf{j} \times \mathbf{B} \tag{20.23}$$

In cylindrical polar coordinates, Eq. (20.23) is satisfied trivially for the z- and ϕ-components. The radial equilibrium condition is

$$\frac{\partial p}{\partial r} = -jB \tag{20.24}$$

Since

$$B = \frac{\mu_0}{r} \int_0^r r' j(r')\, dr' \tag{20.25}$$

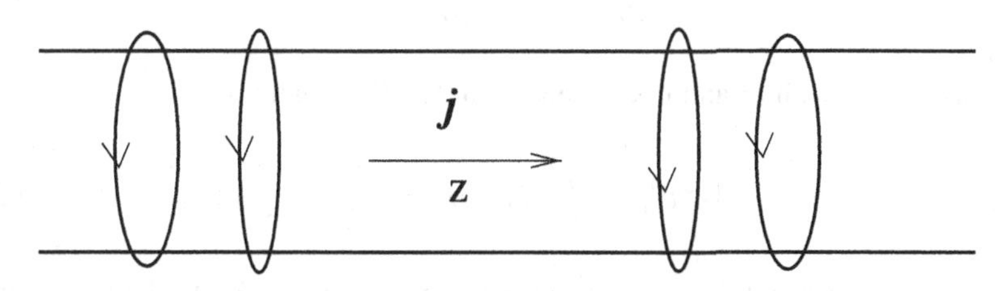

Fig. 20.2 Cylindrical pinch configuration: magnetic field lines in a cylindrical column of magnetic fluid

if the current distribution be given, we can find the pressure as a function of r. As p is to vanish at the boundary this enables us to estimate the current necessary for an equilibrium.

The simplest case is when j is a constant for $0 \leq r \leq R$ and vanishes for $r > R$, where R is the radius of the cylinder. In terms of the total current $I = \pi R^2 j$, we have

$$B = \frac{\mu_0}{r} \int_0^r \frac{I}{\pi R^2} r' dr' = \frac{\mu_0 I r}{2\pi R^2}$$

$$\frac{\partial p}{\partial r} = -\frac{I}{\pi R^2} \frac{\mu_0 I r}{2\pi R^2}$$

which integrates to

$$p = \frac{\mu_0 I^2}{4\pi^2 R^2} \left(1 - \frac{r^2}{R^2}\right)$$

where we have used the condition $p = 0$ at $r = R$. In the other extreme, the current may be confined to a surface layer. Hence dp/dr approaches the form of a δ-function—the pressure remains practically constant over the entire interior, then drops abruptly to zero at the surface. We then have

$$p = -\int jB\, dr = -\frac{\mu_0 I^2}{\pi^2 R^2}$$

As a numerical example, suppose that the temperature of the plasma is 10^6 K and the number density of particles is 10^{26} per m^3, so that the pressure $p \sim 10^9$ N/m^2. The requisite current will then be $I \sim 10^6 R$ A, where the radius R is measured in metres.

An interesting relation connecting the current with the temperature of the plasma can be easily obtained, without introducing any assumption about the distribution of the current. We have from Eqs. (20.24) and (20.25)

$$\frac{dp}{dr} = -\frac{B}{\mu_0 r} \frac{d}{dr}(rB)$$

Multiplying this with r^2 and integrating from 0 to R, we get

$$\left[r^2 p\right]_0^R - \int_0^R 2pr\, dr = -\frac{R^2 B^2}{2\mu_0} \tag{20.26}$$

The boundary terms on the left vanishes since $p = 0$ at $r = R$. iI n be the number density of particles and T the temperature (assumed constant)

$$-\int_0^R 2pr\, dr = -\frac{1}{\pi} \int nk_B T\, 2\pi r\, dr = -\frac{Nk_B T}{\pi}$$

where $N = \int_0^R 2\pi nr\, dr$ is the number of particles per unit length of the cylinder. Finally, we arrive at Bennett's relation

$$2Nk_BT = \frac{R^2B^2\pi}{\mu_0} = \frac{\mu_0 I^2}{4\pi} \tag{20.27}$$

If, for example, $I = 10^6$ A and $N = 10^{19}$ m^{-1}, the temperature is $T \sim 10^8$ K.

20.3 Instability of the Cylindrical Plasma

The cylindrical pinch that we have just considered is basically unstable. We shall not go into a mathematical discussion, but only point out two types of instability.

(i) *The sausage instability.* Suppose that a wavelike change takes place in the surface of the plasma as shown in Fig. 20.3. Since the current is continuous, the magnetic field at the surface would vary inversely as the radius. The field will thus have comparatively higher values at the constrictions and lower values at the bulges. To understand the effect of this, recall that the equilibrium of the pinch is due to the opposing effects of pressure of the fluid and magnetic pressure gradient forces (cf. Eq. (20.14) and the remarks following as well as Eq. (20.1)). The fluid pressure tries to expand the diameter of the pinch while magnetic pressure works in the opposite direction. Hence, the increase of magnetic pressure at the necks would still further accentuate the constrictions, thus bringing about an instability.

(ii) *The kink instability.* The kink instability is associated with the development of curvature of the plasma column, as shown in Fig. 20.4. As a result of the deformation, the magnetic lines of force crowd together on the concave side and become comparatively well separated on the convex side. Thus these changes in the magnetic field bring in forces which tend to increase the curvature, hence the instability.

Both these types of instability are overcome, or hindered at least to some extent, by having an axial magnetic field in the plasma. In case of sausage instability, the

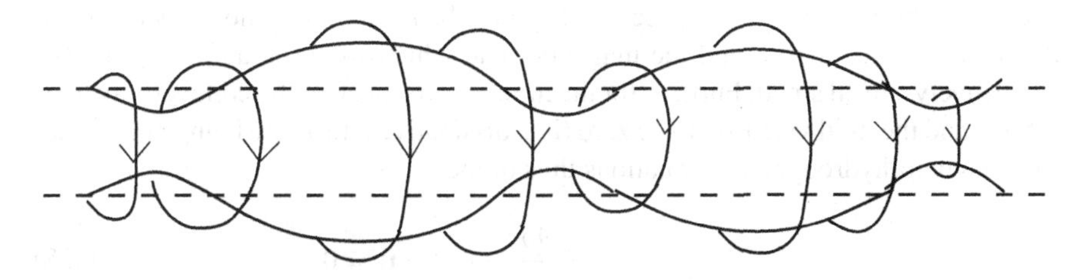

Fig. 20.3 Sausage or neck instability resulting in harmonic variation of beam radius along the beam axis

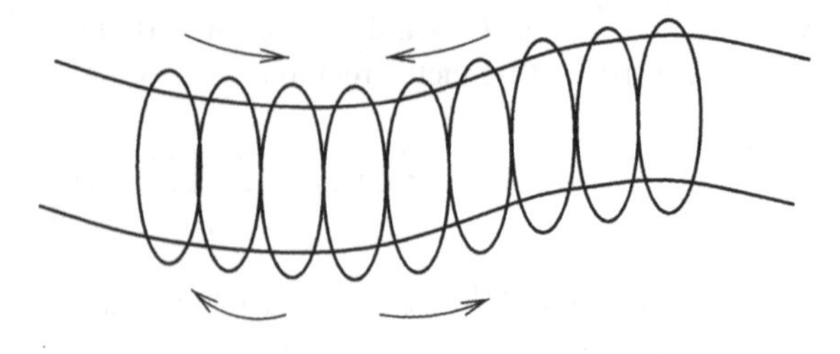

Fig. 20.4 Kink instability of cylindrical plasma

flux $\pi B_z r^2$ being conserved, the axial magnetic field at a neck increases. Thus, the increase δB_z for a decrease of the radius δr is given by $\delta B_z = 2B_z\delta r/r$. Consequently, the corresponding magnetic pressure increases by

$$\delta p = \frac{2B_z^2}{\mu_0}\frac{\delta r}{r}$$

This opposes the effect of the pressure due to the azimuthal magnetic field. The latter pressure increases by $\frac{B_\phi^2 \delta r}{\mu_0 r}$. Hence, if the axial magnetic field be strong enough, i.e., $B_z^2 > \frac{1}{2}B_\phi^2$, one may have stability against sausage deformation. In case of kink deformation, it may be shown that an axial magnetic field controls the disturbances of short wavelengths, but cannot do so if the wavelengths be large.

20.4 The Magneto-Hydrodynamic Waves

We seek solutions of the equations of magneto-hydrodynamics representing waves. The equations of magneto-hydrodynamics are non-linear (as indeed the equations of simple hydrodynamics are), this, therefore, brings in great complexity when compared to pure electromagnetic waves. However, we shall assume that the departures from equilibrium conditions are small enough so as to justify neglecting non-linear terms, thus the equations are treated in a linearized approximation. The equilibrium condition is defined by $\mathbf{u} = 0$, and there is an impressed magnetic field $\mathbf{B}_0$, which is constant both in time and space. In the perturbed state, the fluid velocity $\mathbf{u}$ and the perturbations in p, ρ and the magnetic field which we write as p_1, ρ_1 and $\mathbf{B}_1$, respectively, are all small. Further, for the moment we consider the conductivity to be infinite and the fluid viscosity to be zero (i.e., dissipative effects are being neglected). The magneto-hydrodynamic equations thus of the

$$\frac{\partial \rho}{\partial t} + \rho_0 \nabla \cdot \mathbf{u} = 0 \qquad (20.28)$$

$$\rho_0 \frac{\partial \mathbf{u}}{\partial t} + \nabla p_1 + \frac{1}{\mu_0} \mathbf{B}_0 \times (\nabla \times \mathbf{B}_1) = 0 \tag{20.29}$$

$$\nabla \times (\mathbf{u} \times \mathbf{B}_0) = \frac{\partial \mathbf{B}_1}{\partial t} \tag{20.30}$$

Now with

$$\nabla p_1 = \nabla \left(\frac{\partial p}{\partial \rho} \cdot \rho_1 \right) = C_s^2 \nabla \rho_1$$

where $C_s = \sqrt{\partial p / \partial \rho}$ is the velocity of the compressional waves in the fluid, commonly called the sound velocity, Eq. (20.29) becomes

$$\rho_0 \frac{\partial \mathbf{u}}{\partial t} + \rho_0 C_s^2 \nabla \rho_1 + \frac{1}{\mu_0} \mathbf{B}_0 \times (\nabla \times \mathbf{B}_1) = 0 \tag{20.31}$$

Differentiating the above with respect to t and then using Eqs. (20.28) and (20.30), we get

$$\frac{\partial^2 \mathbf{u}}{\partial t^2} - C_s^2 \nabla(\nabla \cdot \mathbf{u}) + \frac{1}{\mu_0 \rho_0} \mathbf{B}_0 \times [\nabla \times (\nabla \times (\mathbf{u} \times \mathbf{B}_0))] = 0 \tag{20.32}$$

The first two terms on the left are the familiar terms of ordinary fluid dynamics, which result in compressional waves, while the third term represents the influence of the magnetic field. We may consider a monochromatic wave solution (or more precisely, a Fourier component) $\mathbf{u} = \mathbf{u}_0 e^{i(\mathbf{k} \cdot \mathbf{r} - \omega t)}$, where the amplitude $\mathbf{u}_0$ is a constant. Substituting in Eq. (20.32), we get

$$\omega^2 \mathbf{u} + \left(C_s^2 + \frac{B_0^2}{\mu_0 \rho_0} \right) (\mathbf{k} \cdot \mathbf{u}) \mathbf{k} + \frac{\mathbf{k} \cdot \mathbf{B}_0}{\mu_0 \rho_0} ((\mathbf{k} \cdot \mathbf{B}_0)\mathbf{u} - (\mathbf{u} \cdot \mathbf{B}_0)\mathbf{k} - (\mathbf{k} \cdot \mathbf{u})\mathbf{B}_0) = 0 \tag{20.33}$$

A straightforward analysis of Eq. (20.33) is difficult and not very illuminating. We shall consider different cases according to the relative orientation of the velocity vector **u** and the propagation vector **k**.

- **Case I**: The vectors **u** and **k** are parallel.
  Taking the cross product of Eq. (20.33) with **u**, we get

$$(\mathbf{k} \cdot \mathbf{B}_0)(\mathbf{u} \times \mathbf{B}_0) = 0$$

  whence, either **u** is parallel to or normal to $\mathbf{B}_0$. We take the two subcases separately.

  - **Case I (a)**: All three vectors $\mathbf{B}_0$, **u** and **k** are in the same direction. These are longitudinal waves and the phase velocity is C_s, the velocity of compressional waves in the absence of magnetic fields. Indeed, from Eq. (20.30) we see that B_1 vanishes, so that the magnetic field is unperturbed and the current also vanishes.

- **Case I (b)**. The vectors **u** and **k** are in the same direction, which is normal to $\mathbf{B}_0$. The velocity of these waves is given by Eq. (20.33) as

$$v^2 = \frac{\omega^2}{k^2} = C_s^2 + \frac{B_0^2}{\mu_0 \rho_0} \tag{20.34}$$

These waves are non-dispersive. From Eq. (20.30), we have

$$\mathbf{B}_1 = \mathbf{B}_0 |\mathbf{u}| \sqrt{C_s^2 + \frac{B_0^2}{\mu_0 \rho_0}} \tag{20.35}$$

Thus, the perturbation in the magnetic field is confined to the direction of the unperturbed field.

- **Case II**: The vector **u** is normal to **k**.
Equation (20.33) reduces to

$$-\omega^2 \mathbf{u} + \frac{\mathbf{k} \cdot \mathbf{B}_0}{\mu_0 \rho_0} \left((\mathbf{k} \cdot \mathbf{B}_0)\, \mathbf{u} - (\mathbf{u} \cdot \mathbf{B}_0)\, \mathbf{k} \right) = 0$$

or, because of the orthogonality of **u** and **k**

$$\omega = \frac{\mathbf{k} \cdot \mathbf{B}_0}{\sqrt{\mu_0 \rho_0}} \tag{20.36}$$

$$(\mathbf{k} \cdot \mathbf{B}_0)\, (\mathbf{u} \cdot \mathbf{B}_0) = 0 \tag{20.37}$$

Equation (20.37) shows that $\mathbf{B}_0$ is perpendicular to either **k** or **u**. But with $\mathbf{k} \cdot \mathbf{B}_0 = 0$, Eq. (20.36) shows that there is no relative orientation of the velocity vector **u** and the propagation vector **k**. Hence for the propagation of the disturbance, $\mathbf{u} \cdot \mathbf{B}_0 = 0$. Therefore, **u** is normal to the plane containing $\mathbf{B}_0$ and **k** (of course in case the directions of $\mathbf{B}_0$ and **k** coincide, **u** can have any direction in the plane perpendicular to this direction). Equations (20.36) and (20.30) now give

$$v = \frac{\omega}{k} = \frac{B_0 \cos\alpha}{\sqrt{\mu_0 \rho_0}} \tag{20.38}$$

$$\mathbf{B}_1 = \frac{\mathbf{k}}{\omega} \times (\mathbf{u} \times \mathbf{B}_0) = -\frac{k}{\omega} B_0 \cos\alpha\, \mathbf{u} = -\sqrt{\mu_0 \rho_0}\, \mathbf{u} \tag{20.39}$$

where α is the angle between $\mathbf{B}_0$ and **k**. Equation (20.39) shows that the magnetic field perturbation is parallel to **u** and, is therefore normal to **k** and $\mathbf{B}_0$. This perturbation may be called transverse. Further Eq. (20.28) shows that in this case there is no compression of the fluid. This type of transverse magnetic wave was first studied by Alfvén and the waves are called Alfvén waves and the velocity $\frac{B_0}{\sqrt{\mu_0\rho_0}}$ known as Alfvén velocity.

- **Case III**: The vector **u** makes an angle β (where $\beta \neq 0, \frac{\pi}{2}$) with **k**. After taking the cross product with **k**, Eq. (20.33) gives

$$\left(-\omega^2 u + \frac{(\mathbf{k}\cdot\mathbf{B}_0)^2}{\mu_0\rho_0}\right)(\mathbf{u}\times\mathbf{k}) + \frac{(\mathbf{k}\cdot\mathbf{B}_0)(\mathbf{k}\cdot\mathbf{u})}{\rho_0\mu_0}(\mathbf{k}\times\mathbf{B}_0) = 0 \tag{20.40}$$

There are now three possibilities.

- **Case III (a)**: The direction of propagation is perpendicular to the applied field, so $\mathbf{k}\cdot\mathbf{B}_0 = 0$, but $\mathbf{u}\times\mathbf{k} \neq 0$. This gives $\omega = 0$, i.e., there is no propagation of disturbance.
- **Case III (b)**: The direction of propagation is perpendicular to the applied field, so $\mathbf{k}\times\mathbf{B}_0 = 0$. Then

$$v = \frac{\omega}{k} = \frac{B_0}{\sqrt{\mu_0\rho_0}}$$

From Eqs. (20.30) and (20.28), we get

$$\begin{aligned}
\mathbf{B}_1 &= -\frac{1}{\omega}\mathbf{k}\times(\mathbf{u}\times\mathbf{B}_0) = -\frac{1}{\omega}\left((\mathbf{k}\cdot\mathbf{B}_0)\,\mathbf{u} + (\mathbf{k}\cdot\mathbf{u})\,\mathbf{B}_0\right) \\
&= \sqrt{\mu_0\rho_0}\left(-\mathbf{u} + u\cos\beta\,\frac{\mathbf{B}_0}{B_0}\right)\rho_1 \\
&= \rho_0 u\cos\beta\,\frac{k}{\omega} = \frac{\rho_0}{B_0}\,u\cos\beta\,\sqrt{\mu_0\rho_0}
\end{aligned}$$

- **Case III (c)**: The angles between the direction of propagation and $\mathbf{B}_0$ and **u** are α and β, respectively, with none of the two angles being zero or $\pi/2$. This is the most general case. However, Eq. (20.40) shows that the vectors $\mathbf{B}_0\times\mathbf{k}$ and $\mathbf{u}\times\mathbf{k}$ are parallel, i.e., $\mathbf{B}_0$, **u** and **k** are in the same plane. Taking the component in the direction of **k**, from Eqs. (20.33) and (20.40), we get

$$v^2 = \frac{\omega^2}{k^2} = C_s^2 - \frac{B_0^2\sin\alpha\,\sin(\beta-\alpha)}{\mu_0\rho_0\cos\beta} \tag{20.41}$$

$$v^2 = \frac{B_0^2\cos\alpha\,\sin(\beta-\alpha)}{\mu_0\rho_0\sin\beta} \tag{20.42}$$

respectively. From these two equations, we obtain

$$C_s^2 = \frac{B_0^2\sin 2(\beta-\alpha)}{\mu_0\rho_0\sin 2\beta} \tag{20.43}$$

$$C_s^2\left(\frac{B_0^2}{\mu_0\rho_0}\right)\sin^2\alpha = (v^2 - C_s^2)\left(v^2 - \frac{B_0^2}{\mu_0\rho_0}\right) \tag{20.44}$$

The latter shows that the waves will have velocity either greater than both C_s and $B_0/\sqrt{\mu_0\rho_0}$, it will be less than both, the maximum possible value of the wave-velocity being $\sqrt{C_s + \mu_0\rho_0 B_0^2}$.

All the above results have been obtained for small enough disturbances so that the use of linearized equations is justified. Besides, dissipative effects, as well as the displacement current have been neglected. In the following section, we consider the effects of viscosity and electrical resistance. These will cause an irreversible generation of heat and consequent attenuation of the waves.

20.5 Dissipative Effects

Equations (20.29) and (20.30) are now modified to

$$\rho_0\frac{\partial \mathbf{u}}{\partial t} + \nabla p_1 + \frac{1}{\mu_0}\mathbf{B}_0 \times (\nabla \times \mathbf{B}_1) - \eta\,\nabla^2\mathbf{u} = 0 \tag{20.45}$$

$$\frac{\partial \mathbf{B}_1}{\partial t} = \frac{c^2}{\mu_0\sigma}\,\nabla^2\mathbf{B}_1 + \nabla \times (\mathbf{u} \times \mathbf{B}_0) \tag{20.46}$$

where η is the coefficient of viscosity as before. The viscous force expression $\eta\,\nabla^2\mathbf{u}$ is obtained in fluid dynamics under the condition $\nabla \cdot \mathbf{u} = 0$. Therefore, to be consistent we consider the case $\mathbf{u} \cdot \mathbf{k} = 0$. Using this, Eq. (20.45) yields

$$\frac{\partial^2\mathbf{u}}{\partial t^2} + \frac{1}{\mu_0\rho_0}\mathbf{B}_0 \times \left(\nabla \times \frac{\partial \mathbf{B}_1}{\partial t}\right) - \frac{\eta}{\rho_0}\nabla^2\left(\frac{\partial \mathbf{u}}{\partial t}\right) = 0$$

Now with $\mathbf{B}_1 = (\mathbf{B}_1)_0\, e^{i(\mathbf{k}\cdot\mathbf{r}-\omega t)}$ the above equation reduces to

$$-\omega^2\mathbf{u} + \frac{\omega}{\mu_0\rho_0}\mathbf{B}_0 \times (\mathbf{k} \times \mathbf{B}_1) - i\frac{\eta\omega k^2}{\rho_0}\mathbf{u} = 0 \tag{20.47}$$

But from the condition above and Eq. (20.46)

$$\mathbf{B}_1 = -\frac{(\mathbf{k}\cdot\mathbf{B}_0)\,\mathbf{u}}{\omega + i\frac{c^2k^2}{\mu_0\sigma}} \tag{20.48}$$

Using the last two equations and keeping in mind that $\mathbf{u} \cdot \mathbf{k} = 0$, one gets

$$-\omega^2 - \frac{i\eta k^2}{\rho_0}\omega + (\mathbf{k} \cdot \mathbf{B}_0)^2 \mu_0 \rho_0 \left(1 + i\frac{c^2 k^2}{\mu_0 \sigma \omega}\right) = 0 \tag{20.49}$$

$$\mathbf{u} \cdot \mathbf{B}_0 = 0 \tag{20.50}$$

Neglecting the product of the small quantities η and $1/\sigma$, Eq. (20.49) gives

$$\omega^2 + ik^2\xi\omega - \frac{B_0^2 k^2 \cos^2\alpha}{\mu_0 \rho_0} = 0$$

where α is the angle between $\mathbf{B}_0$ and $\mathbf{k}$, and $\xi = \frac{\eta}{\rho_0} + \frac{c^2}{\mu_0 \sigma}$. The solution of this quadratic equation is

$$\omega = -\frac{i}{2}k^2\xi \pm \sqrt{\frac{B_0^2 k^2 \cos^2\alpha}{\mu_0 \rho_0} - \xi^2 k^4}$$

Thus only long waves for which the wavelength exceeds $2\pi\xi\sqrt{\mu_0\rho_0}/B_0$ can propagate, and the imaginary term gives a decay of the intensity as the wave proceeds.[2] Lastly, we note that the velocity is frequency dependent so that these waves are dispersive.

Problems

1. Assuming that there is no charge separation, but taking into account the displacement current, investigate the velocity of waves in different cases. In particular if $\mathbf{u}$ is perpendicular to $\mathbf{B}_0$ and $\mathbf{k}$ parallel to $\mathbf{B}_0$, show that the phase velocity of the waves is $cv/\sqrt{c^2 + v^2}$ where c is the velocity of light in vacuo and v is the Alfven velocity.
2. In a thetatron, the plasma cylinder is field-free but is enclosed in a vacuous region where there is a uniform field in the axial direction. Show how such a field may be obtained in practice and investigate the condition of equilibrium. Is the equilibrium stable, unstable or neutral?
3. A non-viscous liquid is contained in a cylinder in which there is a uniform axial field. The fluid velocity as well as the magnetic field variations are solely in the azimuthal direction. Set up the equations of the problem in cylindrical polar coordinates and show that there are progressive waves proceeding in the axial direction, which suffer attenuation if the conductivity of the liquid be finite.

[2] In any case our considerations have been limited to low frequencies for which there is neither charge separation nor appreciable displacement current.

Chapter 21
Two Component Plasma Oscillations

In case the frequency of collisions is not so high, the continuous fluid approximation is no longer appropriate. Further, for frequencies of oscillation comparable to the plasma frequency, the assumptions of no charge separation and neglect of displacement current are also untenable. One can then adopt the kinetic theory approach of introducing a distribution function and use Boltzmann's equation. If $f(\mathbf{r}, \mathbf{v}, t)\, d\mathbf{r}\, d\mathbf{v}$ denote the number of particles in the spatial volume lying between $\mathbf{r}$ and $\mathbf{r} + d\mathbf{r}$, having velocities in the range $\mathbf{v}$ to $\mathbf{v} + d\mathbf{v}$ at time t, then Boltzmann's equation is

$$\frac{\partial f}{\partial t} + v_i \frac{\partial f}{\partial r_i} + \frac{F_i}{m} \frac{\partial f}{\partial v_i} = \left(\frac{\partial f}{\partial t} \right)_{\text{coll}} \tag{21.1}$$

In the above equation a summation over the repeated index i is to be understood, r_i and v_i being the components of $\mathbf{r}$ and $\mathbf{v}$ so that $i = 1, 2, 3$, F_i is the ith component of the force acting on a particle of mass m and $(\partial f/\partial t)_{\text{coll}}$ represents the change in the distribution due to what are called collisions. In applying th to the case of a plasma, we note that there are invariably at least two components—the electrons and the positive ions. There may of course be more, e.g., neutral particles or different types of ions. Here, we shall consider only two types of particles and for simplicity assume the positive ions to be protons. We assume that the composite distribution function involving all different types of particles may be factored out and expressed as a product of the distribution f functions for each different type. Thus in (21.1) f is the distribution function for either the electrons or the protons but the $(\partial f/\partial t)_{\text{coll}}$ term will involve both the protons and the electrons, for the velocity of an electron (say) may undergo a change by a collision with a proton. For the force F_i we shall assume that we can talk of a electromagnetic field in the plasma and write

$$\mathbf{F} = e\,(\mathbf{E} + \mathbf{v} \times \mathbf{B}) \tag{21.2}$$

A. K. Raychaudhuri, *Classical Theory of Electricity and Magnetism*, Texts and Readings in Physical Sciences 21, https://doi.org/10.1007/978-981-16-8139-4_21

where the electromagnetic field (**E**, **B**) should be self-consistent, i.e., their sources should be given by the distribution functions of the electrons and protons, which are solutions of Eq. (21.1).

The collisions responsible for the $(\partial f/\partial t)_{\text{coll}}$ term may be considered to be a short-range interaction which lasts over an infinitesimal time and brings about an abrupt change of velocity (but not of position as that would require the velocity to blow up). In case of a plasma, although the Coulomb force is a long-range one, one has effectively a short-range force field about any charged particle. This may be seen as follows. A positively charged particle attracts the electrons and repels the positive ions. Thus, in the immediate neighbourhood of a positive ion, there will be an excess of electrons which would shield the field of the positive charge at larger distances. This is how a short-range field results. If the number density of particles of charge e be n, Poisson's equation for the electrostatic potential ϕ is

$$\nabla^2\phi = -\frac{ne}{\epsilon_0} \tag{21.3}$$

The potential energy of a particle of charge $\pm e$ in the field being $\pm\phi e$, the number density of the particles will vary with position according to the Boltzmann statistics: $n = \frac{n_0}{2}\exp(\mp\phi/k_BT) \simeq \frac{n_0}{2}(1 \mp e\phi/k_BT)$ for $\frac{e\phi}{k_BT} \ll 1$, a condition that is usually satisfied. In the above, n_0 is the total number density of ions at $\phi \to 0$, so that the plasma being neutral in the undisturbed state the number density of each type is $n_0/2$. Hence Eq. (21.3) gives

$$\nabla^2\phi = -\frac{1}{\epsilon_0}(n_+e - n_-e) = \frac{n_0\phi e^2}{\epsilon_0 k_B T}$$

Assuming the field to have central symmetry, the potential due to a point charge in a plasma is

$$\phi = \frac{e}{4\pi\epsilon_0 r}e^{-r/D} \tag{21.4}$$

where $D = \sqrt{\frac{\epsilon_0 k_B T}{n_0 e^2}} \sim 69\sqrt{\frac{T}{n_0}}$ m is called the Debye shielding distance. (Debye first deduced the formula in his discussion of the theory of electrolytes.) The distance is usually small, e.g., for $T \sim 10^3$ K and $n_0 \sim 10^{21}/\text{m}^3$, we find $D \sim 7 \times 10^{-8}$ m. Even then there may be quite a number of 'many particle collisions', since even for the above example $n_0 D^3 \sim 0.3$.

Consider now a function $g(\mathbf{v})$ of $\mathbf{v}$ alone, independent of $\mathbf{r}$ and t. Multiplying Eq. (21.1) by $g(\mathbf{v})$ and integrating over the entire range of velocities, we get (in tensor notation)

$$\int g(\mathbf{v})\left(\frac{\partial f}{\partial t} + v_i\frac{\partial f}{\partial r_i} + \frac{F_i}{m}\frac{\partial f}{\partial v_i}\right)d\mathbf{v} = \int g(\mathbf{v})\left(\frac{\partial f}{\partial t}\right)_{\text{coll}} d\mathbf{v} \tag{21.5}$$

The terms in the integral on the left side may be written in physically more significant forms as follows.

$$\int g(\mathbf{v})\frac{\partial f}{\partial t}\,d\mathbf{v} = \frac{\partial}{\partial t}\int g(\mathbf{v})\,f\,d\mathbf{v} = \frac{\partial n}{\partial t}\,\langle g\rangle \tag{21.6}$$

$$\int g(\mathbf{v})\,v_i\frac{\partial f}{\partial r_i}\,d\mathbf{v} = \frac{\partial}{\partial r_i}\int g(\mathbf{v})\,v_i f\,d\mathbf{v} = \frac{\partial}{\partial r_i}\,(n\,\langle g v_i\rangle) \tag{21.7}$$

$$\int F_i\frac{\partial f}{\partial v_i}g(\mathbf{v})\,d\mathbf{v} = \int f\,\frac{\partial(F_i g)}{\partial v_i}\,d\mathbf{v} = -\int f F_i\frac{\partial g}{\partial v_i}\,d\mathbf{v}$$

$$= -n\left\langle F_i\frac{\partial g}{\partial v_i}\right\rangle \tag{21.8}$$

where in (21.8) we have first used the condition that f vanishes sufficiently rapidly for $v_i \to \pm\infty$, and in the second step used the result $\frac{\partial F_i}{\partial v_i} = e\frac{\partial}{\partial v_i}\,(\epsilon_{ikl}v_k B_l) = e\epsilon_{ikl}\delta_{ik}B_l = 0$, since ϵ_{ikl} is the totally antisymmetric symbol. In the above equation, n indicates the number of particles at r for all velocities. In view of Eqs. (21.6–21.8), Eq. (21.5) may be written in the form

$$\frac{\partial n}{\partial t}\,\langle g\rangle + \frac{\partial n}{\partial r_i}\,\langle g v_i\rangle - n\left\langle\frac{F_i}{m}\frac{\partial g}{\partial v_i}\right\rangle = \int g\left(\frac{\partial f}{\partial t}\right)_{\text{coll}} d\mathbf{v} \tag{21.9}$$

The integral on the right gives the change of $n\langle g\rangle$ due to collisions.

In order to obtain the so-called first two moment equations, we take, successively, $g = 1$ and $g = m v_k$. The first equation so obtained is

$$\frac{\partial n}{\partial t} + \frac{\partial n}{\partial r_i}\,\langle v_i\rangle = 0 \tag{21.10}$$

which simply indicates the conservation of the particle number. One can obtain from this the principles of charge and mass conservation. The second moment equation reads

$$\frac{\partial}{\partial t}\rho\langle v_k\rangle + \frac{\partial}{\partial r_i}\rho\langle v_k v_i\rangle - n\langle F_k\rangle = \int m v_k\left(\frac{\partial f}{\partial t}\right)_{\text{coll}} d\mathbf{v} \tag{21.11}$$

where $\rho = nm$ is the mass per unit volume. We introduce a velocity vector $\mathbf{u}$ such that $\langle v_i\rangle = u_i$, where u_i indicates the velocity of the hypothetical continuous fluid made up of the discrete particles. Writing $v_k = u_k + V_k$, where V_k is the 'random' part of the particle velocity so that $\langle V_k\rangle = 0$, we have

$$\langle v_k v_i\rangle = u_k u_i + \langle V_k V_i\rangle$$

$$\frac{\partial}{\partial t}\langle\rho v_k\rangle + \frac{\partial}{\partial r_i}(\rho u_k u_i) + \frac{\partial p_{ik}}{\partial r_i} - n\langle F_k\rangle = \int m v_k\left(\frac{\partial f}{\partial t}\right)_{\text{coll}} d\mathbf{v} \tag{21.12}$$

where $p_{ik} = \rho\langle V_i V_k\rangle$ is the stress tensor that arises from the departure of the particle velocities from the mass velocity. If these velocities be isotropic, as is to be expected in the absence of a preferred direction like the direction of an external magnetic field, $p_{ik} = \frac{1}{3}\rho\langle V^2\rangle\delta_{ik} = p\delta_{ik}$ (say) and we can introduce the temperature of the plasma by the relation $p = nk_BT$, or $\langle\frac{1}{2}mV^2\rangle = \frac{3}{2}k_BT$.

The collision integral on the right of Eq. (21.12) would vanish if there be only one type of particle, for in collisions the total momentum is conserved. In case there are two types, the integral would represent the momentum exchange between the two species and would cancel out if added together for the two species. However, we shall assume that the plasma is so dilute that we can neglect the collision effects—the Boltzmann equation is then known as Boltzmann–Vlasov equation. Then, taking into account Eq. (21.10), Eq. (21.12) becomes

$$\frac{\partial u_k}{\partial t} + u_i\frac{\partial u_k}{\partial r_i} = \frac{e}{m}\left(E_k + (\mathbf{u}\times\mathbf{B})_k\right) - \frac{1}{\rho}\frac{\partial p}{\partial r_k} \tag{21.13}$$

where of course $\mathbf{u}$, p and ρ as well as m would have different values for the electrons and protons.

21.1 Waves in the Plasma

We shall consider two different situations.

Case 1. *Electrostatic waves involving oscillations of both electrons and protons*

We consider oscillations of the electric field unaccompanied by any magnetic field. The equations of motion of the protons and electrons, from (21.13) in terms of the displacement vector $\boldsymbol{\xi}$ (where $\mathbf{u} = \dot{\boldsymbol{\xi}}$), are

$$\ddot{\boldsymbol{\xi}}_1 = -\frac{1}{n_1 m_1}\nabla p_1 + \frac{e}{m_1}\mathbf{E} \tag{21.14}$$

$$\ddot{\boldsymbol{\xi}}_2 = -\frac{1}{n_2 m_2}\nabla p_2 + \frac{e}{m_2}\mathbf{E} \tag{21.15}$$

where subscripts 1 and 2 refer to the protons and electrons, respectively. The equation of continuity (21.10) may be integrated to give

$$n_i = n_0\left(1 - \nabla\cdot\boldsymbol{\xi}_i\right), \qquad i = 1, 2 \tag{21.16}$$

where the constant of integration n_0 is the same for both species, indicating that in the equilibrium condition, the plasma is electrically neutral. In view of Eq. (21.16)

$$\nabla\cdot\mathbf{E} = \frac{e}{\epsilon_0}(n_1 - n_2) = -\frac{e}{\epsilon_0}n_0\,\nabla\cdot\left(\boldsymbol{\xi}_1 - \boldsymbol{\xi}_2\right) \tag{21.17}$$

Taking the divergence of Eqs. (21.14–21.15) and using Eqs. (21.16–21.17), we get

$$\ddot{n}_1 = C_1^2 \nabla^2 n_1 - \nu_1^2 (n_1 - n_2) \tag{21.18}$$
$$\ddot{n}_2 = C_2^2 \nabla^2 n_2 - \nu_2^2 (n_1 - n_2) \tag{21.19}$$

where

$$\nu_i = \sqrt{\frac{n_0 e^2}{\epsilon_0 m_i}} \tag{21.20}$$

$$C_i = \sqrt{\left(\frac{\partial p}{\partial \rho}\right)_i} \tag{21.21}$$

In particular ν_2 is the so-called plasma frequency, which is the frequency of electronic oscillations when the kinetic pressure terms are negligible. While C_i's are the velocities of compressional waves in the neutral plasma and are named the sound velocities.

For a monochromatic wave

$$n_i = A_i e^{i(\mathbf{k}\cdot\mathbf{r} - \omega t)} \tag{21.22}$$

Equations (21.18) and (21.19) give

$$\left(\omega^2 - C_1^2 k^2 - \nu_1^2\right) A_1 + \nu_1^2 A_2 = 0 \tag{21.23}$$
$$\left(\omega^2 - C_2^2 k^2 - \nu_2^2\right) A_2 + \nu_2^2 A_1 = 0 \tag{21.24}$$

For a nontrivial solution of Eqs. (21.23–21.24), we must have

$$\left(\omega^2 - C_1^2 k^2 - \nu_1^2\right)\left(\omega^2 - C_2^2 k^2 - \nu_2^2\right) - \nu_1^2 \nu_2^2 = 0 \tag{21.25}$$
$$\omega^4 - \omega^2\left((C_1^2 + C_2^2)k^2 + \nu_1^2 + \nu_2^2\right) + C_1^2 C_2^2 k^4 + (C_1^2 \nu_2^2 + C_2^2 \nu_1^2)k^2 = 0 \tag{21.26}$$

Equations (21.25–21.26) look complicated, however, it is clear that there will be no root of ω^2 lying between $C_1^2 k^2 + \nu_1^2$ and $C_2^2 k^2 + \nu_2^2$. As $m_1 \gg m_2$, $\nu_2 \gg \nu_1$ and in general $C_2 \gg C_1$ (unless the electron temperature be very low compared to the proton temperature)

$$C_2^2 k^2 + \nu_2^2 \gg C_1^2 k^2 + \nu_1^2$$

Hence if ω_1 and ω_2 be the lower and higher eigenfrequency, i.e., $\omega_1^2 < (C_1^2 k^2 + \nu_1^2)$ and $\omega_2^2 > (C_2^2 k^2 + \nu_2^2)$, respectively. Further, in view of the smallness of ν_1^2, we can replace the inequalities by equalities (see Eq. (21.25))

$$\omega_1^2 = C_1^2 k^2 + \nu_1^2$$
$$\omega_2^2 = C_2^2 k^2 + \nu_2^2 \tag{21.27}$$

From Eq. (21.24), we have

$$\frac{A_1}{A_2} = 1 + \frac{C_2^2 k^2 - \omega^2}{\nu_2^2} \tag{21.28}$$

We may consider three special cases.

1 (a) If the frequency ω be small compared to the plasma frequency ν_2, a condition which is consistent with Eqs. (21.27) with $i = 1$ only, and the wavelength $\lambda = 2\pi/k$ be also large compared with the Debye shielding distance, then $A_1/A_2 = 1$ (i.e., the protons and electrons more or less move together). To see this, consider

$$\frac{C_2 k}{\nu_2} = \frac{2\pi C_2}{\lambda}\sqrt{\frac{m_2 \epsilon_0}{n_0 e^2}} = \frac{2\pi D}{\lambda}\sqrt{\frac{C_2^2 m}{k_B T_2}} \sim \frac{D}{\lambda}$$

where we have used the result that the sound velocity is equal to the root-mean-square velocity of the particles $\sqrt{3k_B T/(2m)}$ to within a factor of the order of unity. Thus low-frequency long waves may proceed without charge separation, as we have already assumed in the study of magneto-hydrodynamic waves.

1 (b) If $C_2 k \gg \nu_2$, i.e., $\lambda \ll D$, and also $C_2 \gg C_1$, then with Eq. (21.27) and (21.28) $A_1/A_2 \gg 1$. These waves are primarily due to proton oscillations and are accompanied by charge separation.

1 (c) The condition $C_2^2 k^2 - \omega^2 + \nu_2^2 = 0$ may be satisfied with Eq. (21.27). The oscillations are primarily electronic and the frequency higher than the plasma frequency.

Case 2. *Waves in which the motion of the protons may be neglected.*

The basic equations now are

$$\frac{\partial n_e}{\partial t} + \nabla \cdot (n_e \mathbf{u}) = 0 \tag{21.29}$$

$$\frac{\partial \mathbf{u}}{\partial t} + (\mathbf{u} \cdot \nabla)\,\mathbf{u} = \frac{e}{m}(\mathbf{E} + \mathbf{u} \times \mathbf{B}) - \frac{1}{m n_e}\nabla p \tag{21.30}$$

$$\nabla \cdot \mathbf{E} = \frac{e}{\epsilon_0}(n_e - n_0) \tag{21.31}$$

$$\mathbf{j} = n_e e \mathbf{u} \tag{21.32}$$

$$\nabla \cdot \mathbf{B} = 0 \tag{21.33}$$

$$\nabla \times \mathbf{E} = -\dot{\mathbf{B}} \tag{21.34}$$

$$\nabla \times \mathbf{B} = \mu_0 \epsilon_0 \dot{\mathbf{E}} + \mu_0 n_e e \mathbf{u} \tag{21.35}$$

In the above equations $\mathbf{u}$ and p refer to the electronic velocity and kinetic pressure, n_0 is the equilibrium number density of either type of particles, therefore, is the number density of protons in the disturbed state as well, and finally, n_e is the number density of electrons in the disturbed state.

We restrict our considerations to small departures from the equilibrium state, which is defined by $\mathbf{E} = c\mathbf{B} = \mathbf{u} = \nabla p = n_e - n_0 = 0$. Writing $n = n_e - n_0$, the linearized equations are

$$\frac{\partial n}{\partial t} + n_0 \nabla \cdot \mathbf{u} = 0 \tag{21.36}$$

$$\frac{\partial \mathbf{u}}{\partial t} - \frac{e}{m}\mathbf{E} + \frac{1}{mn_0}\left(\frac{\partial p}{\partial n}\right)_0 \cdot \nabla n = 0 \tag{21.37}$$

$$\nabla \cdot \mathbf{E} = \frac{ne}{\epsilon_0} \tag{21.38}$$

$$\nabla \times \mathbf{B} = \mu_0 \epsilon_0 \dot{\mathbf{E}} + \mu_0 n_0 e \mathbf{u} \tag{21.39}$$

$$\nabla \times \mathbf{E} = -\dot{\mathbf{B}} \tag{21.40}$$

Equation (21.37) no longer contains the $\mathbf{u} \times \mathbf{B}$ term, as it is nonlinear. This enables us to have what are called electrostatic oscillations.

Case 2(a). *Electrostatic oscillations.*

These are oscillations of the electric field unaccompanied by magnetic fields. This would mean, in view of Eq. (21.40), that the field is irrotational (hence the term electrostatic). Now Eq. (21.39) gives

$$\dot{\mathbf{E}} = -\frac{n_0 e}{\epsilon_0}\mathbf{u} \tag{21.41}$$

which means that the effect of the convection current is exactly annulled by the displacement current. Equation (21.41) shows that the oscillations of $\mathbf{E}$ and $\mathbf{u}$ will be of identical period but opposite in phase. Further for a plane wave $\mathbf{E} = \mathbf{E}_0 \exp i(\mathbf{k} \cdot \mathbf{r} - \omega t)$, the condition $\nabla \times \mathbf{E} = 0$ gives $\mathbf{k} \times \mathbf{E} = 0$, i.e., the electric intensity lies in the direction of propagation. From Eq. (21.41) $\nabla \times \mathbf{u} = 0$, i.e., the velocity vector also is in the direction of propagation and the vorticity vanishes. The waves are thus completely longitudinal. From Eq. (21.36) for a monochromatic plane wave we have

$$n = \frac{1}{\omega} n_0 |\mathbf{k}|\, |\mathbf{u}| \tag{21.42}$$

where we have taken into account the parallelism of $\mathbf{u}$ and $\mathbf{k}$. Thus, from Eqs. (21.41) and (21.42), $\mathbf{E}$, $\mathbf{u}$ and n would obey identical differential equations (however, while the former two are vector fields, n is a scalar field). The differential equation is obtained from Eqs. (21.36), (21.37) and (21.40)

$$\left(\frac{\partial^2}{\partial t^2} - \frac{1}{m}\left(\frac{\partial p}{\partial n}\right)_0 \nabla^2 + \frac{n_0 e^2}{m\epsilon_0}\right)\chi = 0 \tag{21.43}$$

where χ may stand for n or any component of $\mathbf{E}$ or $\mathbf{u}$. With $\chi = \chi_0 e^{-i(\mathbf{k}\cdot\mathbf{r}-\omega t)}$, we get

$$\omega^2 = \frac{n_0 e^2}{m\epsilon_0} + \frac{1}{m}\left(\frac{\partial p}{\partial n}\right)_0 k^2 = \nu_2^2 + \frac{1}{m}\left(\frac{\partial p}{\partial n}\right)_0 k^2 \tag{21.44}$$

where $\nu_2 = \sqrt{\frac{n_0 e^2}{m\epsilon_0}}$ is the plasma frequency. As ω/k, the velocity of the waves, is frequency dependent, we have dispersion. The waves, however, do not carry energy as $\mathbf{E} \times \mathbf{B} = 0$ (hence the name non-radiative waves).

Case 2(a). *General electromagnetic waves.*

We shall no longer take $\mathbf{B} = 0$. The equation for n is still of the form of Eq. (21.43), but for $\mathbf{E}$, from Eqs. (21.39) and (21.40), the equation is now

$$\ddot{\mathbf{E}} = -\frac{n_0 e}{\epsilon_0}\dot{\mathbf{u}} - c^2\nabla(\nabla\cdot\mathbf{E}) + c^2\nabla^2\mathbf{E}$$

Eliminating $\mathbf{u}$ with the help of Eq. (21.37), the above equation gives

$$\begin{aligned}\ddot{\mathbf{E}} &= -\frac{n_0 e}{\epsilon_0}\left(\frac{e}{m}\mathbf{E} - \frac{1}{mn_0}\left(\frac{\partial p}{\partial n}\right)_0 \nabla n\right) - c^2\nabla(\nabla\cdot\mathbf{E}) + c^2\nabla^2\mathbf{E} \\ &= -\frac{n_0 e}{\epsilon_0}\left(\frac{e}{m}\mathbf{E} - \frac{\epsilon_0}{mn_0 e}\left(\frac{\partial p}{\partial n}\right)_0 \nabla(\nabla\cdot\mathbf{E})\right) - c^2\nabla(\nabla\cdot\mathbf{E}) + c^2\nabla^2\mathbf{E}\end{aligned}$$

which may be written as

$$\frac{\partial^2\mathbf{E}}{\partial t^2} - c^2\nabla^2\mathbf{E} + (c^2 - c_s^2)\nabla(\nabla\cdot\mathbf{E}) + \frac{ne^2}{\epsilon_0 m}\mathbf{E} = 0 \tag{21.45}$$

where we have written $c_s^2 = \frac{1}{m}\left(\frac{\partial p}{\partial n}\right)_0$ for sound velocity in the plasma. For a monochromatic plane wave $\mathbf{E} = \mathbf{E}_0 e^{i(\mathbf{k}\cdot\mathbf{r}-\omega t)}$, Eq. (21.45) gives

$$(\omega^2 - c_s^2 k^2)\mathbf{E} + (c^2 - c_s^2)(\mathbf{k}\cdot\mathbf{E})\mathbf{k} - \frac{n_0 e^2}{m\epsilon_0}\mathbf{E} = 0$$

Splitting $\mathbf{E}$ into components $E_{\parallel}$ and $E_{\perp}$, parallel and perpendicular to $\mathbf{k}$, respectively

$$\begin{aligned}\left(\omega^2 - c^2k^2 - \nu_2^2\right) E_{\perp} &= 0 \\ \left(\omega^2 - c_s^2k^2 - \nu_2^2\right) E_{\parallel} &= 0\end{aligned}$$

where, as before, ν_2 stands for the plasma frequency. Thus for longitudinal waves, Eq. (21.44) is still valid while for transverse waves

$$\omega^2 = \nu_2^2 + c^2k^2$$

Again with $\mathbf{B} = \mathbf{B}_0 e^{i(\mathbf{k}\cdot\mathbf{r}-\omega t)}$, Eq. (21.40) gives

$$\mathbf{k} \times \mathbf{E}_0 = \omega \mathbf{B}_0 \tag{21.46}$$

showing that for transverse waves, **B**, **k** and **E** are mutually perpendicular. We have from Eqs. (21.35) and (21.46)

$$\begin{aligned}
\mu_0 n_0 e\, \mathbf{u} &= i\,(\mathbf{k} \times \mathbf{B}) + \frac{i\omega}{c^2}\mathbf{E} \\
&= \frac{i}{\omega}\mathbf{k} \times (\mathbf{k} \times \mathbf{E}) + \frac{i\omega}{c^2}\mathbf{E} \\
&= -\frac{i}{\omega}\mathbf{E}\left(k^2 - \frac{\omega^2}{c^2}\right) \\
&= \frac{i v_2^2}{c^2\omega}\,\mathbf{E}
\end{aligned} \tag{21.47}$$

Equation (21.31) shows that the transverse waves are not associated with charge separation. Also in view of Eq. (21.47) the motion is now vortical.

Chapter 22
The Theory of the Electron

The emission of radiation by an accelerated electron means that if not replenished by external agencies, the accelerated electron would be losing energy. As the electron radiates, it must be experiencing a retarding force which we call the radiation reaction. We would expect from the principle of conservation of energy that the work done by the radiation reaction exactly accounts for the loss of energy by radiation, so that at first sight one may be tempted to write

$$\mathbf{F} \cdot \mathbf{u} + \frac{e^2 \dot{u}^2}{6\pi \epsilon_0 c^3} = 0 \tag{22.1}$$

where **F** denotes the radiation reaction. This equation, however, leads to absurdities as can be easily seen. Split up **F** into parts parallel and perpendicular to **u**. Then from Eq. (22.1)

$$F_{\parallel} = -\frac{e^2 \dot{u}^2}{6\pi \epsilon_0 c^3 u}$$

which diverges if $\dot{u} \neq 0$ and $u \to 0$. This would mean that to start the motion of an electron at rest one requires an infinite force.

The defect in the above discussion is that there is, besides the radiation loss and the change in the kinetic energy of the electron, a change in the field energy which we have not taken into consideration in the energy balance. However, if the motion of the electron be periodic, the field returns to the same state after each period and then the total work done by the radiation reaction during one period must exactly equal the radiation loss during the period. This condition is satisfied if we take

$$\mathbf{F} = \frac{1}{6\pi \epsilon_0} \frac{e^2}{c^3} \ddot{\mathbf{u}}$$

since then the work done by this force is

A. K. Raychaudhuri, *Classical Theory of Electricity and Magnetism*, Texts and Readings in Physical Sciences 21, https://doi.org/10.1007/978-981-16-8139-4_22

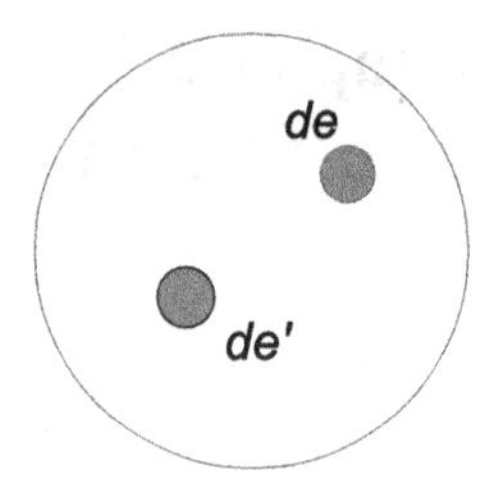

Fig. 22.1 Charge elements in a finite sized electron with spherical symmetry

$$\int_{t_1}^{t_2} \mathbf{F} \cdot \mathbf{u}\, dt = \frac{e^2}{6\pi \epsilon_0 c^3} \left[\dot{\mathbf{u}} \cdot \mathbf{u} \right]_{t_1}^{t_2} - \frac{e^2}{6\pi \epsilon_0 n c^3} \int_{t_1}^{t_2} u^2\, dt$$

The integrated term on the right may be interpreted as the change of electromagnetic field energy—it vanishes if the interval $t_2 - t_1$ is a multiple of the time period, or if either $\dot{u}$ or u vanish at either limit.

Apparently, the radiation reaction comes not from any external agency but from the electron itself. (This is, however, not so in the Wheeler–Feynman theory as we shall see later.) We proceed to build a theory to obtain the radiation reaction term.

We shall consider the electron to be of finite size and have spherical symmetry—the assumption of the finite size raises some difficulties if one introduces the ideas of the special theory of relativity but for the present we shall not get into those. We shall allow for any possible non-uniformity in the distribution of the electric charge within the electron, subject to the restriction of spherical symmetry of course. We shall adopt some approximations, the meaning of which will be clear as we proceed.

Let us work out the interaction between two elements of charge de and de', as shown in Fig. 22.1. The field due to de' at the position of de is given by (see Eq. (17.1)

$$\begin{aligned} dE_k = &\frac{de'}{4\pi \epsilon_0} \frac{\left(R_k - \frac{R u_k}{c}\right)}{\left(R - \frac{R_l u_l}{c}\right)^3} \left[\left(1 - \frac{u^2}{c^2}\right) + \frac{\dot{u}_i R_i}{c^2} \right] \\ &- \frac{de'}{4\pi \epsilon_0} \frac{\dot{u}_k R \left(R - \frac{R_l u_l}{c}\right)}{c^2 \left(R - \frac{R_l u_l}{c}\right)^3} \end{aligned} \tag{22.2}$$

where all the variables pertaining to the source are to be taken at the retarded time t'. Since $\mathbf{R}$ is $\mathbf{r} - \mathbf{r}'$, its time derivatives are $\dot{\mathbf{R}} = -\dot{\mathbf{r}}' = -\mathbf{u}$ and $\ddot{\mathbf{R}} = -\dot{\mathbf{u}}$ etc.

Let us take the instant at which the field is calculated to be $t = 0$. Then the retarded time is $t' = -R/c$. We make a Taylor series expansion to express the variables at the retarded time t' in terms of their values at the time $t = 0$

$$(R_k)_{\text{ret}} = R_k + \dot{R}_k t' + \frac{1}{2}\ddot{R}_k t'^2 + \frac{1}{6}\dddot{R}_k t'^3 + \cdots$$
$$= R_k - u_k t' - \frac{1}{2}\dot{u}_k t'^2 - \frac{1}{6}\ddot{u} t'^3 + \cdots$$
$$(R^2)_{\text{ret}} = R^2 - 2R_k u_k t' - R_k \dot{u}_k t'^2 - \frac{1}{3} R_k \ddot{u}_k t'^3 + \cdots$$

In the above, we have retained only the linear terms in u_k and its derivatives. Therefore, consistent to this order of approximatIon

$$R_{\text{ret}} = R\left(1 - \frac{R_k u_k t'}{R^2} - \frac{\dot{u}_k R_k t'^2}{2R^2} - \frac{\ddot{u}_k R_k t'^3}{6R^2} + \cdots\right)$$

Also, $t' = R_k/c$ differs from R/c by terms involving u's—since we shall ignore terms of higher order in u's etc, these terms in t' would not enter into our calculations. Therefore, we may take $t' = -R/c$, so that

$$(R_k)_{\text{ret}} = R_k + \frac{Ru_k}{c} - \frac{R^2 \dot{u}_k}{2c^2} + \frac{R^3 \ddot{u}_k}{6c^3} + \cdots \tag{22.3a}$$
$$(u_k)_{\text{ret}} = u_k - \frac{R}{c}\dot{u}_k + \frac{R^2}{2c^2}\ddot{u}_k - \frac{R^3}{6c^3}\dddot{u}_k + \cdots \tag{22.3b}$$

Hence

$$(R_k u_k)_{\text{ret}} = R_k u_k - \frac{R R_k \dot{u}_k}{c} + \frac{R^2 R_k \ddot{u}_k}{2c^2} - \frac{R^3 R_k \dddot{u}_k}{6c^3} + \cdots \tag{22.4}$$
$$\left(R_k - \frac{Ru_k}{c}\right)_{\text{ret}} = R_k + \frac{R^2 \dot{u}_k}{2c^2} - \frac{R^3 \ddot{u}_k}{3c^3} + \cdots \tag{22.5}$$
$$\left(R - \frac{R_k u_k}{c}\right)_{\text{ret}} = R + \frac{R R_k \dot{u}_k}{2c^2} - \frac{R^2 R_k \ddot{u}_k}{3c^3} + \frac{R^3 R_k \dddot{u}_k}{8c^4} + \cdots \tag{22.6}$$
$$\left(R - \frac{R_k u_k}{c}\right)^{-3}_{\text{ret}} = \frac{1}{R^3}\left(1 - \frac{3R_k \dot{u}_k}{2c^2} + \frac{R R_k \ddot{u}_k}{c^3} - \frac{3}{8}\frac{3R^2 R_k \dddot{u}_k}{8c^4} + \cdots\right) \tag{22.7}$$

Substituting these in the expression (22.2) for $dE_k(t)$ and keeping in mind that only linear terms in u and its derivatives are being retained, we have, using (22.5) and (22.7), for the first term in $dE_k(t)$

$$\frac{de'}{4\pi\epsilon_0}\frac{\left(R_k - R\frac{u_k}{c}\right)_{\text{ret}}}{\left(R - \frac{R_l u_l}{c}\right)^3_{\text{ret}}}$$
$$= \frac{de'}{4\pi\epsilon_0 R^3}\left(R_k + \frac{R^2\dot{u}_k}{2c^2} - \frac{R^3\ddot{u}_k}{3c^3} - \frac{3R_k R_l \dot{u}_l}{2c^2} + \frac{R R_k R_l \ddot{u}_l}{c^3} - \frac{3R^2 R_k R_l \dddot{u}_l}{8c^4}\right)$$

The second term in of (22.2) is split up into two parts. The first part, using (22.3a), and (22.3b), and the second term are

$$(\text{i}) = \frac{de'}{4\pi\epsilon_0}\frac{R_i R_k}{R^3 c^2}\left(\dot{u}_i - \frac{R\ddot{u}_i}{c} + \frac{R^2\dddot{u}_i}{2c^2}\right)$$
$$(\text{ii}) = \frac{1}{4\pi\epsilon_0}\frac{de'}{Rc^2}\left(\dot{u}_k - \frac{R\ddot{u}_k}{c} + \frac{R^2\dddot{u}_k}{2c^2}\right)$$

Collecting all these parts we have finally

$$dE_k = \frac{de'}{4\pi\epsilon_0}\left[\frac{R_k}{R^3}\left(1 - \frac{R_l\dot{u}_l}{2c^2}\right) - \frac{\dot{u}_k}{2Rc^2} + \frac{2\ddot{u}_k}{3c^3} + O\left(\frac{R\dddot{u}}{c^4}\right)\right] \tag{22.8}$$

The force on the element de due to the element de' will be $(de)(dE_k)$ and the resultant force due to the interaction between different elements will be obtained by integrating over all the elements de and de'. One may wonder why this integral would not vanish for in this double integral we are in reality adding up terms like the force on de due to de' and force on de' due to de. If Newton's third law of motion holds good then these should be equal and opposite and thus cancel each other. However, in the field-theoretic formulation interactions travel with a finite velocity and thus the third law regarding the action and reaction does not hold good in general, hence the resultant self-force does not vanish. Of course if the conditions be static the velocity of interaction plays no essential part and as we shall just now see the self-force does vanish in that case.

The first term in E_k gives rise to the force

$$\frac{1}{4\pi\epsilon_0}\frac{de\,de' R_k}{R^3}$$

which on integration over de and de' vanishes, for R_k changes sign (equality of action and reaction). Next consider the term involving the acceleration vector $\dot{u}_k$. This term is

$$-\frac{1}{8\pi\epsilon_0}\frac{de\,de'}{Rc^2}\left(\frac{R_k R_l \dot{u}_l}{R^2} + \dot{u}_k\right)$$

In case of spherical symmetry, averaging out $R_k R_l$ gives $\frac{1}{3}\delta_{kl}R^2$. If we assume that the acceleration $\dot{u}$ is the same for all elements (hypothesis of rigid electron—a hypothesis not consistent with the special theory of relativity) we get for the resultant from this term

$$-\frac{1}{8\pi\epsilon_0}\frac{4}{3}\iint\frac{\rho\rho'}{R}\dot{u}_k\,dvdv' = -\frac{4\dot{u}_k}{3c^2}U = -\frac{4\dot{u}_k}{3c^2}\times\frac{1}{8\pi\epsilon_0}\iint\frac{\rho\rho'}{R}dv\,dv'$$

where ρ and ρ' are the charge densities at the volume elements dv and dv', R is the distance between the elements and the integrations for dv and dv' extend over the entire volume of the electron. The quantity U is the electrostatic self-energy of the electron. This term in $\dot{u}$ represents a force opposing the acceleration, i.e., an apparent increase of inertia of magnitude $4U/(3c^2)$. It does not agree with the special theory result that an energy U is associated with an inertia U/c^2. The factor 4/3 as distinct from unity apparently indicates that the inertia of the electron should not be interpreted as purely electromagnetic.

The term in $\ddot{u}_k$ is $\frac{1}{6\pi\epsilon_0}\frac{de\,de'\ddot{u}_k}{c^3}$ and on integration gives $\frac{1}{6\pi\epsilon_0}\frac{e^2\ddot{u}_k}{c^3}$. Thus we have recovered the radiation reaction[1] term we had guessed from the energy conservation principle. It is rather remarkable that this is independent of the dimensions of the electron and is thus justifiable even for the point electron.

The next term involving $\dddot{u}_k$ involves R/c^4 and in the limit of a point particle would vanish. However, it is hardly permissible to go to this limit for the expression of self-energy then blows up. However, irrespective of this limit, the term is of small magnitude as it is of the order of $\frac{Re^2}{c^4} \sim 10^{-85}$ (if one takes $R \sim 10^{-15}$ m and $e \sim 1.6 \times 10^{-19}$ C). Its precise evaluation would require ideas about the 'structure of the electron'.

22.1 Critique of the Classical Theory

The fact that inertia comes out as $4U/(3c^2)$, in place of the relativistic expression U/c^2, indicates that there must be non-Maxwellian interactions. Indeed, it is easy to see that a charge distribution cannot be in equilibrium under electromagnetic interactions alone. This follows directly for a static distribution—the force density $\rho\mathbf{E}$ must vanish or $\mathbf{E}$ must vanish but then *boldsymbol*$\nabla\cdot\mathbf{E} = \rho/\epsilon_0$ would also vanish and we have the trivial case of no charge anywhere. If the distribution of charges be stationary rather than static (i.e., there be currents present), the force density $\rho\mathbf{E} + \mathbf{j}\times\mathbf{B}$ must vanish. Thus

$$\mathbf{j}\times\mathbf{B} = -\rho\mathbf{E}$$

[1] To see explicitly that this term gives the form of radiation reaction given in Eq. (22.1). Note that on taking the dot product with the velocity $\mathbf{u}$, we get $\mathbf{F}\cdot\mathbf{u} = -\frac{1}{6\pi\epsilon_0}e^2\ddot{u}_k u_k = -\frac{e^2}{6\pi\epsilon_0}\left(\frac{d}{dt}(\dot{u}_k u_k) - \dot{u}_k^2\right)$. The second term is always positive. The first term averages to zero for periodic motion.

The right-hand side, as we have just seen, cannot vanish in the non-trivial case, so that we have here three non-homogeneous linear equations for the three components of the current vector j. For a non-trivial solution the determinant of the coefficients, i.e.,

$$\begin{vmatrix} 0 & B_z & -B_y \\ -B_z & 0 & B_x \\ B_y & -B_x & 0 \end{vmatrix} = 0$$

must not vanish. However, it is easy to see that the determinant above vanishes identically. In the classical model, $a \sim e^2 / \left(4\pi\epsilon_0 mc^2\right) \sim 10^{-15}$ m appears as the radius of the electron. However, observationally there is no evidence of finite extension of the electron even down to 10^{-17} m. All these go to show the limitations of the theory we have just discussed.

22.2 The Radiation Reaction

We return to the radiation reaction term. Firstly, it gives the rather awkward result that if $\ddot{u}$ vanishes, then even with $\dot{u}$ non-vanishing, the radiation reaction vanishes. Does it indicate that a charge moving with uniform acceleration would not radiate? (Note that in arriving at the expression of radiation reaction from the energy conservation principle, we had to consider either a periodic motion, or a motion in which either u or $\dot{u}$ vanishes at both limits of integration. In the case of uniform acceleration, neither of these conditions hold good.) In a classical paper Born investigated the field due to a charged particle in uniform accelerated motion. His results may be given in cylindrical coordinates (t, ρ, ϕ, z), the non-zero components are

$$E_\rho = \frac{e}{2\pi\epsilon_0} \frac{\alpha^2 \rho z}{\left(4\alpha^2\rho^2 + (\alpha^2 + t^2 - \rho^2 - z^2)^2\right)^{3/2}}$$

$$E_z = -\frac{e}{\pi\epsilon_0} \frac{\alpha^2\left(\alpha^2 + t^2 - \rho^2 - z^2\right)}{\left(4\alpha^2\rho^2 + (\alpha^2 + t^2 - \rho^2 - z^2)^2\right)^{3/2}}$$

$$B_\phi = \frac{\mu_0 e}{2\pi} \frac{\alpha^2 \rho t}{\left(4\alpha^2\rho^2 + (\alpha^2 + t^2 - \rho^2 - z^2)^2\right)^{3/2}}$$

where we have written $\xi^2 = 4\alpha^2\rho^2 + (\alpha^2 + t^2 - \rho^2 - z^2)^2$ and the trajectory of the particle is $z = \sqrt{\alpha^2 + t^2}$ (units chosen are Gaussian with the velocity of light taken to be unity). Since at the instant at which the velocity of the charged particle vanishes, i.e., at $t = 0$, we have $\mathbf{B} = 0$, it was argued that there is no radiation flux at this instant anywhere. Again there is nothing peculiar with this instant, for the velocity may be made to vanish at any instant whatsoever by a Lorentz transformation—hence the conclusion of no radiation by a uniformly accelerated charge. However, a close examination shows that the Born field is not quite consistent with the ideas of the special theory of relativity, as the field apparently 'travels' with a velocity exceeding the light velocity. Besides, as pointed out by Das (*J. Phys.* **A13**, 529, 1980) simple vanishing of **B** at some space-time points is not sufficient to indicate the absence of radiation. Although there still persists some controversy on the point, it seems that the charge in uniform acceleration does in fact radiate. Therefore, one feels that the validity of the radiation reaction term has some limitations.

Another puzzling feature of the radiation reaction term is the existence of the so-called runaway solutions: The equation of motion of a free electron is

$$m\dot{\mathbf{u}} = \frac{e^2}{6\pi\epsilon_0 c^3}\ddot{\mathbf{u}}$$

One solution obviously corresponds to uniform motion, i.e., $\dot{\mathbf{u}} = 0$. But there is also the solution $\dot{u} = A\exp(\alpha t)$, where $\alpha = 6\pi\epsilon_0 mc^3/e^2$. According to this equation, the electron velocity goes on increasing exponentially and the characteristic time is $\alpha^{-1} \sim 10^{-23}$ s. These runaway solutions are obviously in conflict with our experience.

22.3 The Wheeler–Feynmann Absorber Theory of the Radiation Reaction

The theory of radiation reaction that we have presented makes the force appear as a self-force and the deduction proceeds with the idea of a finite electron. Both these undesirable features are absent in the Wheeler–Feynmann theory. According to them, the intrinsic field due to any charge is time symmetric, i.e., equal to $\frac{1}{2}\left(F_{\text{ret}} + F_{\text{adv}}\right)$, where F_{ret} and F_{adv} represent, respectively, the retarded and advanced fields. Thus in the absence of any other charges, the isolated charge would not radiate. The Wheeler–Feynman theory considers the accelerated charge placed in a medium containing other charges and the medium is a 'complete absorber'. This means that outside there is no resultant field, i.e.,

$$\frac{1}{2}\sum_k \left(F_{\text{ret}}^{(k)} + F_{\text{adv}}^{(k)}\right) = 0 \quad \text{(outside the absorber)}$$

where the superscript k indicates the kth particle and the sum extends over all particles (including, in particular, the particle 1 in whose radiation reaction we are interested). The vanishing of the above sum everywhere outside and at all times indicate

$$\sum_k F_{\text{ret}}^{(k)} = \sum_k F_{\text{adv}}^{(k)} = 0 \quad \text{(outside)}$$

This is because the retarded sum indicates an outgoing wave and the advanced sum an incoming wave, so they cannot completely destroy each other, except when they separately vanish. Hence for a completely absorbing medium

$$\frac{1}{2}\sum_{k}\left(F_{\text{ret}}^{(k)} - F_{\text{adv}}^{(k)}\right) = 0 \quad \text{(outside)}$$

But the above is a singularity-free solution of Maxwell's equation and as it always vanishes outside, it must be zero inside also and at all times. Thus follows the very powerful and simple property of the completely absorbing medium

$$\frac{1}{2}\sum_{k}\left(F_{\text{ret}}^{(k)} - F_{\text{adv}}^{(k)}\right) = 0 \quad \text{(everywhere)}$$

The total field acting on particle 1 is

$$\frac{1}{2}\sum_{k\neq 1}\left(F_{\text{ret}}^{(k)} + F_{\text{adv}}^{(k)}\right) = \sum_{k\neq 1} F_{\text{ret}}^{(k)} + \frac{1}{2}\left(F_{\text{ret}}^{(1)} - F_{\text{adv}}^{(1)}\right) - \frac{1}{2}\sum_{k}\left(F_{\text{ret}}^{(k)} - F_{\text{adv}}^{(k)}\right)$$

Of the above expression, the third term vanishes on the complete absorber hypothesis and the second term is easily shown to be the radiation reaction term.

Chapter 23
Special Theory of Relativity and Electromagnetism

Einstein was puzzled by the fact that while there existed a symmetry in electromagnetic phenomena so far as observations are concerned, the theoretical explanations lacked that symmetry. The example that he cited was the electromagnetic induction that takes place when there is a relative motion between a magnet and a closed coil. Observationally only the relative motion is important but in explaining the observations, when the magnet moves we say that because $\dot{\mathbf{B}} \neq 0$, there originates an electric field $\nabla \times \mathbf{E} = -\dot{\mathbf{B}}$ whose line integral does not vanish. When the coil moves, we invoke the Lorentz force to explain the observed current. More formally, one may say that the equations of Maxwell do not remain valid when one makes a Galilean transformation to a relatively moving frame. That this is the case was shown very clearly by the experiments of Michelson and Morley who found that the velocity of light is isotropic in all frames although they are in relative motion.

To Einstein we owe two simple postulates from which the Lorentz transformation formulae linking the coordinates in one frame with those in another frame in uniform relative motion can be obtained. The postulates are usually expressed in the following form.

I. *Postulate I:* The basic laws of nature will be of identical form in all inertial frames so that there is nothing to distinguish between any two such frames in uniform relative motion. Put in other words, physically one cannot talk significantly of absolute rest or absolute velocity.
II. *Postulate II:* The velocity of light is the same in all inertial frames and in all directions.

As is easy to see, the two postulates require a revision of the ideas of simultaneity and time measurements in relatively moving frames. Consider two frames O and O′ in uniform motion relative to each other along the x-axis, with the primed frame moving with respect to the unprimed frame with a speed u along the positive x-axis; see Fig. 23.1. When the origins of the two frames are coincident, a light signal is emitted from

A. K. Raychaudhuri, *Classical Theory of Electricity and Magnetism*, Texts and Readings in Physical Sciences 21, https://doi.org/10.1007/978-981-16-8139-4_23

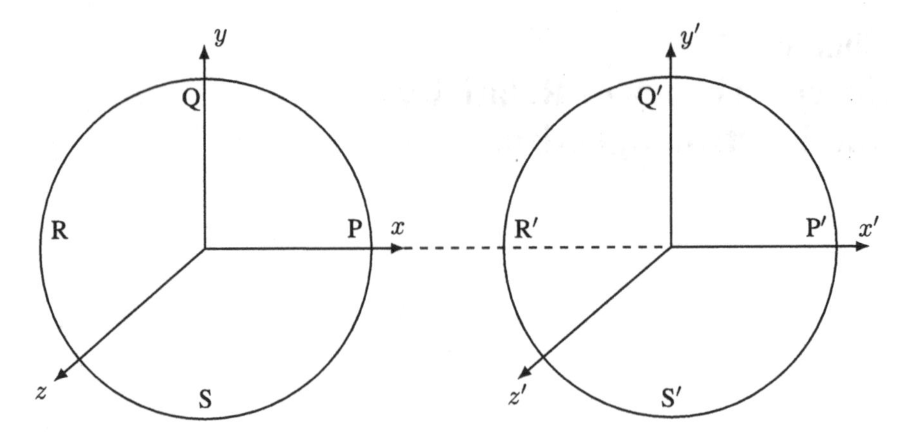

Fig. 23.1 Wavefronts of light emitted by two sources in uniform relative motion

the origin. According to the second postulate, in both the frames the light will spread out isotropically with the same velocity, say c. Thus, the wavefronts in the two frames appear to be the sphere $PQRS$ and $P'Q'R'S'$. The light reaches the points P' and R' at the same instant t' according to the observer in the primed frame, whereas light reaches the points P and R at the same instant t according to the unprimed frame. If $t = t'$, the light has reached P and R' simultaneously—an apparently absurd thing. One has thus to sacrifice the invariance of time measurements in the two frames. In a similar manner one has to allow for a change of length, i.e., distance measurements in going from one frame to another.

23.1 The Lorentz Transformation Formulae

The first question that may puzzle the reader is this—if time and length measurements are different in the two frames, what is the meaning of the second postulate—is it simply a prescription for correlating the time measurements in the two frames? It is not exactly so—implicit in the first postulate is the idea that because of the equivalence of different inertial frames, the characteristic times for natural processes in these frames are identical, e.g., the half-life of a radioactive material at rest in the unprimed frame as measured in the unprimed frame will be identical with the half-life of the same material at rest in the primed frame as measured in the same frame. In a similar manner the length measurements may be standardized by taking the wavelength of a spectral line with the source at rest. However in the deduction that we are going to give, we are not going to use these ideas. Instead we are making the assumption that the velocity of the primed frame as measured by the unprimed frame is the same as the velocity of the unprimed frame as measured by the primed frame in magnitude. Thus, the equality of velocity of light means that it bears the same ratio to this velocity in the two frames.

From the first postulate, we can say that a free particle will move with uniform velocity in either frame. The coordinates of a particle moving with uniform velocity are related by an equation of the form

$$ax + by + cz + dt + e = 0, \qquad a, b, c, d, e \text{ are constants}$$

On transformation, it must assume the form

$$a'x' + b'y' + c'z' + d't' + e' = 0, \qquad a', b', c', d', e' \text{ are constants}$$

As this holds for arbitrary values of the constants, the transformation relations must be linear. A consideration of rotational symmetry demands that the coordinates y and z which are normal to the direction of the relative velocity will not undergo any change on transformation, i.e., $y = y'$ and $z = z'$. Therefore, the transformations are of the form

$$\begin{aligned} x' &= \alpha_1 x + \beta_1 t, & t' &= \gamma_1 x + \delta_1 t \\ dx' &= \alpha_1 dx + \beta_1 dt, & dt' &= \gamma_1 dx + \delta_1 dt \end{aligned}$$

so that

$$\frac{dx'}{dt'} = \frac{\alpha_1 \frac{dx}{dt} + \beta_1}{\gamma_1 \frac{dx}{dt} + \delta_1}$$

Now if $\frac{dx}{dt} = 0$ then $\frac{dx'}{dt'} = -u$. Also if $\frac{dx}{dt} = u$ and $\frac{dx'}{dt'} = 0$, hence

$$\beta_1 = -\delta_1 u \tag{23.1}$$

$$\beta_1 = -\alpha_1 u \tag{23.2}$$

respectively, which imply

$$\alpha_1 = \delta_1 \tag{23.3}$$

Using the above equations

$$x' = \alpha_1 (x - ut), \qquad t' = \gamma_1 x + \alpha_1 t$$

Let us now use the group property. Physically this means that if A is moving uniformly relative to B along the x-direction and B moves uniformly relative to C along the same direction, then between A and C also there is a uniform motion along the x-direction. Hence, the combination of two Lorentz transformations with velocities in the same direction will be a Lorentz transformation with a velocity in the very same direction. Thus

$$x' = \alpha_1(x - u_1 t), \qquad t' = \gamma_1 x + \alpha_1 t \qquad \text{(from A to B)}$$
$$x'' = \alpha_2(x' - u_2 t), \qquad t'' = \gamma_2 x' + \alpha_2 t' \qquad \text{(from B to C)}$$

which combine to

$$\begin{aligned} x'' &= \alpha_2\left(\alpha_1(x - u_1 t) - u_2(\gamma_1 x + \alpha_1 t)\right) \\ &= \alpha_2(\alpha_1 - u_2\gamma_1)x - \alpha_2\alpha_1(u_1 + u_2)t \qquad (23.4) \\ t'' &= \gamma_2(\alpha_1 x - \alpha_1 u_1 t) + \alpha_2(\gamma_1 x + \alpha_1 t) \\ &= (\gamma_2\alpha_1 + \gamma_1\alpha_2)x + \alpha_1(-u_1\gamma_2 + \alpha_2)t \qquad (23.5) \end{aligned}$$

In order to qualify as a Lorentz transformation, transformations (23.4) and (23.5) must obey the relations corresponding to Eqs. (23.1)–(23.3). Hence $\alpha_2 u_2 \gamma_1 = \alpha_1 u_1 \gamma_2$, or

$$\frac{\gamma_2}{\alpha_2 u_2} = \frac{\gamma_1}{\alpha_1 u_1}$$

The left-hand side is a function of u_2 alone while the right-hand side that of u_1 alone. As u_1 and u_2 are independent, this requires them to be simply a universal constant. Thus $\gamma_1 = \frac{\alpha_1 u_1}{k^2}$, where k^2 is constant. Hence

$$x' = \alpha(x - ut), \qquad t' = \alpha\left(t - \frac{ux}{k^2}\right) \qquad (23.6)$$

where we have dropped the subscript, however, Here $\alpha = \alpha(u)$ depends on u. The reciprocal transformation is $x = \alpha(-u)(x' + ut')$, which should be compared to Eq. (23.6)

$$x = \frac{x' + ut'}{\alpha(u)\left(1 - \frac{u^2}{k^2}\right)}$$

to get

$$\alpha(u)\,\alpha(-u) = (1 - u^2/k^2)^{-1}$$
$$\beta(u)\,\beta(-u) = 1$$

where we have defined $\beta(u) = \alpha(u)\sqrt{1 - u^2/k^2}$. The function $\beta(u)$ cannot depend on the direction of u, for then the space will not be isotropic. Thus $\beta(u) = \beta(-u) = \pm 1$. The constant k has the dimension of velocity and defines an invariant velocity. If we take the limit k to infinity, the transformations reduce to Galilean form. With the second postulate of special relativity the velocity of light is an invariant. Hence in the transformation formulae of special relativity, k is to be identified with c, the velocity of light. The resulting transformation is known as Lorentz transformation. Thus, the transformation formulae finally read

$$x' = \frac{x - ut}{\sqrt{1 - u^2/c^2}} \qquad y' = y \tag{23.7}$$

$$z' = z \qquad t' = \frac{t - ux/k^2}{\sqrt{1 - u^2/c^2}} \tag{23.8}$$

This can be expressed in matrix form, by taking $(x_0, x_1, x_2, x_3) = (ct, x, y, z)$ as

$$x_i' = \sum_{k=0}^{3} \alpha_i{}^k x_k$$

where

$$\alpha = ||\alpha_i{}^k|| = \begin{pmatrix} \frac{1}{\sqrt{1-\frac{u^2}{c^2}}} & -\frac{u/c}{\sqrt{1-\frac{u^2}{c^2}}} & 0 & 0 \\ -\frac{u/c}{\sqrt{1-\frac{u^2}{c^2}}} & \frac{1}{\sqrt{1-\frac{u^2}{c^2}}} & 0 & 0 \\ 0 & 0 & 1 & 0 \\ 0 & 0 & 0 & 1 \end{pmatrix} \tag{23.9}$$

which defines the Lorentz transformation. In this transformation, $c^2t^2 - x^2 - y^2 - z^2$ is an invariant. This may be compared with the situation in Galilean transformations where t and $x^2 + y^2 + z^2$ are separately invariant.

23.2 Covariant and Contravariant Vectors and Tensors

Written in terms of infinitesimal differences of the coordinates, we have the invariant

$$ds^2 = c^2dt^2 - dx^2 - dy^2 - dz^2 \tag{23.10}$$

while in three-dimensional Euclidean space with rectangular Cartesian coordinates, the corresponding invariant is

$$ds^2 = dx^2 + dy^2 + dz^2 \tag{23.11}$$

The similarity between Eqs. (23.10) and (23.11) is obvious—only there are four variables instead of three in Eq. (23.10) and the coefficient of dt^2 is $-c^2$ instead of unity.

One may reduce Eq. (23.10) to the form of Eq. (23.11) by introducing a complex variable $\tau = ict$. This was the origin of the imaginary time coordinate which was rather popular in the past and one used to say that the three space and the time coordinates fused together to form a four-dimensional pseudo-euclidean space (pseudo because of the appearance of an imaginary variable). With the advent of the general theory of relativity, where also an invariant exists in the form of a homogeneous quadratic expression in the coordinate differentials, i.e., $ds^2 = g_{ik} dx^i dx^k$ (recall the summation convention), the use of imaginary time became more or less obsolete. Instead one talked in terms of metric tensor g_{ik}, which has the following form in special theory of relativity[1]

$$g_{ik} = \begin{pmatrix} 1 & 0 & 0 & 0 \\ 0 & -1 & 0 & 0 \\ 0 & 0 & -1 & 0 \\ 0 & 0 & 0 & -1 \end{pmatrix}$$

It is necessary to introduce two types of transformation laws in the definition of vectors. Suppose that there are a set of a numbers A^l (i.e., $A^1, A^2, \ldots, A^n$) where n is the dimensionality of the space we are considering. Then if a coordinate transformation is made from x^is to $\bar{x}^i$s, the relation between the coordinate differentials may be written as

$$d\bar{x}^i = \frac{\partial \bar{x}^i}{\partial x^k} dx^k \quad \text{or equivalently} \quad dx^i = \frac{\partial x^i}{\partial \bar{x}^k} d\bar{x}^k \tag{23.12}$$

If owing to this transformation the A^i s change to $\bar{A}^i$ following the same transformation law, i.e.,

$$\bar{A}^i = \frac{\partial \bar{x}^i}{\partial x^k} A^k \qquad \text{or} \qquad A^i = \frac{\partial x^i}{\partial \bar{x}^k} \bar{A}^k \tag{23.13}$$

then the A^i's are said to constitute a *contravariant* vector. Thus in ordinary three-dimensional space, (in Galilean transformations) the velocity vector $v^i \equiv \left(\frac{dx}{dt}, \frac{dy}{dt}, \frac{dz}{dt}\right)$, the acceleration vector $\frac{d^2x^i}{dt^2}$, the force vector $(m\frac{d^2x^i}{dt^2})$ are all contravariant vectors. In the four-dimensional space-time of special relativity, however, dx^i/dt, d^2x^i/dt^2 etc. do not qualify as vectors because t is a coordinate and not an invariant now. One may, however, use the invariant ds defined by Eq. (23.10) to define the velocity vector as $\left(c\frac{dt}{ds}, \frac{dx}{ds}, \frac{dy}{ds}, \frac{dz}{ds}\right)$. Thus for a particle at rest we shall now say that the velocity components are $(1, 0, 0, 0)$. The 'time component' of the velocity can never vanish, for that would require stopping the flux of time.

The other type of transformation law defines a so-called *covariant vector*. We shall give an illustration to obtain this law. Consider the set of quantities $\left(\frac{d\phi}{dx}, \frac{d\phi}{dy}, \frac{d\phi}{dz}\right)$ (in general $\frac{d\phi}{dx^i}$ in an n-dimensional space), where ϕ is an invariant or a scalar function. Then under a coordinate transformation these will transform to $\frac{\partial \phi}{\partial \bar{x}^i}$, where

[1] The original text by the author uses $x_1 = x$, $x_2 = y$, $x_3 = z$ and $x_4 = t$ as the four coordinates. However, this leads to equations which are at variance with other standard texts. In order to avoid such confusion, this version uses $x_0 = ct$, $x_1 = x$, $x_2 = y$ and $x_3 = z$. The same reason has also necessitated a change of the metric signature to $(+ - - -)$ leading the invariant $dx_0^2 - dx_1^2 - dx_2^2 - dx_3^2$.

$$\bar{A}_i = A_k \frac{\partial x^k}{\partial \bar{x}^i} \quad \text{or} \quad A_i = \bar{A}_k \frac{\partial \bar{x}^k}{\partial x^i} \tag{23.14}$$

Note that the convention is of using a superscript for contravariant vectors and a subscript for covariant vectors. However, as we are calling both the objects as vectors, there should be some relation between them—put in other words given the contravariant components of a vector, how are we to find its covariant components? The rule is to use the metric tensor g_{ik}. The components A_i and A^i are related by $A_i = g_{ik}A^k$ and $A^i = g^{ik}A_k$ where g^{ik} is the reciprocal matrix of g_{ik}. They satisfy the relation $g^{ik}g_{kj} = \delta^i_j$, where δ^i_k is the Kroenecker symbol, $\delta^i_k = 1$ if $i = k$ and 0 if $i \neq k$. In matrix notation, the Kroenecker symbol is just the unit matrix.

It is easy to check that, in Euclidean space, in transformation from one rectangular Cartesian coordinate system to another, there is no difference between the two transformation laws, hence there is no point in talking of covariance and contravariance. However, even in Euclidean space if one considers curvilinear coordinates or oblique coordinates, the two transformation laws are no longer identical. Thus in oblique Cartesian coordinates, one may talk of contravariant components as those obtained by drawing parallels to the coordinate axes, while the covariant components are obtained by dropping perpendiculars on the coordinate axes. We illustrate the case of two dimensions in Fig. 23.2, here OM and ON are the contravariant components of OP, while OM' and ON' are its covariant components.

It will be noted that $(OP)^2 \neq (OM)^2 + (ON)^2$ or $(OM')^2 + (ON')^2$, but rather it is equal to $OM \cdot OM' + ON \cdot ON'$. We generalize this result to define the *norm* or magnitude of a vector A^i as $\sqrt{|A^iA_i|}$. Defined in this way the norm of a vector is a scalar. The reason that we have put the modulus sign in the definition of the norm is that A^iA_i may be of either sign—this is due to the fact that in the diagonalized metric tensor components there are elements of both signs. In case A^iA_i is positive with the metric (23.10), we call the vector time-like, if it is negative, the vector is called space-like, whereas if it be zero, without all the components vanishing, we call it a null vector. Thus the velocity vector dx^i/ds is null in case of light, time-like in case of

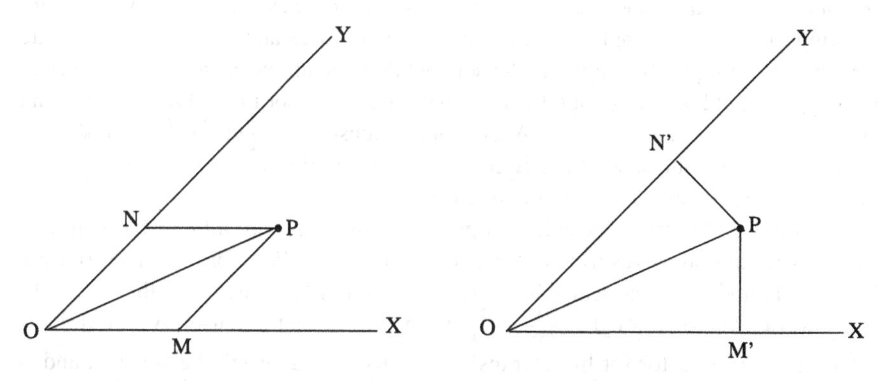

Fig. 23.2 Contravariant and covariant components of a vector in a non-orthogonal system of coordinates

ordinary particles and space-like in case of tachyons which are hypothetical particles having velocity exceeding the velocity of light. The scalar formed from two vectors A^i and B^i, namely $A^i B_i$, is called the scalar product of the two vectors—when this vanishes we say that the vectors are orthogonal, or normal, to one another. A rather curious consequence of this definition is that a null vector is normal to itself.

We next go to define tensors of rank two. In a n-dimensional space, it will be a set of n^2 quantities designated by two indices each of which obeys the transformation law for a vector. Thus if A^{ik} obeys the transformation rule

$$\bar{A}^{ik} = A^{lm} \frac{\partial \bar{x}^i}{\partial x^l} \frac{\partial \bar{x}^k}{\partial x^m}, \qquad A^{ik} = \bar{A}^{lm} \frac{\partial x^i}{\partial \bar{x}^l} \frac{\partial x^k}{\partial \bar{x}^m} \tag{23.15}$$

we say that A^{ik} is a contravariant tensor of rank two. A similar definition (with appropriate modification) is given for a covariant tensor—we may also have a mixed tensor A^i_k, where the superscript obeys the contravariant and the subscript the covariant transformation rule. The rule for passing from a covariant to a contravariant index is the same as we have given for vectors. Thus

$$A_{ik} = g_{il} A^l_k = g_{il} g_{km} A^{lm}$$

It is easy to check that g_{ik}, δ^i_k and g^{ik} are all tensors of rank two. The scalar $g_{ik} A^{ik} = A^i_i$ is called the *spur* or the *trace* of the tensor. An important property of a tensor of rank two is its symmetry property—it is important because the symmetry property is maintained in a coordinate transformation. Thus if $A_{ik} = A_{ki}$, the tensor is called symmetric. The transformed tensor $\bar{A}_{ik}$, from its definition,

$$\bar{A}_{ik} = A_{lm} \frac{\partial x^l}{\partial \bar{x}^i} \frac{\partial x^m}{\partial \bar{x}^k} = A_{ml} \frac{\partial x^l}{\partial \bar{x}^i} \frac{\partial x^m}{\partial \bar{x}^k} = \bar{A}_{ki} \tag{23.16}$$

is also symmetric. In case $A_{ik} = -A_{ki}$, we call the tensor antisymmetric (or skew-symmetric). This is also a coordinate invariant property. The metric tensor g_{ik}, the moment of inertia tensor $I_{ik} = \sum m(r^2 \delta_{ik} - r_i r_k)$ (for these examples we are considering three-dimensional euclidean space with a Cartesian rectangular coordinate system) are examples of symmetric tensors, while we shall see presently that the electromagnetic field is represented by an antisymmetric tensor in the four-dimensional space-time of special relativity. A symmetric tensor has $\frac{1}{2}n(n+1)$ independent components—in our space-time it comes to ten, while the antisymmetric tensor has $\frac{1}{2}n(n-1)$ components, i.e., in our case six.

The metric (23.10) is specially simple in that the g_{ik}'s are all constant—in this case, restricting ourselves to linear transformations, $\partial A_i / \partial x^k$ form a tensor of rank two and its trace, i.e., the scalar $\partial A^m / \partial x^m$, is called the divergence of the vector A^i. The curl of a vector defined as $\frac{\partial A_i}{\partial x^k} - \frac{\partial A_k}{\partial x^i}$ is an antisymmetric tensor. Again, if A^{ik} is a tensor $\frac{A^{ik}}{\partial x^k}$ is a vector for linear transformations, as may easily be verified, and is called the divergence of the tensor A^{ik}.

Before going further, we introduce a notation which will make writing simpler: a comma followed by an index will indicate the partial derivative with respect to that coordinate, e.g., $A_{i,k} = \frac{\partial A_i}{\partial x^k}$.

23.3 Variation of Mass with Velocity

An important change in the older concepts brought about by the special theory of relativity is the dependence of the mass of a particle on its velocity. Let us first make clear what we mean by mass. From a combination of Newton's second and third laws of motion, we get the conservation of momentum principle

$$\sum_{(i)} m_{(i)} \frac{dx^{\mu}_{(i)}}{dt_{(i)}} = \text{constant} \tag{23.17}$$

where the subscript i within braces labels the particle, the greek subscript μ runs over the four coordinates ($x_0 = ct, x_1 = x, x_2 = y, x_3 = z$). The equation corresponding to the zeroth coordinate reads $\sum m_{(i)} = \text{constant}$, giving the conservation of mass. The above equation defines mass—we shall call 'mass' the coefficients of the velocities in the conservation relation, which we expect will hold in all inertial systems because of the principle of relativity, namely, the equivalence of all inertial frames. If there is a system S′ moving relatively to S with a uniform velocity we shall have the transformation relations

$$dx^{\mu} = \alpha^{\mu}{}_{\nu}\, dx'^{\nu}$$

where $\alpha^{\mu}{}_{\nu}$ is the appropriate Lorentz transformation matrix (cf. 23.9). Also because of (23.14)

$$\frac{ds_{(i)}}{dt_{(i)}} = \sqrt{c^2 - v^2_{(i)})} = c\sqrt{1 - \frac{v^2_{(i)}}{c^2}}$$

Hence

$$\frac{dx^{\mu}_{(i)}}{dt_{(i)}} = \frac{dx^{\mu}_{(i)}}{ds_{(i)}} \frac{ds_{(i)}}{dt_{(i)}} = \alpha^{\mu}{}_{\nu} \frac{dx'^{\nu}}{ds_{(i)}} c\sqrt{1 - \frac{v^2_{(i)}}{c^2}}$$

Note that $\alpha^{\mu}{}_{\nu}$ depends only on the relative velocity between S and S′, therefore, is independent of the subscript i so we are justified in taking it outside the summation. Thus, Eq. (23.17) becomes

$$\alpha^{\mu}{}_{\nu} \sum_{(i)} m_{(i)} \sqrt{1 - \frac{v^2_{(i)}}{c^2}} \frac{dx'^{\nu}_{(i)}}{ds_{(i)}} = \text{constant} \tag{23.18}$$

If we multiply (23.18) by the matrix α'^{σ}_{μ}, which is reciprocal to the matrix $\alpha^{\mu}{}_{\nu}$ (this reciprocal matrix exists because determinant of $\alpha^{\mu}{}_{\nu}$ does not vanish, in fact the reciprocal matrix is merely the Lorentz transformation matrix from S to S′, i. e., $dx'^{\mu} = \alpha'^{\mu}{}_{\nu}\, dx^{\nu}$) we get

$$\sum_{(i)} m_{(i)} \sqrt{1 - \frac{v_{(i)}^2}{c^2}}\, \frac{dx'^{\sigma}_{(i)}}{ds_{(i)}} = \text{constant}$$

$$\text{or,} \quad \sum_{(i)} m_{(i)} \sqrt{\frac{1 - v_{(i)}^2/c^2}{1 - v'^2_{(i)}/c^2}}\, \frac{dx'^{\sigma}}{dt} = \text{constant}$$

Recalling our definition of mass, the mass $m'_{(i)}$ in S′ is, therefore

$$m'_{(i)} = m_{(i)} \sqrt{\frac{1 - v_i^2/c^2}{1 - v'^2_i/c^2}}$$

$$\text{or,} \quad m'_{(i)} \sqrt{1 - \frac{v'^2_i}{c^2}} = m_{(i)} \sqrt{1 - \frac{v_i^2}{c^2}}$$

This holds for all inertial frames, hence

$$m_{(i)} \sqrt{1 - \frac{v_i^2}{c^2}} = m_{(i)0}$$

where $m_{(i)0}$ is the mass in the frame in which the velocity of the particle vanishes—this is called the rest mass of the particle. Finally the mass of a particle at velocity v is

$$m = \frac{m_0}{\sqrt{1 - \frac{v^2}{c^2}}} \tag{23.19}$$

Our definition of mass fixes it up to a multiplicative constant, for we may multiply Eq. (23.17) by a constant without altering its meaning. The formula (23.19) is obtained on the implicit assumption that this constant has been fixed by choosing the same unit of mass in different inertial frames.

The conservation of momentum cum conservation of mass principle may now be written as

$$\sum_i m_{(i)0} \frac{dx^{\mu}_{(i)}}{ds_{(i)}} = \text{constant} \tag{23.20}$$

This is a vector equation in four dimensions. The beauty of vector and tensor equations is that they retain their form unaltered in coordinate transformations. Formally the special theory of relativity is an assertion that the basic laws of nature must be Lorentz covariant, i.e., maintain their form unaltered in a Lorentz transformation. We shall now put the Maxwell equations into a tensor form and thus demonstrate their Lorentz covariance.

23.4 Tensor Form of the Equation of Electromagnetism

First, we take note of all the following points.

1. Maxwell's equations are expected to remain valid, at least so long as the velocities involved are small compared to c.
2. In a transformation involving only the space coordinates (i.e., between frames at rest with respect to one another) we can introduce two vectors **E** and **B**, each of which has three 'components'. However, in the special theory of relativity a vector has four components, hence **E** and **B** are apparently not vectors in the special theory of relativity.
3. The Lorentz force expression $e\,(\mathbf{E} + \mathbf{v} \times \mathbf{B})$ indicates that in a Lorentz transformation the electric and magnetic fields do not remain independent entities.
4. In an inversion of space coordinates alone, if the equations of Maxwell are to remain valid, while the vector **E** appears to be a polar vector, the magnetic field **B** appears as an axial vector. Again it is a well-known result that an axial vector in three dimensions is in reality an antisymmetric tensor of rank two.

All these suggest that we should bring in an antisymmetric tensor in four dimensions to represent the electromagnetic field. Interestingly, as we have already seen, such a tensor has six independent components while the electromagnetic field also requires the same number of variables. We shall make the following ansatz.

An antisymmetric tensor $F^{\alpha\beta}$ exists such that we may make the following identifications:

$$\begin{aligned} F^{01} &= -F^{10} = -E_x/c \\ F^{02} &= -F^{20} = -E_y/c \\ F^{03} &= -F^{30} = -E_z/c \end{aligned} \tag{23.21}$$

and

$$\begin{aligned} F^{12} &= -F^{21} = -B_z \\ F^{23} &= -F^{32} = -B_x \\ F^{31} &= -F^{13} = -B_y \end{aligned} \tag{23.22}$$

The covariant components of $F^{\mu\nu}$ are easily written down using $F_{\mu\nu} = g_{\mu\alpha} g_{\nu\beta} F^{\alpha\beta}$.

$$\begin{aligned} F_{01} &= -F^{01},\ F_{02} = -F^{02},\ F_{03} = -F^{03} \\ F_{12} &= F^{12},\ \ F_{23} = F^{23},\ \ F_{31} = F^{31} \end{aligned} \tag{23.23}$$

Expressed as matrices, these tensors have the following form.

$$F^{\mu\nu} = \begin{pmatrix} 0 & -E_x/c & -E_y/c & -E_z/c \\ E_x/c & 0 & -B_z & B_y \\ E_y/c & B_z & 0 & -B_x \\ E_z/c & -B_y & B_x & 0 \end{pmatrix} \tag{23.24}$$

$$F_{\mu\nu} = g_{\mu\alpha}g_{\nu\beta}F^{\alpha\beta} = \begin{pmatrix} 0 & E_x/c & E_y/c & E_z/c \\ -E_x/c & 0 & -B_z & B_y \\ -E_y/c & B_z & 0 & -B_x \\ -E_z/c & -B_y & B_x & 0 \end{pmatrix} \tag{23.25}$$

We now assert that the basic equations of electromagnetism are the tensor equations

$$F^{\mu\nu}{}_{,\mu} = \mu_0 J^\nu = \mu_0\rho_0 v^\nu \tag{23.26}$$

$$F_{\mu\nu,\sigma} + F_{\nu\sigma,\mu} + F_{\sigma\mu,\nu} = 0 \tag{23.27}$$

where ρ_0 is the *proper charge density*, i.e., the charge density in the frame in which the charge is at rest, and v^ν are the components of the four velocity. Let us write out Eqs. (23.26) and (23.27) in terms of the components of **E** and **B** using Eqs. (23.21–23.23). We get from Eq. (23.26)

$$\frac{\partial E_x}{\partial x} + \frac{\partial E_y}{\partial y} + \frac{\partial E_z}{\partial z} = \frac{c^2\mu_0\rho}{\sqrt{1-v^2/c^2}} \tag{23.28a}$$

$$\begin{aligned} \frac{\partial B_z}{\partial y} - \frac{\partial B_y}{\partial z} &= \frac{1}{c^2}\frac{\partial E_x}{\partial t} + \mu_0\rho v^x \\ &= \frac{1}{c^2}\frac{\partial E_x}{\partial t} + \frac{\mu_0\rho}{\sqrt{1-v^2/c^2}}\frac{dx}{dt} \end{aligned} \tag{23.28b}$$

$$\begin{aligned} \frac{\partial B_x}{\partial z} - \frac{\partial B_z}{\partial x} &= \frac{1}{c^2}\frac{\partial E_y}{\partial t} + \mu_0\rho v^y \\ &= \frac{1}{c^2}\frac{\partial E_y}{\partial t} + \frac{\mu_0\rho}{\sqrt{1-v^2/c^2}}\frac{dy}{dt} \end{aligned} \tag{23.28c}$$

$$\begin{aligned} \frac{\partial B_y}{\partial x} - \frac{\partial B_x}{\partial y} &= \frac{1}{c^2}\frac{\partial E_z}{\partial t} + \mu_0\rho v^z \\ &= \frac{1}{c^2}\frac{\partial E_z}{\partial t} + \frac{\mu_0\rho}{\sqrt{1-v^2/c^2}}\frac{dz}{dt} \end{aligned} \tag{23.28d}$$

while from Eq. (23.27) we find

$$\frac{\partial B_x}{\partial y} + \frac{\partial B_y}{\partial y} + \frac{\partial B_z}{\partial z} = 0 \tag{23.29a}$$

$$\frac{\partial E_z}{\partial y} - \frac{\partial E_y}{\partial z} = -\frac{\partial B_x}{\partial t} \tag{23.29b}$$

$$\frac{\partial E_x}{\partial z} - \frac{\partial E_z}{\partial x} = -\frac{\partial B_y}{\partial t} \tag{23.29c}$$

$$\frac{\partial E_y}{\partial x} - \frac{\partial E_x}{\partial y} = -\frac{\partial B_z}{\partial t} \tag{23.29d}$$

Thus, at least formally, we have recovered Maxwell's equations, if we take the charge density to be given by $\rho_0/\sqrt{1-v^2/c^2}$, where v is the velocity of the charge in the particular frame. Obviously, ρ_0 is the charge density in the frame in which the charge is at rest. The total charge of the system is given by the volume integral

$$\int \frac{\rho_0}{\sqrt{1-v^2/c^2}}\, dV$$

and since $dV/\sqrt{1-v^2/c^2}$ is an invariant under the Lorentz transformation, we get the result that the total charge in any system is Lorentz invariant—this may be contrasted with the dependence of mass on velocity that we have already investigated. Both are the results of the basic requirement of Lorentz covariance of the laws of physics.

However, we have still to justify the identifications (23.21) and (23.22) by showing that they are consistent with the empirical transformation rules, at least up to terms in the first order in v/c. Considering a primed frame moving with velocity v in the x-direction, and using the tensor transformation formulae (23.16), we obtain

$$\begin{aligned} E_{x'} = -cF_{1'0'} &= -cF_{10}\frac{\partial x^1}{\partial x^{1'}}\frac{\partial x^0}{\partial x^{0'}} - cF_{01}\frac{\partial x_0}{\partial x^{1'}}\frac{\partial x^1}{\partial x^{0'}} \\ &= -cF_{10}\left(\frac{1}{1-v^2/c^2} - \frac{v^2}{c^2}\frac{1}{1-v^2/c^2}\right) \\ &= -cF_{10} = E_x \end{aligned} \tag{23.30a}$$

$$\begin{aligned} E_{y'} = cF_{2'0'} &= cF_{12}\frac{\partial x^1}{\partial x^{1'}} + cF_{02}\frac{\partial x^0}{\partial x^{0'}} \\ &= -cB_z\frac{v/c}{\sqrt{1-v^2/c^2}} + E_y\frac{1}{\sqrt{1-v^2/c^2}} \\ &= \frac{E_y - vB_z}{\sqrt{1-v^2/c^2}} \end{aligned} \tag{23.30b}$$

$$E_{z'} = -cF_{3'0'} = -cF_{30}\frac{\partial x^0}{\partial x_{0'}} - cF_{31}\frac{\partial x^1}{\partial x^{0'}}$$

$$= E_z \frac{1}{\sqrt{1 - v^2/c^2}} + B_y \frac{v}{\sqrt{1 - v^2/c^2}}$$

$$= \frac{E_z + vB_y}{\sqrt{1 - v^2/c^2}} \tag{23.30c}$$

$$B_{x'} = F_{3'2'} = F_{32} = B_x \tag{23.30d}$$

$$B_{y'} = F_{1'3'} = F_{13} \frac{\partial x^1}{\partial x^{1'}} + F_{03} \frac{\partial x^0}{\partial x^{1'}} = \frac{B_y + \frac{v}{c^2} E_z}{\sqrt{1 - v^2/c^2}} \tag{23.30e}$$

$$B_{z'} = F_{2'1'} = F_{21} \frac{\partial x^1}{\partial x^{1'}} + F_{20} \frac{\partial x^0}{\partial x^{1'}} = \frac{B_z - \frac{v}{c^2} E_y}{\sqrt{1 - v^2/c^2}} \tag{23.30f}$$

Excepting terms of order v^2/c^2, the above relations agree with the classical transformation formulae. Eq. (23.30) may be compactly represented as

$$E'_{\parallel} = E_{\parallel}$$

$$E'_{\perp} = \frac{E_{\perp} + \mathbf{v} \times \mathbf{B}_{\perp}}{\sqrt{1 - v^2/c^2}}$$

$$B'_{\parallel} = B_{\parallel}$$

$$B'_{\perp} = \frac{B_{\perp} - \frac{1}{c^2} \mathbf{v} \times \mathbf{E}_{\perp}}{\sqrt{1 - v^2/c^2}}$$

Of some interest are the invariants of the electromagnetic field. By direct calculation, we find

$$\begin{aligned} |\mathbf{E}|^2 - c^2|\mathbf{B}|^2 &= E_x^2 + E_y^2 + E_z^2 - c^2(B_x^2 + B_y^2 + B_z^2) \\ &= E_{x'}^2 + E_{y'}^2 + E_{z'}^2 - c^2(B_{x'}^2 + B_{y'}^2 + B_{z'}^2) \\ &= |\mathbf{E}'|^2 - c^2|\mathbf{B}'|^2 \end{aligned} \tag{23.31}$$

and

$$\begin{aligned} \mathbf{E}' \cdot \mathbf{B}' &= E_{x'} B_{x'} + E_{y'} B_{y'} + E_{z'} B_{z'} \\ &= E_x B_x + E_y B_y + E_z B_z \\ &= \mathbf{E} \cdot \mathbf{B} \end{aligned} \tag{23.32}$$

It is well known that for a pure radiation field, both $|\mathbf{E}|^2 - c^2|\mathbf{B}|^2$ and $\mathbf{E} \cdot \mathbf{B}$ vanish. We now see that these characteristics persist independent of the velocity of the observer. In other words, an electromagnetic field which appears to be a pure radiation field to any observer will appear to be such to any other relatively moving observer.

In case the electric and magnetic fields are orthogonal in one frame so that $\mathbf{E} \cdot \mathbf{B} = 0$, one may reduce the field to a simple electric or magnetic field by a Lorentz transformation. Suppose $|\mathbf{E}|^2 > c^2|\mathbf{B}|^2$, we take the y-and z-directions along $\mathbf{E}$ and $\mathbf{B}$, respectively. If we now transform to a frame moving along the negative x-direction (i.e., the direction orthogonal to both $\mathbf{E}$ and $\mathbf{B}$) with a velocity $v = c^2|\mathbf{B}|/|\mathbf{E}|$ (such a velocity being permissible because $c|\mathbf{B}| < |\mathbf{E}|$) we get the field components in the primed frame (cf. Eq. (23.30) as

$$\mathbf{E}' = \left(0, E_y\sqrt{1 - v^2/c^2}, 0\right)$$
$$\mathbf{B}' = (0, 0, 0)$$

Thus, the field is now a simple electric field in the y'-direction. Similarly if $c^2B^2 > E^2$, one may reduce to a simple magnetic field by transforming to a frame moving with velocity $|\mathbf{E}|/|\mathbf{B}|$.

The situation is more complicated in case $\mathbf{E} \cdot \mathbf{B} \neq 0$. In this case even after a Lorentz transformation both the electric and magnetic fields must persist otherwise $\mathbf{E} \cdot \mathbf{B}$ would vanish contrary to its invariant character. One can, however, reduce the field to an electric and magnetic field, both in the same direction. Suppose in one frame

$$\mathbf{E} = (0, E_y, 0), \qquad \mathbf{B} = (0, B_y, B_z)$$

are the non-zero components of the fields. After transforming to a frame moving along the x-axis (i.e., perpendicular to the plane containing $\mathbf{E}$ and $\mathbf{B}$) we get

$$\mathbf{E}' = \left(0, \frac{E_y - vB_z}{\sqrt{1 - v^2/c^2}}, \frac{vB_y}{\sqrt{1 - v^2/c^2}}\right)$$
$$\mathbf{B}' = \left(0, \frac{B_y}{\sqrt{1 - v^2/c^2}}, \frac{B_z - \frac{v}{c^2}E_y}{\sqrt{1 - v^2/c^2}}\right)$$

The electric and magnetic fields will be in the same direction if $vB_y/(E_y - vB_z) = (B_z - (v/c^2)E_y)/B_y$, or

$$v^2 - \frac{E^2 + c^2B^2}{E_yB_z} v + c^2 = 0$$

The quadratic equation determines two values of v whose product is equal to c^2—the only value for which $\frac{v}{c} < 1$ is an acceptable solution of the problem.

The above examples go to show that except in the case of the pure radiation fields (also called null fields sometimes) the Poynting vector at any point may always be made to vanish by a Lorentz transformation. However, the velocity involved may not be the same at all points. Let us give a few examples.

(i) If the electric and magnetic fields be uniform, $\mathbf{v}$ will be the same at all points—this is the case of crossed uniform fields very often considered in the experimental study of charged particles.
(ii) In case of a charge moving with uniform velocity $\mathbf{u}$, we have seen that $\mathbf{B} = \mathbf{u} \times \mathbf{E}$. Obviously a transformation with velocity $\mathbf{u}$, which introduces the rest frame of the charge, makes the Poynting vector vanish at all points.
(iii) In case of a charged particle in non-uniform motion, $\mathbf{B} = \mathbf{n} \times \mathbf{E}$, where $\mathbf{n}$ is the unit vector along the line joining the field point and the source point (i.e.,

the retarded position). As **n** is variable, the Lorentz transformation to make the Poynting vector vanish would involve different velocities at different points. In other words, a single Lorentz transformation cannot remove the flux of radiation everywhere.

The above discussion opens up the possibility of defining radiation fields in an unambiguous manner.

The relations (23.21) and (23.22) are non-tensorial relations—this is not strange for the definitions of the electric and magnetic fields are frame-dependent. The electric intensity in any frame is defined by the force on an infinitesimal charge *at rest in that frame*. Similarly, the definition of the magnetic field involves the velocity of the charge *in that frame*. Keeping these in mind one can introduce tensorial formulae for E^μ and B^μ involving a velocity vector v^μ.

$$E_\nu = F_{\mu\nu} v^\mu \tag{23.33a}$$

$$B_\nu = \frac{1}{2} \epsilon_{\nu\mu\lambda\sigma} F^{\mu\lambda} v^\sigma \tag{23.33b}$$

where $\epsilon_{\nu\mu\lambda\sigma}$ is a completely antisymmetric tensor with $\epsilon_{0123} = 1$. By complete antisymmetry we mean that any single interchange of indices changes the sign of the component without changing its magnitude, i.e., $\epsilon_{1023} = -\epsilon_{0123} = +\epsilon_{0132} = \cdots = -1$. We leave it to the reader to check that under these prescriptions, $\epsilon_{\nu\mu\lambda\sigma}$ is a tensor of rank 4. With $v^\mu = \delta^\mu_0$, we recover the relations (23.21) and (23.22). In a relatively moving frame we get the formulae (23.30). It may be noted that Eqs. (23.33a)–(23.33b) lead to

$$E_\mu v^\mu = B_\mu v^\mu = 0$$

indicating that the vectors E^μ and B^μ have no component along the world line of the observer. Thus each has only three non-trivial components. The equation of motion of a particle of charge e and rest mass m_0 may now be written in the Lorentz covariant form

$$m_0 \frac{d^2 x^\mu}{ds^2} = eE^\mu = eF^{\nu\mu} v_\nu \tag{23.34}$$

The symmetry of the Maxwell equations with respect to electric and magnetic fields may be brought into relief by introducing the tensor $\widetilde{F}^{\mu\nu}$, called the dual of the tensor $F^{\mu\nu}$ and defined by

$$\widetilde{F}_{\mu\nu} = \frac{1}{2} \epsilon_{\mu\nu\alpha\beta} F^{\alpha\beta}$$

Equations (23.27) and (23.33b) may now be written as

$$\widetilde{F}^{\mu\nu}_{,\mu} = 0$$
$$B_\mu = \widetilde{F}_{\mu\nu} v^\nu$$

Equation (23.27) is the necessary and sufficient conditions for $F_{\mu\nu}$ to be the curl of a vector

$$\widetilde{F}_{\mu\nu} = A_{\mu,\nu} - A_{\nu,\mu}$$

so that from (23.21) and (23.22)

$$\frac{E_x}{c} = \frac{\partial A_x}{\partial t} - \frac{\partial A_0}{\partial x} \qquad \begin{aligned} B_x &= \frac{\partial A_z}{\partial y} - \frac{\partial A_y}{\partial z} \\ B_y &= \frac{\partial A_x}{\partial z} - \frac{\partial A_z}{\partial x} \\ B_z &= \frac{\partial A_y}{\partial x} - \frac{\partial A_x}{\partial z} \end{aligned} \tag{23.35}$$

These formulae agree with Eqs. (24.5) and (24.6), if we identify the space components of A_μ with the vector potential introduced there and the time component[2] A_0 with ϕ. Thus, the scalar and vector potentials of the non-relativistic theory unify to make a four-vector potential A_μ. The field tensor $F_{\mu\nu}$, as usual, is not altered by a gauge transformation

$$A_\mu \to A_\mu + \frac{\partial \psi}{\partial x^\mu}$$
$$F_{\mu\nu} \to F_{\mu\nu}$$

for a scalar function ψ.

Problems

1. Show that the charge conservation principle follows directly from the field equations.
2. Work out the expressions of the invariants $F_{\mu\nu}\widetilde{F}^{\mu\nu}$ and $F_{\mu\nu}F^{\mu\nu}$ in terms of the components of the electric and magnetic fields **E** and **B**.
3. Obtain the field due to a charged particle moving with a uniform velocity in an inertial frame, by a Lorentz transformation of the field due to the particle at rest.

[2] If the zeroth variable is taken to be ct, then $A_0 = \phi/c$ in SI units.

Chapter 24
Variational Principle Formulation of Maxwell's Equations and Lagrangian Dynamics of Charged Particles in Electromagnetic Fields

A convention has developed of deriving the basic equations of physics from a variational principle—it has the advantage of giving directly a conserved energy momentum complex. The procedure is to select a scalar Lagrangian L involving usually the field variables and their first order spatial and time derivatives. More precisely in field theories, the function to be selected is the Lagrangian density $\mathcal{L}$ and hence the Lagrangian is

$$L = \iiint \mathcal{L}\, dx\, dy\, dz$$

One then postulates, in analogy with Hamilton's principle in mechanics, that the action integral $\int L\, dt$ is an extremum for arbitrary infinitesimal variations of the field variables subject to the condition that these variations vanish at the boundary of the domain of integration.

$$\delta \iiint \mathcal{L}\, dx\, dy\, dz\, dt = 0 \tag{24.1}$$

The empirical part of the procedure consists in selecting the scalar function $\mathcal{L}$.

For the electromagnetic field, the scalars that we have met with are $\tilde{F}^{\alpha\beta} F_{\alpha\beta}$ and $F^{\alpha\beta} F_{\alpha\beta}$. When written in terms of **E** and **B**, apart from numerical constants which play no part in the variation principle, these are $\mathbf{E} \cdot \mathbf{B}$ and $E^2 - c^2 B^2$, respectively. Of these, $\tilde{F}^{\alpha\beta} F_{\alpha\beta}$ cannot serve as the Lagrangian density as it vanishes unless both **E** and **B** exist, therefore, it would not admit the existence of simple electric or magnetic fields (if used as the Lagrangian density). Thus, we take $F^{\alpha\beta} F_{\alpha\beta}$ as the Lagrangian density of our variational principle

$$\delta \iiint F^{\alpha\beta} F_{\alpha\beta}\, dx\, dy\, dz\, dt = 0 \tag{24.2}$$

A. K. Raychaudhuri, *Classical Theory of Electricity and Magnetism*, Texts and Readings in Physical Sciences 21, https://doi.org/10.1007/978-981-16-8139-4_24

Using the relation

$$F_{\alpha\beta} = A_{\alpha,\beta} - A_{\beta,\alpha} \tag{24.3}$$

our field variables are now the components of the vector A_μ and the variational principle yields

$$\begin{aligned}
\iiint \frac{\partial \mathcal{L}}{\partial A_{\mu,\nu}} \delta(A_{\mu,\nu})\, dx\, dy\, dz\, dt &= \iiint \frac{\partial \mathcal{L}}{\partial A_{\mu,\nu}} \frac{\partial}{\partial x^\nu} \delta A_\mu\, dx\, dy\, dz\, dt \\
&= -\iiint \frac{\partial}{\partial x^\nu} \left(\frac{\partial \mathcal{L}}{\partial A_{\mu,\nu}} \right) \delta A_\mu\, dx\, dy\, dz\, dt \\
&= 0
\end{aligned} \tag{24.4}$$

The last step has been obtained by integration by parts—the integrated part involves δA_μ at the boundary, which vanishes due to the condition of variational principle. As δA_μ's are arbitrary and independent, from (24.4) we have

$$\frac{\partial}{\partial x^\nu} \left(\frac{\partial \mathcal{L}}{\partial A_{\mu,\nu}} \right) = 0 \tag{24.5}$$

$$\text{or,} \quad F^{\mu\nu}{}_{,\nu} = 0 \tag{24.6}$$

where we have used Eqs. (24.2) and (24.3) as well as

$$\begin{aligned}
\frac{\partial \mathcal{L}}{A_{\mu,\nu}} &= g^{\alpha\lambda} g^{\beta\sigma} \frac{\partial}{A_{\mu,\nu}} \left(F_{\lambda\sigma} F_{\alpha\beta} \right) \\
&= g^{\alpha\lambda} g^{\beta\sigma} \frac{\partial}{A_{\mu,\nu}} \left((A_{\lambda,\sigma} - A_{\sigma,\lambda})(A_{\alpha,\beta} - A_{\beta,\alpha}) \right) \\
&= 4F^{\mu\nu}
\end{aligned}$$

Equation (24.6)above is just the Maxwell's equation in the source-free case. In case there exists a source, i.e., a current density vector J^μ, we can form the scalar $J^\mu A_\mu$ and take as our Lagrangian density the expression

$$\begin{aligned}
\mathcal{L} &= -\frac{1}{4\mu_0} F^{\mu\nu} F_{\mu\nu} + J^\mu A_\mu \\
&\boxed{= -\frac{1}{4\pi} \left(F^{\mu\nu} F_{\mu\nu} - \frac{1}{c} J^\mu A_\mu \right)} \quad \text{(in Gaussian units)}
\end{aligned}$$

In the above expression, the constants have been chosen to agree with the SI system of units we have been using. The variational principle now gives

$$\iiint \left(\frac{\partial}{\partial x^\nu} \frac{\partial \mathcal{L}}{\partial A_{\mu,\nu}} - J^\mu \right) \delta(A_\mu)\, dx\, dy\, dz\, dt = 0$$

$$\text{or,} \qquad F^{\mu\nu}{}_{,\mu} = \mu_0 J^\nu \tag{24.7}$$

It is somewhat instructive to note the relation between gauge invariance and charge conservation. With the gauge transformation

$$A_\mu \to A_\mu + \frac{\partial \Psi}{\partial x^\mu}$$

the Lagrangian density changes to

$$\mathcal{L} = -\left[\frac{1}{4\mu_0} F^{\mu\nu} F_{\mu\nu} - J^\mu \left(A_\mu + \frac{\partial \Psi}{\partial x^\mu} \right) \right] \tag{24.8}$$

Considering that Ψ is now an additional variable—in fact an ignorable 'coordinate' in the language of Lagrangian mechanics, the variational principle leads to

$$\iiint \left[\left(\frac{\partial}{\partial x^\nu} \frac{\partial \mathcal{L}}{\partial A_{\mu,\nu}} - J^\mu \right) \delta A_\mu - J^\mu \delta \left(\frac{\partial \Psi}{\partial x^\mu} \right) \right] dx\, dy\, dz\, dt = 0$$

The additional term involving Ψ gives

$$\iiint \frac{\partial J^\mu}{\partial x^\nu} \delta \Psi\, dx\, dy\, dz\, dt = 0$$

on integration by parts. Hence the variational principle with the new Lagrangian gives, along with (24.7), the additional relation

$$\frac{\partial J^\mu}{\partial x^\mu} = 0 \tag{24.9}$$

which represents the conservation of charge.

The canonical stress energy tensor is defined by

$$T^\alpha_{(\mathrm{can})\beta} = \left(\frac{\partial \mathcal{L}}{A_{\mu,\alpha}} \right) A_{\mu,\beta} - \mathcal{L}\, \delta^\alpha_\beta$$

so that in the present case

$$T^\alpha_{(\mathrm{can})\beta} = \frac{1}{\mu_0} \left(-F^{\mu\alpha} A_{\mu,\beta} + \frac{1}{4} \delta^\alpha_\beta F^{\mu\nu} F_{\mu\nu} \right) \tag{24.10}$$

This canonical tensor obeys the conservation principle $T^\alpha_{(\mathrm{can})\beta,\alpha} = 0$ if $J^\mu = 0$. However, the canonical tensor is not symmetric. This would not be consistent with angular momentum conservation. The symmetric electromagnetic energy stress tensor is obtained by adding an additional tensor which is itself divergence free. In the present case this is $\frac{1}{4}\delta^\alpha_\beta F^{\mu\nu} F_{\mu\nu}$, thus our electromagnetic stress energy tensor finally reads

$$T^\alpha_\beta = -\frac{1}{\mu_0}\left(F^{\mu\alpha} F_{\mu\beta} - \frac{1}{4}\delta^\alpha_\beta F^{\mu\nu} F_{\mu\nu}\right) \tag{24.11}$$

The symmetric tensor involves the field tensor $F_{\mu\nu}$ only, as distinct from the potential vector and it is therefore gauge invariant. In case the field is not source free, the divergence of (24.11) yields

$$T^\alpha_{\nu,\mu} = J_\alpha F^\alpha_\nu \tag{24.12}$$

which leads to the equation of motion (23.29).

The components of T^μ_ν in terms of $\mathbf{E}$ and $\mathbf{B}$ may be written down readily with the help of Eqs. (23.20)–(23.22). Thus

$$\begin{aligned} T^0_0 &= \frac{1}{2\mu_0} B^2 + \frac{\epsilon_0}{2} E^2 \\ T_{ik} &= \frac{1}{\mu_0}\left(\frac{1}{c^2} E_i E_k + B_i B_k - \frac{1}{2}\left(B^2 + \frac{1}{c^2}E^2\right)\delta_{ik}\right) \\ T_{0i} &= -\epsilon_0 (\mathbf{E} \times \mathbf{B})_i \end{aligned}$$

where the subscripts i and k refer to space coordinates and 0 refers to the time coordinate. Hence the ten components of the tensor T^μ_ν are made up of six components of the Maxwell stress tensor, the three components of the momentum vector (the Poynting vector) and the energy density.

24.1 Lagrangian and Hamiltonian of a Charged Particle in an Electromagnetic Field

The Lagrangian for a free particle in the special theory of relativity is taken as

$$-m_0 c^2 \sqrt{1 - \frac{v^2}{c^2}}$$

The following ideas go to make this choice. The variational principle for the free particle motion is

$$-m_0 c^2 \delta \int \sqrt{1 - \frac{v^2}{c^2}}\, dt = -m_0 c^2 \delta \int ds = 0$$

Thus it is obviously Lorentz covariant. One may wonder about the negative sign—indeed so long as we use the work 'extremal' in the enunciation of our variational principle, either sign leads to the same result. Only whereas in one case the extremal is a minimum, in the other case it is a maximum. However, we demand that the relativistic expression should go over to the expression used in pre-relativistic mechanics for small enough velocities, namely $L = \frac{1}{2}mv^2$. This leads to the choice of the negative sign.

The Lagrangian of the discrete particle may be expressed as a volume integral of Lagrangian density by introducing the Dirac delta function. However, in our present discussion we need not go into that.

The complete variational principle would involve three terms. First, that due to the electromagnetic field alone. Secondly, the Lagrangian of the free particle, finally a term involving the interaction between the charged particle and the electromagnetic field. This last term will be

$$L_{\text{int}} = ev^{\mu} A_{\mu} \frac{ds}{dt} = ev^{\mu} A_{\mu} \sqrt{1 - \frac{v^2}{c^2}}$$

Hence, leaving out for the moment the part due to the field itself, the Lagrangian for the particle is

$$L = -m_0 c^2 \sqrt{1 - \frac{v^2}{c^2}} + ev^{\mu} A_{\mu} \sqrt{1 - \frac{v^2}{c^2}}$$

Corresponding to the ith space coordinate, the Lagrangian momentum p_i is

$$p_i = \frac{\partial L}{\partial v_i} = \frac{m_0 v_i}{\sqrt{1 - v^2/c^2}} + eA_i$$

In the above equations v_i represents dx^i/dt and not dx^i/ds. It is to be noted that the canonical Lagrangian momentum differs from the Newtonian definition of product of mass and velocity even for rectangular Cartesian coordinates.

One may construct the Hamiltonian of the particle

$$\begin{aligned} H &= p_i v_i - L \\ &= p_i v_i + m_0 c^2 \sqrt{1 - \frac{v^2}{c^2}} + e\phi - ev_i A_i \\ &= \frac{m_0 c^2}{\sqrt{1 - v^2/c^2}} + e\phi \end{aligned}$$

It is interesting to note that although the momenta differ from the Newtonian momenta, the Hamiltonian is just the total energy.

Index

A. K. Raychaudhuri, *Classical Theory of Electricity and Magnetism*, Texts and Readings in Physical Sciences 21, https://doi.org/10.1007/978-981-16-8139-4

Printed in the United States
by Baker & Taylor Publisher Services